BASIC
MATHEMATICS

REVISION AND PRACTICE FOR
GCSE AND STANDARD GRADE

Iolanthe E. Gordon

BASIC MATHEMATICS

REVISION AND PRACTICE FOR GCSE AND STANDARD GRADE

MICHAEL WARDLE

Senior Lecturer in Mathematics Education
University of Warwick

Adviser for SCE Standard Grade: Dr Robin Waterston,
Madras College, St Andrews

MACMILLAN

First published 1987   Reprinted once
This edition first published 1990

Published by
MACMILLAN EDUCATION LTD
Houndmills, Basingstoke, Hampshire RG21 2XS
and London
Companies and representatives
throughout the world

Printed in Hong Kong

British Library Cataloguing in Publication Data
Wardle, M.E. (Michael Ernest), 1939–
Basic mathematics for GCSE and standard grade.–New ed.
1. Mathematics. Questions & answers. For schools
I. Title
510
ISBN 0–333–51741–5

10   9   8   7   6   5   4   3   2
00   99 98 97 96 95 94 93 92 91

This book is dedicated to RIMP without whom it
would not have been possible. The author would
also like to thank Caroline Evans and Thomas
Hardy at Macmillan for the encouragement and
help given during the production of the book; Jack
Forster of Warwick University who provided the
initial impetus when collaborating on its
predecessor, Countdown to GCSE: Mathematics
(Jack Forster also provided the material for the
GCSE practice papers in this book) and Dr Robin
Waterston who acted as consultant for this edition
and provided the standard Grade exam papers at
the end of the book.

Contents

Preface

This book provides a complete revision course for students aiming at grades C, D, E, F and G of GCSE, or General or Foundation levels of the Standard Grade SCE, examinations in mathematics. The questions have been based entirely on the topics within lists 1 and 2 of the National Criteria, and the revised arrangements as published by the Scottish Examination Board in 1987.

The questions have been designed to develop confidence and to provide ample practice. Particular emphasis is given to the use and application of the topics in everyday life and examples and reminders are given where appropriate.

GCSE

Candidates aiming at the higher grades (C, D, E) would be required to cover all the material in the book. This also applies to candidates aiming at grades A and B, where these topics account for between 50% and 70% of the total marks (although such candidates would need to cover supplementary and extension material defined by the appropriate board in addition to this). Candidates aiming at the lower grades (E, F, G) need only cover those parts of the book relevant to list 1 of the core syllabus. A copy of the core syllabus has been included in the book as a check list.

STANDARD GRADE

Candidates aiming at the General level (grades 3 and 4) would be required to cover all the material in the book except sections 4.18–4.21, whilst those aiming at the Foundation level (grades 5 and 6) would only be expected to cover those parts relevant to their syllabus. For convenience, copies of the content of the two syllabuses have been included in the book as a check list. At each level the lower of the two grades reflects a satisfactory overall standard of performance for those candidates who obtain about 55%. The higher of the two grades is intended to reflect a high overall standard of performance for those who obtain about 75%.

For convenience, the book is divided into four main sections covering:
1 Basic computation
2 Everyday arithmetic
3 Geometry and measures
4 Graphs, algebra and statistics.

These sections are then sub-divided into double-page spreads, each one concentrating on a separate topic. This will allow the reader to work on any particular aspect of the syllabus as the need arises. The topics may be tackled in any order.

At the end of the four main sections there are:
- **short oral tests** – which are designed to develop abilities in mental arithmetic. These should be read out by another person and answered either on paper or orally, allowing 15–30 seconds per question.
- **problems and investigations** – to provide experience in a variety of more open-ended situations. The introduction to the book presents the idea of an investigation, and provides hints for tackling them.
- **fact sheets** – which summarise key pieces of information relevant to each section, and provide reminders of some of the key mathematical techniques and formulae.

At the end of the book there are:
- **practice exam papers**. Three practice papers have been provided for GCSE levels 1 and 2. Candidates aiming at the higher grades would be expected to tackle papers 2 and 3, whilst those aiming at the lower grades would tackle papers 1 and 2. Two practice papers have been provided for each of the Standard Grade levels. One of these is aimed at Knowledge and Understanding, whilst the other is aimed at Reasoning and Applications.
- **full answers** – to all questions (including hints for investigations).
- **mathematical vocabulary** – this includes a list of terms, as specified in the SCE regulations, which may be used in the examination without explanation. It is therefore important that candidates should be familiar with the meaning of these terms.

GCSE Core Syllabus

LEVEL 1

(for those aiming at Grades E, F and G)

- Whole numbers: odd, even, prime, square
- Factors, multiples, idea of square root
- Directed numbers in practical situations
- Vulgar and decimal fractions and percentages; equivalences between these forms in simple cases; conversion from vulgar to decimal fractions with the help of a calculator
- Estimation and approximation to obtain reasonable answers
- The four rules applied to whole numbers and decimal fractions
- Language and notation of simple vulgar fractions in appropriate contexts, including addition and subtraction of vulgar (and mixed) fractions with simple denominators
- Elementary ideas and notation of ratio
- Percentage of a sum of money
- Scales, including map scales
- Elementary ideas and applications of direct and inverse proportion
- Common measures of rate
- Efficient use of an electronic calculator; application of appropriate checks of accuracy
- Measures of weight, length, area, volume and capacity in current units
- Time: 24 hour and 12 hour clock
- Money, including the use of foreign currencies
- Personal and household finance, including hire purchase, interest, taxation, discount, loans, wages and salaries
- Profit and loss, VAT
- Reading of clocks and dials
- Use of tables and charts
- Mathematical language used in the media
- Simple change of units including foreign currency
- Average speed
- Cartesian coordinates
- Interpretation and use of graphs in practical situations, including travel graphs and conversion graphs
- Drawing graphs from given data
- The use of letters for generalised numbers

- Substitution of numbers for words and letters in formulae
- The geometrical terms: point, line, parallel, right angle, acute and obtuse angles, perpendicular
- Bearings
- Similarity
- Measurement of lines and angles
- Angles at a point
- Enlargement
- Vocabulary of triangles, quadrilaterals and circles; properties of these figures directly related to their symmetries
- Angle properties of triangles and quadrilaterals
- Simple solid figures
- Use of drawing instruments
- Reading and making scale drawings
- Perimeter and area of rectangle and triangle
- Circumference of circle
- Volume of cuboid
- Collection, classification and tabulation of statistical data
- Reading, interpreting and drawing simple inferences from tables and statistical diagrams
- Construction of bar charts and pictograms
- Measures of average and the purpose for which they are used
- Probability involving only one event

LEVEL 2

(for those aiming at Grades C, D and E)

This syllabus includes all the Level 1 topics, together with the following additional topics.
- Natural numbers, integers, rational and irrational numbers
- Square roots
- Common factors, common multiples
- Conversion between vulgar and decimal fractions and percentages
- Standard form
- Approximation to a given number of significant figures or decimal places
- Approximate limits of accuracy
- The four rules applied to vulgar (and mixed) fractions

- Expression of one quantity as a percentage of another
- Percentage change
- Proportional division
- Constructing tables for given functions which include expressions of the form: $ax + b$, $a \times 2$, a/x $(x \neq 0)$ where a and b are integral constants
- Drawing and interpretation of related graphs; idea of gradient
- Transformation of simple formulae
- Basic arithmetic processes expressed algebraically
- Directed numbers
- Use of brackets and extraction of common factors
- Positive and negative integral indices
- Simple linear equations in one unknown
- Congruence
- Angles formed within parallel lines
- Properties of polygons directly related to their symmetries
- Angle in a semicircle; angle between the tangent and radius of a circle
- Angle properties of regular polygons
- Practical applications based on simple locus properties

- Area of parallelogram
- Area of circle
- Volume of cylinder
- Results of Pythagoras
- Sine, cosine and tangent for acute angles
- Application of these to the calculation of a side or an angle of a right-angled triangle
- Histogram with equal intervals
- Construction and use of pie charts
- Construction and use of simple frequency distributions
- Simple combined probabilities

Note:

The practice exam papers later in this book have been graded so that Papers 1 and 2 draw from the material in the Level 1 core syllabus, whilst Paper 3 draws from the material in the Level 2 core syllabus.

It is expected that candidates aiming at Grades E, F and G would attempt Papers 1 and 2, whilst those aiming at Grades C, D and E would attempt Papers 2 and 3.

SCE Standard Grade Syllabuses

FOUNDATION LEVEL

Number

- Whole numbers: addition, subtraction, multiplication and division
- Decimals: addition, subtraction, multiplication and division, rounding to the nearest unit
- Fractions: as an operator, normally confined to fractions with unit numerators
- Percentages: finding the percentage of a quantity
- Equivalence of percentages and fractions
- Reading and locating decimal numbers on scales to the nearest hundredths
- Reading and locating directed numbers such as temperatures on scales
- Ratio: in form of 1 to 5 or 3 to 1
- Rate: miles per gallon, production per hour
- Mean of ungrouped data

Money

- Interest: simple interest for whole year
- Calculations involving money in appropriate contexts: home and personal budgeting, income, savings, shopping bills, fuel bills, hire purchase, profit and loss, discount, VAT, bonus, overtime, commission

Measures

- Estimating: length, weight, area, volume and angle
- Measuring to a reasonable degree of accuracy: length, area, volume, weight, angle, time and temperature
- The interrelationships among the units of measure
- Telling and recording the time: 12-hour and 24-hour clock: second, minute, hour, day, week, month, year, leap year and their interrelationships
- Calculating time intervals: within a given half day on a 12-hour clock, a whole day on a 24-hour clock

Shape

- Interpreting and constructing scale drawings: maps and plans, scales in words
- Enlarging and reducing figures: halving and doubling
- Using scale drawings to solve problems, ratio of sides of similar rectangles
- Cartesian coordinates: plotting points and determining coordinates in the first quadrant
- Recognising and naming 3-D shapes: cube, pyramid, cylinder, cuboid, cone and sphere
- Drawing and recognising nets of cube and cuboid
- Calculating perimeters of rectilinear figures
- Finding areas: of rectangles, squares, right-angled triangles. Irregular figures by counting squares
- Finding volumes: of cubes, cuboids. Other solids by counting cubes
- Angle: measuring and drawing, degrees in right-angle, straight-angle, half-turn, full-turn
- Properties of square, rectangle and triangle: sides, angles, diagonals and sum of angles
- Circle: properties of radius and diameter and their relationship
- Using above properties to calculate angles and lengths of sides
- Line symmetry: recognising and drawing simple symmetrical figures
- Bearings: 8 points of compass, 3-figure bearings, measuring bearing from A to B

Relationships

- Number patterns: extending simple ones
- Evaluating formulae expressed in words, constructing formulae from a table
- Direct proportion
- Extracting data from: pictographs, bar graphs, line graphs and pie charts, trends
- Constructing pictographs, bar graphs and line graphs from given data with a given scale
- Extracting data from tables: up to 2 categories, trends
- Constructing tables from data

- Reading structured diagrams and flow charts with up to 2 or 3 decision boxes
- Codes

GENERAL LEVEL

This syllabus includes all the Foundation Level topics, together with these additional topics.

Number

- Rounding to a given number of decimal places
- Integers: reading and locating heights above and below sea level, coordinates, addition and subtraction in everyday contexts
- Fractions: equivalence, addition, subtraction, multiplication, conversion to decimal
- Expression of one quantity as a percentage of another
- Square roots
- Use of index notation, including scientific notation
- Ratio: splitting a quantity in a given ratio
- Speed, distance, time: calculating one given the other two

Money

- Simple interest for fractions of a year
- Exchange rates, insurance premiums

Measures

- Measuring length, weight, area, volume and angles to a required degree of accuracy
- Interpreting tolerance in a recorded measurement
- Calculating time intervals over midnight, or midday on a 12-hour clock

Shape

- Using scales expressed as a ratio, scaled line or representative fraction
- Constructing triangles: given 2 sides and the included angle, 1 side and 2 angles, or 3 sides
- Ratio of sides in similar right-angled triangles
- Using coordinates in all four quadrants: $y = ax + b$ as the equation of a straight line
- Recognising and drawing nets of: pyramid, cylinder and triangular prism
- Circle: circumference and area
- Finding areas of: triangle (given base and height), kite, parallelogram, rhombus, composite figures

- Finding surface areas of: cube, cuboid, cylinder and triangular prism
- Finding volumes of: cylinder and triangular prism
- Properties of: isosceles and equilateral triangles, kite, parallelogram and rhombus
- Circle: relationship between tangent and radius, angle in semi-circle
- The theorem of Pythagoras
- Rotational and line symmetry: recognising and drawing symmetrical figures
- Bearings: plotting B given the distance and bearing from A, plotting C given the bearings from A and B
- Trigonometry: right-angled triangles using sine, cosine and tangent
- Gradient of slope, angles of elevation and depression

Relationships

- Generalising simple number patterns
- Use of collecting like terms, brackets and factors in simple algebraic expressions
- Evaluating a formula expressed in symbols
- Solving simple linear equations and inequalities
- Inverse proportion, direct variation, graphs for variation
- Pie charts: interpretation and use of proportionality between size of sector and angle at centre
- Constructing graphs where the scale may not be given
- Graphs in context, including the intersection of two speed–time graphs
- Extracting data from tables with up to 5 categories

Note:

The practice exam papers later in this book have been graded so that the first two draw from the material in the Foundation Level syllabus, whilst the second two draw from the material in both syllabuses.

It is expected that Foundation Level candidates aiming at Grade 6 would get about 55% on their exam papers whilst those aiming at Grade 5 would need to get about 75%. In a similar way General Level candidates would be expected to get about 55% for a Grade 4 or 75% for a Grade 3 on their papers.

WHAT ARE INVESTIGATIONS?

In an investigation you are given a starting point and you are expected to explore different avenues for yourself.
Usually, having done this, you will be able to make some general statements about the situation.

AN EXAMPLE OF AN INVESTIGATION

The shape below is made of matchsticks.
How many are there?

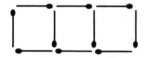

Now try this shape.

Is the number of matchsticks connected to the number of squares in the shape?
Investigate!

There are a number of stages which can help you in tackling an investigation.

Stage 1   Getting started

In the first shape there are ten matches and three squares.
In the second shape there are seven matches and two squares.
Can I see any connection?
Well, there are three more matches in the first shape.

Now let's be more systematic.

Stage 2   Getting some results systematically

Let's start at the beginning and make a table.

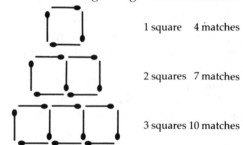

1 square   4 matches

2 squares   7 matches

3 squares 10 matches

It seems that the number of matches goes up by three each time I add another square.

Stage 3   Making some predictions

With five squares there should be 16 matches.
I wonder if this is true.

Yes! 16 matches!

With six squares there should be 19 matches.

Yes! 19 matches!

I wonder if this always works.

Stage 4   Making some generalisations

Each time I add another square I use three more matches. Can I justify this?
Let's look at the shapes again.
Each time I add another square

I need three more matches.

Now can I find a rule for the number of matches in any of these shapes?

Stage 5   Can we find a rule?

Let's look at the results in another way.

1	square	4 matches
2	squares	4 + 3 matches
3	squares	4 + 3 + 3 matches
4	squares	4 + 3 + 3 + 3 matches

.

.

10   squares   4 + (3 + 3 + . . . + 3) matches
←—— nine 3s ——→

.

.

20   squares   4 + (3 + 3 + . . . + 3) matches
← nineteen 3s →

So the rule is
4 + one less than the number of squares, times 3.

Stage 6   Stating the rule using algebra
(if appropriate)

n squares   4 + (n − 1) × 3 matches
or $3n$ + 1   matches

4 + (3n − 3)

= 3n + 1

Let's check:

4 squares 3 × 4 + 1 = 13 matches

Yes I've cracked it!

'The number of matches is three times the number of squares, plus one.'

Or, n squares take $3n$ + 1 matches.

Stage 7   I wonder what happens if . . .
(exploring further)

What about trying the same thing with triangles?

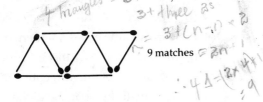

4 Triangles = 3 + 2 + 2 + 2
3 + three 2's
= 3 + (n − 1) × 2
9 matches = 2n
∴ 4△ = (2 × 4) + 1
= 9

or what about . . . ?

Now have a go at this one for yourself.
Start with the triangle matchstick shape and try to use each stage below with these hints.

Stage 1   4 triangles   9 matches
3 triangles   *?* matches
7

Stage 2

Number of triangles	1	2	3	4	5
Number of matches	*3*	*5*	*7*	*9*	*11*

Stage 3   6 triangles ? *13*
7 triangles ? *15*

Stage 4   Each time you add another triangle you add . *2* . more matches.

Stage 5   1 triangle   3 matches
2 triangles   3 + *2* matches
3 triangles   3 + *2* + *2* matches

So it's 3 + *2* . times 2 matches.

Stage 6 n triangles   3 + (n − 1) × 2 matches
Let's check this result for four triangles   3 + (n − 1) × 2 =

So n triangles take *2n + 1* matches.

Stage 7   What about shapes made from cubes?

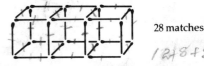

28 matches

12 + 8 + 8
= 28

Summary

1 □ =   12
2 ″   12 + 8
3 ″   12 + 8 + 8

∴ N□ = 12 + (n − 1) × 8
= (8n + 4)

1   If you are to have any hope of making any generalisations at all, it is very important to have some results to look at.

2   Recording systematically is an absolutely crucial stage of any investigation.

3   Try to make some observations about your findings or state the generalisations in your own words.

4   Do try 'what happens if'. You are allowed to choose your own rules or constraints.

1.1 *Number patterns*

1 Look at the pattern of dots below. How many dots will there be in the next pattern? Explain your answer.

2 Look at the set of shapes below. How many lines will there be in the next shape? Explain your answer.

3 Look at the set of squares below. How many small squares will there be in the next shape? Explain your answer.

4 Find: 1   =
           1 + 3   =
           1 + 3 + 5   =
           1 + 3 + 5 + 7   =
           1 + 3 + 5 + 7 + 9 =

What sort of numbers do you get for your answers?

5 (a) What answer do you get if you add together the first six odd numbers?

   (b) Write down the sum of the first ten odd numbers.

6 Look at the pattern of dots below.

   (a) How many dots will there be in the next pattern? Explain your answer.

   (b) How many dots will there be in each of the next two patterns?

You should recognise the results in question **4** as **square numbers**:   1, 4, 9, 16, 25, . . .

The numbers in the pattern in question **6** are called **triangle numbers**:   1, 3, 6, 10, 15, . . .

7 (a) Write down the next two triangle numbers.

   (b) Find: 1 + 3 =
                3 + 6 =
                6 + 10 =
               10 + 15 =

        What sort of numbers are these?

   (c) Do you think this is always true? Try adding the next two triangle numbers. What about the next two?

8 Find: 1   =
           1 + 2 + 1   =
           1 + 2 + 3 + 2 + 1   =
           1 + 2 + 3 + 4 + 3 + 2 + 1 =

What sort of numbers do you get?

9 (a) Find: 1 + 2 + 3 + 4 + 5 + 4 + 3 + 2 + 1 =

   (b) How is your answer connected with the middle number?

   (c) Write down the answer for:
        1 + 2 + 3 + 4 + 5 + 6 + 5 + 4 + 3 + 2 + 1 =
        How did you work it out?

10 Look at the cube patterns.

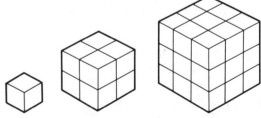

   (a) How many little squares are there on:
        (i) the 2nd cube   (ii) the 3rd cube?

   (b) How many little cubes are there in:
        (i) the 2nd cube   (ii) the 3rd cube?

   (c) Make a sketch of the next cube.

   (d) How many little squares will there be?

   (e) How many little cubes will there be?

11 Look at the set of shapes below.
How many lines will there be in the next shape? Explain your answer.

12 Write down the next two numbers in each set.
   (a) 4, 8, 12, 16, 20, _, _, · · ·
   (b) 4, 7, 10, 13, 16, _, _, · · ·
   (c) 4, 9, 14, 19, 24, _, _, · · ·

In **12**(a) above, the numbers are multiples of four.
They go up in fours, so the next two numbers are 24 and 28.

In (b) the numbers go up in threes, so the next two numbers are 19 and 22.

---

13 For each part of question **12** find:
   (a) the 8th number   (b) the 10th number
   (c) the 20th number   (d) the nth number.

14 Write down the next two numbers in:
   (a) 6, 12, 18, 24, 30, _, _, · · ·
   (b) 9, 18, 27, 36, 45, _, _, · · ·
   (c) 3, 7, 11, 15, 19, _, _, · · ·
   (d) 5, 8, 11, 14, 17, _, _, · · ·

15 For each part of question **14** find:
   (a) the 8th number   (b) the 10th number
   (c) the 20th number   (d) the nth number.

16 Write down the next two numbers in:
   (a) 3, 6, 12, 24, 48, _, _, · · ·
   (b) 1, 3, 9, 27, 81, _, _, · · ·
   (c) 128, 64, 32, 16, 8, _, _, · · ·
   (d) 128, 123, 118, 113, 108, _, _, · · ·

17 For each part of question **16** find:
   (a) the 8th number   (b) the 10th number.

18 Fill in the missing numbers:
   (a) 5, 10, _, 20, _, · · ·
   (b) _, 14, 21, _, 35, · · ·
   (c) 2, 9, 16, _, _, · · ·
   (d) _, 9, 16, 25, _, · · ·
   (e) 1, 3, 6, _, 15, _, · · ·

19 Write down the next two numbers:
   (a) $1, \frac{1}{2}, \frac{1}{4}, \frac{1}{8}, -, -, \cdots$
   (b) 0.1, 0.4, 0.7, 1.0, _, _, · · ·
   (c) 100, 10, 1, 0.1, _, _, · · ·
   (d) $\frac{1}{3}, \frac{2}{3}, 1\frac{1}{3}, 2\frac{2}{3}, -, -, \cdots$
   (e) 0.7, 1.4, 2.1, 2.8, _, _, · · ·
   (f) 3.5, 4.7, 5.9, 7.1, _, _, · · ·

20 (a) You have two parents and four grandparents.
   How many great-great-great-grandparents did you have?
   (b) Make up a table to show the number of your ancestors in each generation.

Description	Generations back	Number in generation
self	0	1
parents	1	2
grandparents	2	4
great-grandparents	3	?
etc.		

   (c) If you were to go back ten generations, how many ancestors did you have in that generation?

21 (a) Find: 1^3 =
   $1^3 + 2^3$ =
   $1^3 + 2^3 + 3^3$ =
   $1^3 + 2^3 + 3^3 + 4^3 =$
   (b) What sort of numbers do you get for your answers?
   (c) Can you say what the next result will be?

22 In a knock-out tennis tournament there are 64 players.
   How many matches will there be in the:
   (a) 1st round
   (b) 2nd round
   (c) 5th round?

23 In a round-robin competition each team plays each of the other teams.
   (a) Find how many matches will be played for different numbers of teams.
   (b) Are your results in (a) always triangle numbers? Can you explain why this so?

1.2 *Multiples, factors and primes*

The **multiples** of 3 are 3, 6, 9, 12 and 15 etc.
The **multiples** of 5 are 5, 10, 15, 20 and 25 etc.

15 and 30 are **common multiples** of 3 and 5.
15 is the **lowest common multiple** of 3 and 5.

---

1 Use the numbers from 1 to 100.
  (a) Write down the multiples of:
    (i) 4 (ii) 6 (iii) 7.
  (b) Write down the common multiples of:
    (i) 4 and 6 (ii) 4 and 7 (iii) 6 and 7.
  (c) Write down the lowest common multiple of:
    (i) 4 and 6 (ii) 4 and 7 (iii) 6 and 7.
  (d) What is the lowest common multiple of 4, 6 and 7?

The diagram below shows the multiples of 6 and the multiples of 8.

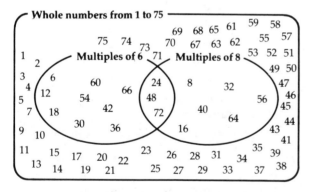

24, 48 and 72 are common multiples of 6 and 8.

---

2 (a) Draw a diagram like the one above to show the multiples of 5 and the multiples of 7 which are less than 75.
  (b) Use your diagram to write down the common multiples of 5 and 7.
  (c) What is the lowest common multiple of 5 and 7?

3 Find the lowest common multiple of:
  (a) 48 and 72 (b) 36, 48 and 60.

4 (a) Write down the first six multiples of 5.
  (b) What can you say about the units digits?

5 (a) Write down the first 12 multiples of 4.
  (b) What can you say about the pattern of numbers made by the units digits?

6 (a) Write down any multiple of 9.
  (b) What is the sum of the digits?

$$6 \times 9 \ = 54 \quad 5 + 4 = 9$$
$$11 \times 9 = 99 \quad 9 + 9 = 18 \quad 1 + 8 = 9$$

  (c) Is this true for all multiples of 9?

7 John visits his grandmother every two days, whilst Jane visits her every three days. If John and Jane both visit her on Sunday, on which day do they next both visit her?

8 Bill always goes to the Chinese restaurant on a Saturday night. Sally works there as a waitress on every fourth night. She serves Bill one Saturday. How many weeks will go by before she could serve him again?

9 Sara has two alarm clocks, one of which gains one hour per day. If she sets them both to the correct time, how long will it be until they next both show the same time?

10 Two salespeople visit a firm on a Monday. One of them re-visits every sixth working day, the other every eighth working day. Assuming everyone works a five day week, how long will it be before both visit the firm on the same day again? What day of the week will this be?

11 In a 10 000 km race one runner is taking a steady 64 seconds for each lap, whilst another takes 72 seconds for each lap. After how many laps will the faster runner overtake the slower one?

12 What difference would it make in question **11** if the faster runner took only 63 seconds for each lap?

The **factors** of 15 are 1, 3, 5 and 15.
The **factors** of 21 are 1, 3, 7 and 21.

1 and 3 are the **common factors** of 15 and 21.
3 is the **highest common factor** of 15 and 21.

---

13 (a) Write down all the factors of:

(i) 8   (ii) 12   (iii) 18.

(b) Write down the common factors of:

(i) 8 and 12   (ii) 8 and 18   (iii) 12 and 18.

(c) What is the highest common factor of:

(i) 8 and 12   (ii) 8 and 18   (iii) 12 and 18?

(d) What is the highest common factor of 8, 12 and 18?

The diagram below shows the factors of 24 and the factors of 30.

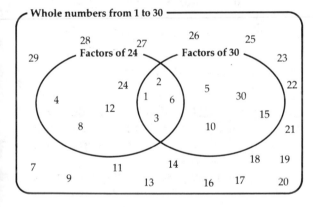

1, 2, 3 and 6 are the common factors of 24 and 30. 6 is the highest common factor of 24 and 30.

---

14 (a) Draw a diagram like the one above to show the factors of 36 and the factors of 48.

(b) Use your diagram to write down the common factors of 36 and 48.

(c) What is the highest common factor of 36 and 48?

15 Find the highest common factor of:
(a) 48 and 72   (b) 36, 48 and 60.

16 Use your calculator to find the factors of:
(a) 256   (b) 2187   (c) 576   (d) 2700.

17 Write down the factors of:
(a) 5   (b) 11   (c) 29   (d) 37   (e) 47.

Each of the numbers in question **17** has exactly two factors, itself and one.
A number which has exactly two factors is called a **prime number**.

---

18 Copy and complete the graph. It shows how many factors each of the numbers has.

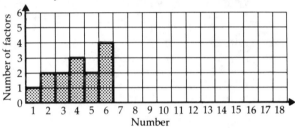

19 (a) Write down for numbers up to 60:

(i) the prime numbers

(ii) the multiples of 6.

(b) How are the prime numbers related to the multiples of 6?

(c) Is this also true for prime numbers greater than 60?

20 (a) Using only £1 coins, or notes all of the same value, in how many ways is it possible to make up a total of:

(i) £20   (ii) £50   (iii) £100?

(b) Which notes can't you use in (ii)?

21 A farmer has 24 cows and 30 sheep. He divides these equally among his sons.
(a) How many sons could he have?

(b) What is the greatest number of sons that he could have?

(c) What difference would it make to your answers if the farmer had 40 cows and 72 sheep?

22 (a) Write down all the factors of 360.

(b) Why do you think 30°, 45°, 60°, 90°, 120° and 180° are the most common special angles?

1.3 *Place value, addition and subtraction*

	Th	H	T	U
7352 is	7	3	5	2
	seven thousands	three hundreds	five tens	two units

seven thousand three hundred and fifty two

1   Write down the value of the ringed number:
(a) 2 ⑤ 3   (b) ⑤ 37   (c) 2 ⑤ 37   (d) ⑤ 237

2   Write using words:
(a) 465   (b) 506   (c) 3270   (d) 4042   (e) 9003

3   Write using numbers:
(a) three hundred and twenty-five
(b) two hundred and six
(c) five thousand and two
(d) twelve hundred and three
(e) eleven thousand eleven hundred.

4   What must you do to the first number to get the second number?
(a) 237 → 287   (b) 359 → 159
(c) 1173 → 1273   (d) 478 → 471
(e) 582 → 732   (f) 4613 → 3713

Th	H	T	U	→	Th	H	T	U
	7	3	5	× 10	7	3	5	0
		7	3	× 100	7	3	0	0
			7	× 1000	7	0	0	0
	4	1	0	÷ 10			4	1
8	4	0	0	÷ 100			8	4

5   How many times is the first three larger than the second three?
(a) 313   (b) 4373   (c) 3397   (d) 3435   (e) 3013

6   Write down:
(a) 253 × 10   (b) 76 × 100
(c) 4370 ÷ 10   (d) 7700 ÷ 10

7   What must you do to the first number to get the second number?
(a) 43 → 430   (b) 6500 → 65   (c) 730 → 7300
(d) 500 → 50   (e) 605 → 6050   (f) 8000 → 8

8   Find:
(a)   17   (b)   43   (c)   217   (d)   526   (e)   869
    +32   +58   +412   +217   +523

9   Find:
(a) 9 + 17   (b) 24 + 36 + 67
(c) 213 + 456 + 729

10   Find a quick way of writing down:
(a) 7 + 3 + 9   (b) 37 + 13 + 9
(c) 77 + 23 + 9   (d) 88 + 45 + 12
(e) 85 + 37 + 15 + 63

11   253 people saw the first night of the school play. 437 saw the second night. How many people saw the play on these two nights?

12   Jo scored 97, 36 and 182 in three innings. How many runs did he score altogether?·

13   On the first two Saturdays in March, 2376 people and 3117 people saw Dunkley United play at home. On the last two Saturdays in March, 2853 people and 2641 people saw the same team play away.
(a) How many people saw the home games?
(b) How many more saw the away games?

14   Attendance figures on the three days of the ABA Games were 21 762, 32 453 and 35 762.
(a) How many attended the games altogether on these three days?
(b) How many more would have been needed on each day to reach a daily target of 40 000?

15   The sales figures for cars in the first three months of the year were 127 000, 135 000 and 142 000 respectively.
(a) How many cars were sold altogether?
(b) By how far did this total fall short of a target of 500 000 cars?

Three subtractions are shown below. Make sure you understand how to do them.

$$\begin{array}{cc} {}^{7\ 1}\\ \cancel{8}5 \\ -27 \\ \hline 58 \end{array} \qquad \begin{array}{cc} {}^{2\ 1}\\ \cancel{3}47 \\ -53 \\ \hline 294 \end{array} \qquad \begin{array}{cc} {}^{2\ 9\ 1}\\ \cancel{3}\cancel{0}7 \\ -59 \\ \hline 248 \end{array}$$

Note: $85 = 70 + 15$ $307 = 290 + 17$
          $347 = 200 + 140 + 7$

---

16 Find:

(a) $\begin{array}{r} 48 \\ -25 \\ \hline \end{array}$ (b) $\begin{array}{r} 48 \\ -29 \\ \hline \end{array}$ (c) $\begin{array}{r} 543 \\ -121 \\ \hline \end{array}$ (d) $\begin{array}{r} 175 \\ -58 \\ \hline \end{array}$ (e) $\begin{array}{r} 516 \\ -47 \\ \hline \end{array}$

17 Find:

(a) $87 - 35$ (b) $63 - 29$ (c) $726 - 513$

(d) $250 - 45$ (e) $243 - 71$ (f) $564 - 298$

18 Find a quick way of writing down the answer:

(a) $100 - 25$ (b) $1000 - 250$

(c) $100 - 73$ (d) $1000 - 427$

(e) $1023 - 623$ (f) $1647 - 847$

19 Copy and complete:

(a) $\begin{array}{r} 283 \\ -\Box \\ \hline 152 \end{array}$ (b) $\begin{array}{r} 716 \\ -\Box \\ \hline 681 \end{array}$ (c) $\begin{array}{r} 238 \\ -\Box \\ \hline 109 \end{array}$ (d) $\begin{array}{r} 1468 \\ -\Box \\ \hline 843 \end{array}$

20 Find:

(a) $46 + 53 - 72$

(b) $57 + 23 - 42 - 17$

(c) $63 - 42 + 21$

(d) $112 - 67 + 41 - 78$

21 When 19 is added to a number the result is 43.
What is the original number?

22 When 23 is subtracted from a number the result is 38.
What is the original number?

23 253 people saw the first night of the new play. 437 saw the second night.
How many more people saw the play on the second night than on the first?

24 Last year G. Boycott scored 1253 runs. This year he scored 1827 runs.
How many more runs did he score this year?

25 Dunkley United Football ground has a capacity of 40 000.
This week 28 362 people watched the game. By how many did the crowd fall short of the full capacity?

26 Wembley stadium has a capacity of 100 000. On the last night of the greyhound races 87 629 people passed through the turnstiles. How many more people would have been needed to fill the stadium?

27 When Jane flew to Sydney in Australia via Hong Kong, the distance was 9857 miles. On the return flight she went via New York and the distance was 12 391 miles.
How much further was the return journey?

28 On the first three days at the Wimbledon tennis championships the attendance figures were 32 561, 35 492 and 38 327 respectively. By how much did the total attendance for these three days exceed 100 000?

29 If a mistake had been made in question **28**, and the third day's attendance figures were only 28 327, by how much did the total attendance fall short of the 100 000?

30 Mrs Sharp went shopping at Tesco's where she could use a £2 voucher if she spent more than £20.
The items she bought cost £7.75, £5.95, £3.50 and £2.95.
What was her final bill?

31 What numbers must go in the circles if each number in a square is the sum of the two circled numbers on either side of it?

(a)   (b)

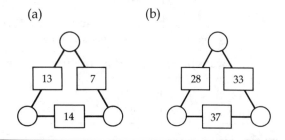

1.4 *Multiplication and division*

Three **multiplications** are shown below.
Make sure you understand how to do them.

$3 \times 47 = 3 \times 40 + 3 \times 7$ $23 \times 47 \leftarrow$ 47
$\qquad = 120 + 21$ $\times 23$
$\qquad = 141$
$\qquad\qquad\qquad\qquad\qquad\qquad\qquad$ 141 (3×47)
$\qquad\qquad\qquad\qquad\qquad\qquad\qquad$ 940 (20×47)

$20 \times 47 = 20 \times 40 + 20 \times 7$
$\qquad = 800 + 140$ $\qquad\qquad\qquad$ 1081
$\qquad = 940$ $\qquad\qquad$ so $23 \times 47 = 1081$

---

1 Find:
  (a) 2×53 (b) 3×61 (c) 4×37
  (d) 28×5 (e) 17×6

2 Find:
  (a) 37   (b)   42   (c)   51   (d) 129   (e) 253
  $\times 9$ $\quad\times 13$ $\quad\times 27$ $\quad\times 8$ $\quad\times 41$

3 Find:
  (a) 16×53 (b) 75×21 (c) 89×132
  (d) 256×32

4 Which of the following are true?
  (a) $27 \times 20 = (27 \times 2) \times 10$
  (b) $27 \times 20 = (27 \times 10) \times 2$
  (c) $27 \times 20 = (27 \times 4) \times 5$
  (d) $27 \times 20 = (27 \times 5) \times 4$

5 Use the ideas of question **4** to write 21×30
   in as many different ways as you can.

6 Which of the following are true?
  (a) $32 \times 15 = 16 \times 30$
  (b) $32 \times 15 = (32 \times 5) \times 3$
  (c) $32 \times 15 = 8 \times 60$
  (d) $32 \times 15 = (8 \times 4) \times 15$

7 Use the ideas of question **6** to write 24×35
   in as many different ways as you can.

Quick ways of finding certain multiplications
are shown below.
  $24 \times 5\ = 12 \times 10 = 120$
  $15 \times 28 = 30 \times 14 = 420$

8 Find:
  (a) 36×5 (b) 5×48 (c) 15×14
  (d) 26×15 (e) 42×35

9 How many hours are there in a week?

10 Sally is building up her collection of compact
   discs. She has found a shop where they cost
   only £7 each.
   How much would she have to pay for:
   (a) 8 discs   (b) 24 discs   (c) 75 discs?

11 John likes collecting stamps. Each set comes
   in a packet of 25 stamps and costs 30p.
   (a) How many stamps will he get if he buys:
      (i) 4 sets   (ii) 12 sets   (iii) 15 sets?
   (b) How much will he have to pay for:
      (i) 4 sets   (ii) 12 sets   (iii) 15 sets?

12 Mrs Gooding has been asked to buy some
   china for the PTA.
   You can buy tea services for six people,
   costing £24. Each set contains eighteen
   pieces of china.
   (a) How many people can she cater for with:
      (i) 5 sets   (ii) 12 sets   (iii) 24 sets?
   (b) How many pieces of china will there be
      altogether in:
      (i) 3 sets   (ii) 8 sets   (iii) 25 sets?
   (c) How much will it cost her to buy:
      (i) 3 sets   (ii) 10 sets   (iii) 25 sets?

13 Every week Jane Drinkwater orders 37 pints
   of milk.
   How many will she have had delivered in:
   (a) 6 weeks   (b) 15 weeks   (c) 26 weeks?

14 If milk costs 25p a pint, how much will it cost
   Jane for:
   (a) 6 weeks   (b) 30 weeks   (c) 52 weeks?

15 Mrs Fleet owns the local newsagents shop.
   Each day except Sunday she sells 57 copies of
   the *Daily Express*.
   How many copies will she sell in:
   (a) 1 week   (b) 4 weeks   (c) 10 weeks?

16 If Mrs Fleet makes 3p profit on each paper she sells, how much will she make in:
(a) 1 week (b) 20 weeks (c) 48 weeks?

Two **divisions** are shown below.
Make sure you understand how to do them.

$$264 \div 12 \qquad 338 \div 13$$

$$20 \times 12 = 240$$
$$2 \times 12 = 24$$
so $22 \times 12 = 264$
so $264 \div 12 = 22$

$$\begin{array}{r} 026 \\ 13\overline{)338} \\ 260 \;(20 \times 13) \\ \hline 78 \\ 78 \;\;(6 \times 13) \\ \hline 00 \end{array}$$

so $338 \div 13 = 26$

---

17 Find:
(a) $96 \div 6$ (b) $48 \div 12$ (c) $108 \div 4$
(d) $312 \div 6$ (e) $732 \div 12$

18 Find:
(a) $5\overline{)165}$ (b) $12\overline{)312}$
(c) $23\overline{)736}$ (d) $17\overline{)2091}$

19 Find:
(a) $371 \div 7$ (b) $848 \div 16$
(c) $1575 \div 21$ (d) $11\,748 \div 89$

20 Which of the following are true?
(a) $960 \div 6 = (960 \div 3) \div 2$
(b) $960 \div 15 = (960 \div 5) \div 3$
(c) $786 \div 6 = (786 \div 2) \div 3$
(d) $786 \div 15 = (786 \div 2) \div 3$

21 Use the ideas of question **20** to write $288 \div 12$ in as many different ways as you can.

22 Which of the following are true?
(a) $135 \div 15 = 270 \div 30$
(b) $144 \div 8 = (144 \div 2) \div 4$
(c) $312 \div 16 = 156 \div 8$
(d) $150 \div 25 = 600 \div 100$

23 Use the ideas of question **22** to write $490 \div 35$ in as many different ways as you can.

Quick ways of finding certain divisions are shown below.
$$350 \div 25 = 1400 \div 100 = 14$$
$$448 \div 28 = 112 \div 7 = 16$$

---

24 Find:
(a) $235 \div 5$ (b) $450 \div 25$
(c) $345 \div 15$ (d) $364 \div 14$

25 (a) How many weeks are there in 301 days?
(b) How many days are there in 192 hours?

26 Sally is given £56 for her 18th birthday.
(a) How many £7 compact discs could she buy?
(b) How many blank cassettes each costing £1.75 could she buy instead?

27 John is given £6 to spend on stamps for his collection. Each packet of 25 stamps costs 30p.
(a) How many packets of stamps can he buy altogether?
(b) How many extra stamps would this add to his collection?

28 Mr Barratt employs 57 labourers equally at three building sites.
How many will work at each site?

29 Austin Rover have 312 new Austin Minis to share equally among 12 distributors.
How many will each get?

30 During a four week period milkman Fred sold 1288 pints of milk.
How many, on average, did he sell:
(a) each week (b) each day?

31 Dunkley United play 16 home games during the football season.
If the total attendance at home matches for the whole season was 211 856, what was the average attendance per game?

32 When Uncle Jo visited England from America, he covered 4935 miles in his hired car during a 21 day holiday.
(a) What was his average daily mileage?
(b) If the hire firm charged 12p per mile, what was Jo's:
(i) total bill (ii) average daily cost?

33 Farmer Giles has 5850 chickens. He does not like to put more than 234 chickens in any one run.
(a) How many runs must he have?
(b) If he spends £25 a week on grain for each run, what is his weekly grain bill?

1.5 *Using arithmetic*

1   Mrs Collins drove 375 miles last week and 463 miles this week.
How many miles did she drive altogether?

2   Mr Read the newsagent sold 4352 papers last month and 3758 the month before.
How many more papers did Mr Read sell last month?

3   Last week Farmer Giles sold 465 trays of a dozen eggs.
How many eggs did he sell altogether?

4   Each of Farmer Giles' customers bought exactly fifteen trays.
How many customers did he have?

5   Jo Moneypenny is an insurance salesman who works a six day week. He made 504 calls during a 12 week period.
How many calls, on average, did he make:
(a) per week   (b) per day?

6   The Hillbilly Blues had a remarkable run scoring 83 baskets during each of 21 consecutive basketball matches.
How many baskets did they score in all?

7   In 1986, Mr Moneypenny did 11 440 miles in his car. In 1987, he clocked up 14 404 miles.
By how much did his 1987 mileage exceed his 1986 mileage?

8   What was Mr Moneypenny's average weekly mileage:
(a) in 1986   (b) in 1987?

9   It costs Mr Moneypenny 6p per mile for petrol to run his car.
How much less did he spend on petrol in 1986?

10 What is the total number of days in four consecutive years?

11 Potatoes are sold in 56 lb bags. The Easy-Eating fish and chip shop bought 35 bags.
How many pounds of potatoes did they buy altogether?

12 Easy-Eating use 98 lb of potatoes each night.
How many nights did the potatoes last them?

13 Farmer Giles decides to keep a record of the number of eggs he collects each day. This is shown below.

Mon	Tue	Wed	Thu	Fri	Sat	Sun
75	63	57	83	71	89	66

(a) How many eggs does he collect during the whole week?

(b) How many more eggs did he collect during the last three days of the week, than during the first three days?

(c) If Farmer Giles sells each egg for 5p how much does he make:
(i) on each day of the week
(ii) for the whole week?

14 The Carefree Motoring company keeps a record of its secondhand car sales during a four week period.

Week 1	Week 2	Week 3	Week 4
327	168	492	285

(a) What was the total number of cars sold during this four week period?

(b) By how much did this figure fall short of their target of 1400 cars?

(c) How many fewer cars did they sell in the first two weeks than in the last two weeks?

(d) If they made a profit of £50 on each car they sold, how much profit did they make:

(i) each week

(ii) over the four weeks?

(e) Assuming the salespeople work a six day week, how many cars, on average, were sold per day during this four week period?

15 In the Copy-It-All office the number of photocopies made each month is recorded.

Jan	Feb	March	April	May	June
3217	4135	2156	2297	3158	4723

(a) By how much did the highest monthly total exceed the lowest monthly total?

(b) How many copies were made altogether?

(c) By how much did this total fall short of 20 000 copies?

(d) What was the average monthly number of copies made?

(e) If the first 3000 copies per month are free and the cost of each of the remainder is $1\frac{1}{2}$p, how much did the office spend on photocopying:

   (i) each month

   (ii) over the six months?

16 Find the cost of buying:

(a) three pairs of tights at 99p each

(b) five boxes of toffees at 99p each.

17 Make a ready reckoner to show the cost of buying any number from one to ten articles each costing 99p.

18 (a) Copy and complete the ready reckoner below which shows the cost of buying articles at £1.99, £2.99, etc.

	Number									
	1	2	3	4	5	6	7	8	9	10
£1.99										
£2.99										
£3.99										
£4.99										
£5.99										
£6.99										
£7.99										
£8.99										
£9.99										

(b) Use your completed ready reckoner to find the cost of buying:

   (i)   3 blouses at £7.99

   (ii) 7 ties at £2.99

   (iii) 5 books at £4.99

   (iv) 9 torches at £1.99.

19 (a) Using your ready reckoner from question 18, find how many shirts, each costing £5.99, you can buy for:

   (i) £17.97   (ii) £29.95   (iii) £47.92

(b) Can you see a quick way of checking your answers, from the total cost?

20 In July 1986, the exchange rate for dollars was: £1 = $1.45.

(a) Copy and complete the ready reckoner which shows the number of dollars you could buy for £1 up to £10, and £10 up to £100.

	£0	£1	£2	£3	£4	£5	£6	£7	£8	£9
£0		1.45	2.90							
£10	14.50	15.95	17.40							
£20	29.00									
£30										
£40										
£50										
£60										
£70										
£80										
£90										
£100										

(b) Use your ready reckoner to find the number of dollars you can buy for:

   (i) £50   (ii) £80   (iii) £41   (iv) £76
   (v) £123

(c) Find how much you will have to pay for:

   (i) $58   (ii) $72.50   (iii) $44.95   (iv) $82.65

21 Mrs Singh has her groceries delivered by the local shop. Part of her order is shown below.

Item	Item cost	Number	Total cost
Cheese	£1.40	2	
Butter	42p	4	
Yoghurt	17p	6	
Gold	38p	1	
Milk	35p	12	
Eggs	8p	18	

(a) Copy and complete the table above.

(b) What is the cost of all these groceries?

(c) Mrs Singh repeats this order each week but only pays at the end of the month. What is her total bill going to be?

1.6 *Fractions*

1 For each of the diagrams below, write down:
(a) the fraction that is shaded
(b) the fraction that is not shaded.

(i)   (ii)   (iii)

2 For each of the diagrams below, write down:
(a) the fraction that is shaded
(b) the fraction that is dotted
(c) the fraction that is not shaded.

(i) 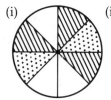 (ii) 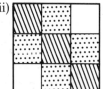 (iii)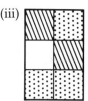

3 (a) Copy the diagrams below.
(b) In each shade in one third of the sections.

(i) 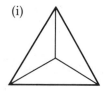 (ii)   (iii)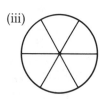

4 On each of the diagrams below two thirds have been shaded.
(a) Use another fraction to describe the shaded sections.

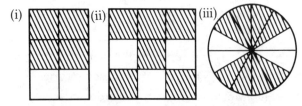

(b) Describe the unshaded sections of each diagram using two different fractions.

5 Copy and complete:
(a) $\frac{2}{3} = \frac{\square}{6}$ (b) $\frac{3}{4} = \frac{\square}{12}$
(c) $\frac{1}{8} = \frac{5}{\square}$ (d) $\frac{4}{7} = \frac{8}{\square}$

6 Write each of these as a simpler fraction:
(a) $\frac{2}{8}$ (b) $\frac{6}{9}$ (c) $\frac{12}{16}$
(d) $\frac{15}{25}$ (e) $\frac{24}{30}$ (f) $\frac{27}{60}$

Fractions can only be added or subtracted if they are of the same type.

$\frac{1}{5} + \frac{3}{5} = \frac{4}{5}$ $\frac{5}{9} + \frac{8}{9} = \frac{13}{9} = 1\frac{4}{9}$ $\frac{5}{6} - \frac{1}{6} = \frac{4}{6} = \frac{2}{3}$

---

7 Find and then write in its simplest form:
(a) $\frac{2}{7} + \frac{3}{7}$ (b) $\frac{3}{8} - \frac{1}{8}$
(c) $\frac{3}{4} + \frac{3}{4}$ (d) $\frac{7}{10} + \frac{9}{10}$

Fractions of different types can be combined if they are *first* changed to the *same* type.

$\frac{3}{4} - \frac{1}{2} = \frac{3}{4} - \frac{2}{4} = \frac{1}{4}$ $\frac{3}{5} + \frac{7}{10} = \frac{6}{10} + \frac{7}{10} = \frac{13}{10} = 1\frac{3}{10}$

---

8 Find and then write in its simplest form:
(a) $\frac{1}{4} + \frac{1}{2}$ (b) $\frac{2}{3} - \frac{1}{6}$
(c) $\frac{3}{4} - \frac{5}{8}$ (d) $\frac{3}{10} + \frac{2}{5}$

It may be necessary to change both fractions before you can add or subtract.

$\frac{2}{3} + \frac{1}{5} = \frac{2 \times 5}{15} + \frac{1 \times 3}{15} = \frac{10}{15} + \frac{3}{15} = \frac{13}{15}$

---

9 Find and then write in its simplest form:
(a) $\frac{1}{2} + \frac{1}{3}$ (b) $\frac{3}{4} + \frac{2}{3}$
(c) $\frac{1}{4} - \frac{1}{5}$ (d) $\frac{3}{5} - \frac{2}{7}$

10 Copy and complete:
(a) $1\frac{1}{2} = \frac{\square}{2}$ (b) $2\frac{1}{3} = \frac{\square}{3}$
(c) $1\frac{5}{8} = \frac{\square}{8}$ (d) $3\frac{3}{4} = \frac{\square}{4}$

11 First change each mixed number to a fraction and then write in its simplest form:
(a) $1\frac{3}{4} + 2\frac{1}{2}$ (b) $3\frac{2}{3} - 1\frac{5}{6}$
(c) $2\frac{2}{5} + 1\frac{1}{8}$ (d) $3\frac{3}{4} - 1\frac{7}{8}$

12 The diagram below shows three circles with two-thirds shaded on each.

How many complete circles could you make with the shaded sections?

13 How many squares and parts of squares can you make with the shaded sections in the diagrams below?

14 Find:
(a) $3 \times \frac{2}{3}$ (b) $3 \times \frac{3}{4}$ (c) $5 \times \frac{2}{5}$
(d) $4 \times \frac{2}{3}$ (e) $7 \times \frac{5}{8}$

15 (a) Copy the squares below.
(b) Shade three-quarters of each square.
(c) How many squares and parts of squares can you make with the shaded sections?

16 (a) What is one-quarter of twenty?
(b) Now write down $\frac{3}{4}$ of 20.

17 (a) What is one-eighth of fifty-six?
(b) Now write down:
(i) $\frac{3}{8}$ of 56
(ii) $\frac{5}{8}$ of 56
(iii) $\frac{7}{8}$ of 56

18 Find:
(a) $\frac{1}{4} \times 8$ (b) $\frac{2}{3} \times 12$ (c) $\frac{3}{4} \times 5$
(d) $\frac{2}{5} \times 9$ (e) $\frac{5}{8} \times 48$

19 (a) What is one-quarter of a half?
(b) What is one-fifth of a half?
(c) Now write down:
(i) $\frac{3}{4}$ of $\frac{1}{2}$
(ii) $\frac{2}{5}$ of $\frac{1}{2}$
(iii) $\frac{4}{5}$ of $\frac{1}{2}$

20 (a) Copy the diagram below.

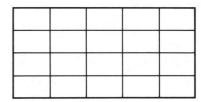

(b) Show, using dots, two-fifths of the sections.
(c) Now shade in three-quarters of the dotted sections.
(d) What fraction of the whole shape is now shaded?

21 Find:
(a) $\frac{1}{2}$ of $\frac{3}{4}$ (b) $\frac{2}{5}$ of $\frac{3}{4}$
(c) $\frac{1}{2} \times \frac{3}{5}$ (d) $\frac{2}{3} \times \frac{4}{5}$

22 Find:
(a) $\frac{3}{4} \times 72$ (b) $\frac{2}{3} \times 51$ (c) $\frac{4}{5} \times 1\frac{2}{3}$
(d) $\frac{5}{8} \times 4\frac{4}{7}$ (e) $2\frac{1}{2} \times 1\frac{3}{4}$

23 (a) How many halves are there in three whole ones?
(b) How many quarters are there in five whole ones?
(c) How many thirds are there in two and two-thirds?
(d) How many three-quarters are there in six whole ones?

24 Find:
(a) $3 \div \frac{1}{2}$ (b) $5 \div \frac{1}{4}$ (c) $2\frac{2}{3} \div \frac{1}{3}$
(d) $6 \div \frac{3}{4}$ (e) $5 \div \frac{2}{5}$

25 (a) Say which of the following are true.
(b) For any that are not true, write down the correct answer.
(i) $\frac{2}{3} + \frac{3}{5} = \frac{19}{15}$
(ii) $\frac{3}{7} \times \frac{5}{8} = \frac{8}{15}$
(iii) $2\frac{2}{3} \div \frac{3}{4} = \frac{8}{9}$

1.7 *Using fractions*

1 There are five school days in the week. On what fraction of the days of the week do you not go to school?

2 There are 32 pupils in Winston's class. 24 of these are girls.
(a) What fraction of the class are girls?
(b) What fraction of the class are boys?

3 In Stanley's school three-fifths of the pupils are boys.
What fraction are girls?

4 There are 15 men and 21 women in the Downside tennis club.
(a) What fraction of the club are:
(i) men (ii) women?
(b) Write your answers for (a) in their simplest form.

5 The table below shows how the pupils in Winston's class come to school.

Means of transport	Number of students
Bus	8
Bicycle	12
Car	4
Walk	8

(a) What fraction of the class:
(i) come by bus
(ii) come by bicycle
(iii) come by car
(iv) walk to school?
(b) Write each fraction in its simplest form.
(c) What is the total of your four fractions?

6 Pato adores sweets. He buys $\frac{1}{8}$ kg on his way to school and another $\frac{1}{4}$ kg on his way home each day.
(a) What fraction of a kilogram of sweets does he buy altogether?
(b) How many more must he buy to make 1 kg in total?

7 Mrs Hadlee buys two bags of rice, one of which weighs $\frac{3}{4}$ kg, the other weighs $\frac{5}{8}$ kg.
(a) How much rice has she bought altogether?
(b) If she uses $\frac{1}{2}$ kg for the family meal, how much does she have left?
(c) How much more will she need to have 1 kg?

8 Saleem's dog eats $\frac{2}{5}$ kg of meat each night. How much does the dog eat in:
(a) 20 nights (b) 4 weeks
(c) one year?

9 Jo jogs $4\frac{1}{2}$ km every morning. How far does he run in:
(a) one week (b) February (c) one year?

10 Pato spends $\frac{1}{2}$ of his pocket money on sweets and $\frac{1}{3}$ on comics. He saves the rest.
(a) How much of his pocket money does Pato:
(i) spend (ii) save?
(b) How long will it take Pato to save an amount equal to his weekly pocket money?

11 Anne's father drives her $\frac{1}{5}$ of her way to work. She then travels $\frac{3}{4}$ of the way by bus. She walks the remainder of her journey.
(a) For what fraction of her journey does she have to walk?
(b) If her total journey is five miles, how far does she:
(i) walk (ii) go by car
(iii) go by bus?

12 Mr Peters has no children so he decides to leave his money to his nieces, Anne, Beni and Carla. Anne gets $\frac{2}{5}$, and Beni gets $\frac{1}{3}$.
(a) What fraction of the money does Carla receive?
(b) If Mr Peters left £3000, how much would each niece receive?

13 (a) Three cakes are divided equally among five people. How much will each receive?

(b) If the cakes had been divided among six people, how much less would each receive?

14 Sally earns £132 a week. Her wages are spent as shown in the table below.

Rent	Food	Travel	Savings	Others
$\frac{1}{4}$	$\frac{1}{3}$	$\frac{1}{6}$	$\frac{1}{12}$	?

(a) How much does Sally spend on each of these five items?

(b) What fraction of her wages is spent on other things?

15 Mr Soo uses half of a sheet of blockboard for some cupboards. He is then able to use two-thirds of what is left for the shelving inside the cupboards.
What fraction of the sheet is wasted?

16 James works part-time in the local store. He saves one-quarter of his wages towards his summer holiday.
He spends two-fifths of what is left of his wages on music tapes.
(a) What fraction of his total wages does he spend on tapes?

(b) If he earns £60 per week, how much does he spend on tapes?

17 Pato decides to eat three-fifths of his sweets today and two-thirds of what are left tomorrow.
What fraction will be left then?

18 A cask of whisky holds 60 litres.
(a) How many $\frac{1}{2}$ litre bottles can be filled from the cask?

(b) How many $\frac{3}{4}$ litre bottles can be filled from the cask?

(c) How many $1\frac{1}{2}$ litre bottles can be filled from the cask?

19 I measured a particular book and found it was $2\frac{1}{2}$ cm thick.
(a) What length of shelf will be needed for nine books of this size?

(b) How many books of this size will fit onto a shelf which is four metres long?

20 A sheet of plastic panelling is $1\frac{1}{2}$ cm thick.
(a) Forty-eight of these sheets are stacked on top of each other.
How tall will the stack be?

(b) How many sheets can be stacked between shelves which are 54 cm apart?

21 Josie is making rich fruit cakes. Each cake needs $\frac{2}{5}$ kg of fruit.
(a) How much fruit will she need to make four of these cakes?

(b) If she has $5\frac{1}{5}$ kg of fruit available, how many cakes can she make?

22 Mrs Jacobs saves £25 a week. If this is one-fifth of her weekly wage, what does she earn each week?

23 Mr Soo used one quarter of a bag of cement when mending his drain. The remainder of the cement weighed 18 kg.
(a) What did the bag weigh originally?

(b) What would five-sixths of the bag weigh?

24 A jug holds $1\frac{1}{2}$ litres when it is $\frac{2}{3}$ full.
(a) How many litres does it hold when full?

(b) How many litres does it hold when it is three-quarters empty?

25 Tania makes dresses. Each dress requires $2\frac{3}{4}$ metres of material.
(a) What length of material will she need for a dozen dresses?

(b) How many dress lengths can she get from a roll of 38.5 metres of material?

26 Jamie completed the first quarter of his project in January. He completed the next third in February and another fifth in March.
(a) What fraction of Jamie's project was still left to complete at the end of March?

(b) During April Jamie managed to do another sixth.
How much is there still left to do?

(c) By what fraction had Jamie exceeded one half at this stage?

(d) What is $\frac{1}{2} + \frac{1}{3} + \frac{1}{4} + \frac{1}{5} + \frac{1}{6}$?

1.8 *Decimals*

$0.1 = \frac{1}{10}$ $0.01 = \frac{1}{100}$ $7.35 = 7 + \frac{3}{10} + \frac{5}{100}$

1 Write as a fraction:
(a) 0.3 (b) 1.7 (c) 0.03
(d) 3.07 (e) 5.43

2 Write as a decimal:
(a) $\frac{3}{10}$ (b) $\frac{7}{100}$ (c) $\frac{23}{100}$
(d) $7\frac{3}{10}$ (e) $2\frac{7}{100}$

$0.6 = \frac{6}{10} = \frac{3}{5}$ $0.05 = \frac{5}{100} = \frac{1}{20}$ $0.75 = \frac{75}{100} = \frac{3}{4}$

3 Write as a fraction:
(a) 0.8 (b) 0.06 (c) 0.15
(d) 0.25 (e) 2.16

$\frac{3}{5} = \frac{6}{10} = 0.6$ $\frac{1}{25} = \frac{4}{100} = 0.04$ $\frac{3}{20} = \frac{15}{100} = 0.15$

4 Write as a decimal:
(a) $\frac{1}{5}$ (b) $\frac{1}{4}$ (c) $\frac{7}{20}$
(d) $\frac{13}{20}$ (e) $\frac{3}{25}$ (f) $1\frac{23}{25}$

Some important decimals

$0.5 = \frac{1}{2}$ $0.2 = \frac{1}{5}$ $0.25 = \frac{1}{4}$ $0.75 = \frac{3}{4}$ $0.125 = \frac{1}{8}$

	T	U	·	t	h
27.35 is	2	7	·	3	5
	two tens	seven units		three tenths	five hundredths

5 Write down the value of the ringed number:
(a) 2.⑤3 (b) 5.3⑦ (c) 25.③7 (d) 75.2③

6 Write using words:
(a) 46.5 (b) 5.06 (c) 32.70 (d) 90.06

7 What must you do to the first number to get the second number?
(a) $23.7 \rightarrow 28.7$ (b) $1.59 \rightarrow 1.79$
(c) $1.73 \rightarrow 1.78$ (d) $4.78 \rightarrow 4.71$
(e) $5.82 \rightarrow 6.32$ (f) $46.17 \rightarrow 46.23$

H	T	U	·	t	h	$\rightarrow$	H	T	U	·	t	h
7	3	5	·			÷ 10	7	3	·	5		
7	3	5	·			÷ 100		7	·	3	5	
		5	·	8		× 10		5	8	·		
		0	·	5	8	× 100		5	8	·		

8 How many times is the first three larger than the second three?
(a) 3.13 (b) 4.33 (c) 3.39 (d) 34.35 (e) 30.13

9 Write down:
(a) 25.3×10 (b) 7.6×100 (c) $43 \div 10$
(d) $7.7 \div 10$

10 What must you do to the first number to get the second number?
(a) $4.3 \rightarrow 43$ (b) $65 \rightarrow 6.5$ (c) $73 \rightarrow 0.73$
(d) $5.6 \rightarrow 0.56$ (e) $600 \rightarrow 0.6$ (f) $0.85 \rightarrow 850$

11 Find:
(a) $\begin{array}{r} 1.7 \\ +3.2 \\ \hline \end{array}$ (b) $\begin{array}{r} 4.3 \\ +5.8 \\ \hline \end{array}$ (c) $\begin{array}{r} 2.17 \\ +4.12 \\ \hline \end{array}$ (d) $\begin{array}{r} 5.26 \\ +2.17 \\ \hline \end{array}$ (e) $\begin{array}{r} 0.69 \\ +0.23 \\ \hline \end{array}$

12 Find:
(a) $2.4 + 3.6 + 6.7$ (b) $2.13 + 4.56 + 7.67$

13 Find:
(a) $\begin{array}{r} 3.7 \\ -1.2 \\ \hline \end{array}$ (b) $\begin{array}{r} 5.3 \\ -4.8 \\ \hline \end{array}$ (c) $\begin{array}{r} 4.17 \\ -4.12 \\ \hline \end{array}$ (d) $\begin{array}{r} 5.26 \\ -2.17 \\ \hline \end{array}$ (e) $\begin{array}{r} 0.63 \\ -0.29 \\ \hline \end{array}$

14 Find:
(a) $12.4 - 3.6 - 6.7$ (b) $12.24 - 4.56 - 7.67$

15 Mrs Patel buys three shirts costing £7.25, £6.15 and £5.50.
(a) How much did she spend altogether?
(b) How much change did she get from £20?

16 Mrs Asid had £15.49 in her purse.
(a) If she spent £4.23 in the Post Office, how much did she have left?
(b) If she then spent another £7.58 at the market, how much did she have left?

17 Find the perimeter of each of these shapes.

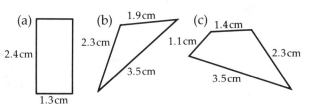

18 Julie needed a piece of copper pipe which was 42.5 cm long, but the piece she had was 61.8 cm long.
(a) How much did she have to cut off?
(b) If she actually cut off a piece which was 18.9 cm, how much too long was what she had left?

19 Find the perimeters of these shapes.

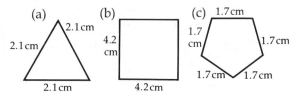

20 Find:
(a) 2×4.3 (b) 3×5.1 (c) 4×7.3
(d) 5×9.7

21 Find the length of the side of a square of perimeter:
(a) 4.4 cm (b) 8.4 cm (c) 12.8 cm
(d) 13.2 cm

22 Find the length of the side of an equilateral triangle of perimeter:
(a) 3.3 cm (b) 6.9 cm (c) 13.2 cm
(d) 14.7 cm

23 Find:
(a) $4.8 \div 4$ (b) $7.2 \div 3$ (c) $9.5 \div 5$
(d) $25.2 \div 6$

24 (a) Find the cost of:
(i) eight packets of crisps at 12p each
(ii) a dozen eggs at 6p each
(iii) twenty chocolate bars at 17p each.
(b) Now write down:
(i) 8×0.12 (ii) 12×0.06
(iii) 20×0.17

25 Find: (a) 8×0.25 (b) 6×0.32 (c) 12×0.09

26 Saleem is stocktaking and wants to know how much his stock is worth.
(a) Find the cost of:
(i) 3 shirts at £7.50 each
(ii) 4 pairs of trainers at £8.15 each
(iii) 5 pairs of tights at £1.35 each
(iv) 6 pairs of jeans at £13.85 each
(v) 7 jackets at £24.95 each.
(b) Find the total value of Saleem's stock.

27 Six friends want to go youth hostelling. Each night costs £2.57 per person. Find the cost for the whole group of:
(a) one night (b) one week.

The area of the rectangle below is shown in two different ways.

$(1\,cm = 10\,mm$ and $1\,cm^2 = 100\,mm^2)$

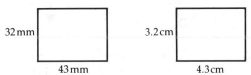

Area $= 32 \times 43\,mm^2$ Area $= 3.2 \times 4.3\,cm^2$
$\quad\quad = 1376\,mm^2$ $= 13.76\,cm^2$

So $3.2 \times 4.3 = \frac{32}{10} \times \frac{43}{10} = \frac{1376}{100} = 13.76$

---

28 (a) Find the area of a rectangle with sides of 21 mm and 34 mm.
(b) Write down the area of a rectangle with sides of 2.1 cm and 3.4 cm.

29 (a) Find the area of a rectangle with sides of 47 mm and 53 mm.
(b) Write down the area of a rectangle with sides of 4.7 cm and 5.3 cm.

30 Find the area of a rectangle with sides of:
(a) 1.2 cm and 2.3 cm
(b) 2.4 cm and 3.7 cm.

31 (a) Use your calculator to find:
(i) 47×53 (ii) 56×73 (iii) 82×91
(b) Now write down:
(i) 4.7×5.3 (ii) 5.6×7.3 (iii) 8.2×9.1

1.9 *Using decimals*

1   Anne is training in the pool. She takes 27.2 seconds for the first length and 32.5 seconds for the second length.
  (a) What is her total time for two lengths?
  (b) How many seconds inside one minute is her combined time?

2   David is practising for the high jump. On his first jump he clears 156.3 cm, whilst on his second he clears 163.8 cm.
  (a) What is the total of his two jumps?
  (b) By how much does the second jump exceed the first?

3   Sali is buying drawing materials. He spends £1.23 on paper, £2.34 on inks and £4.58 on a new pen.
  (a) How much does he spend altogether?
  (b) How much change will he get from a £10 note?

4   Kimo is trying out a new recipe. She needs 0.25 kg of sugar, 0.67 kg of flour, 0.33 kg of butter and 1.24 kg of rice.
  (a) How much do the ingredients weigh altogether?
  (b) What weight of salt must she add if she knows the total weight of the recipe comes to 2.5 kg?

5   Yusuf is making a fruit punch. The recipe requires 0.80 litres of pineapple juice, 0.25 litres of orange juice, 0.12 litres of lemon juice and 0.06 litres of grenadine.
  (a) How many litres does this come to so far?
  (b) When he has mixed these ingredients he adds lemonade. If he wants to make 1.5 litres altogether how much lemonade must he add?

6   Mr Jay is using a copper pipe 66.7 cm long. He cuts off 23.4 cm for the hot tap and 26.9 cm for the cold tap.
  (a) How much does he cut off altogether?
  (b) How much is left of the original length?

7   The lap times of Steve Cram in a 1500 metres race are:
    54.8 seconds, 56.7 seconds, 53.6 seconds and 47.5 seconds.
  (a) By how much is the fastest lap time better than the slowest lap time?
  (b) What is his total time for the race?
  (c) How far under 3 minutes 40 seconds is he?
  (d) How far over 3 minutes 30 seconds is he?

8   A concrete mix requires 6.7 kg of cement, 20.1 kg of sand and 33.5 kg of gravel.
  (a) If the sand and cement come in 25 kg bags, how much of each will be left over?
  (b) If 100 kg of gravel are delivered, how much is left over?
  (c) What is the total weight of the mix?
  (d) How far short of 100 kg is the total weight?
  (e) What is the weight of the leftover ingredients?

9   Grania cuts five dress lengths from a 20 metre length of material.
  (a) If each dress length is 2.65 metres, how much material does she use altogether?
  (b) How much material is left on the roll?

10   Sue buys a five litre bottle of fruit punch to split up as presents for her friends. She uses smaller bottles which hold.
     0.75 l, 0.66 l, 0.5 l, 0.33 l and 0.05 l.
  (a) How much more does the largest bottle hold than the smallest bottle?
  (b) How much will she need to fill all of these bottles?
  (c) How much of the five litres will be left over?

11   Find how much these bottles hold.
  (a) four of 0.75 l   (b) six of 0.66 l
  (c) ten of 0.5 l   (d) twelve of 0.33 l
  (e) twenty of 0.05 l   (f) thirty of 1.25 l

12 (a) Copy and complete the table.

Item	Item cost	Number	Total cost
Biro	£0.12	30	
Crayon	£0.15	20	
Pencil	£0.09	12	
Pen	£2.75	8	
Rubber	£0.40	7	
Ruler	£0.65	4	
Sellotape	£0.43	2	

   (b) What is the total cost of the items listed above?

13 Tania wants to make tapes for the house team.
   How long will each tape be if she makes:
   (a) four from a 68 cm length of tape
   (b) five from a 56.5 cm length of tape
   (c) three from a 52.5 cm length of tape
   (d) six from a 96.24 cm length of tape?

14 (a) What is the perimeter of a square with a side of 36 mm?
   (b) Write down the perimeter of a square of side 3.6 cm.

15 (a) What is the area of a square with a side of 12 mm?
   (b) Write down the area of a square of side 1.2 cm.

16 (a) What is the perimeter of a rectangle with sides 27 mm and 72 mm?
   (b) Write down the perimeter of a rectangle with sides 2.7 cm and 7.2 cm.

17 (a) What is the area of a rectangle with sides 25 mm and 64 mm?
   (b) Write down the area of a rectangle with sides 2.5 cm and 6.4 cm.

18 The length of a rectangle is 8 cm and its area is 60.8 cm^2.
   What is the width of the rectangle?

Use your calculator for these questions:

---

19 Find the area of a square of side 12.5 cm.

20 Find the cost of buying twenty-five books if each costs £5.95.

21 A car can travel 52.3 miles in one hour. How far can it travel in 3.5 hours?

22 A car averages 47.3 miles per gallon. How many miles can the car travel on 7.25 gallons?

23 A car travels 347.1 miles and uses 6.5 gallons.
   How many miles can it travel per gallon?

24 A rectangular room is 4.5 metres long and 3.5 metres wide.
   If carpet costs £6.95 per square metre, how much will it cost to carpet the room?

25 A lawn is 23.4 metres long and 15.9 metres wide. A bag of lawn dressing will treat 12 square metres and costs £4.35.
   Find:
   (a) the area of the lawn
   (b) the number of bags of lawn dressing needed to treat the whole lawn
   (c) the cost of treating the lawn.

26 In March 1987 you could buy $1.62 for a £1.
   (a) How many dollars could be bought for:
      (i) £20 (ii) £120 (iii) £365 (iv) £7.50?
   (b) How much would it have cost to buy:
      (i) $20 (ii) $120 (iii) $729 (iv) $5.67?

27 Raj borrowed some money from his father to buy a music centre.
   They agreed that he would pay £5.15 each week for a year.
   (a) How much did Raj have to pay altogether?
   (b) If the music centre actually cost £250, how much did Raj have to pay his father for borrowing the money?

28 A cuboid has a length of 7.3 cm, breadth 4.9 cm and height 6.1 cm.
   Find:
   (a) the area of each face
   (b) the total surface area
   (c) the volume of the cuboid.

29 A cube has one edge of length 3.5 cm.
   Find:
   (a) the surface area
   (b) the volume of the cube.

1.10 *Using your calculator*

Use your calculator, for all the questions on the next two pages, unless you are specifically told not to.

1 Find out which of the following is incorrect.
  (a) $713 + 519 = 1212$ (b) $123 \times 517 = 63\,591$
  (c) $571 - 326 = 255$ (d) $676 \div 52 = 13$

2 (a) Find the total bill for three articles costing £3.62, £5.97 and £1.45.
  (b) How much change would you have from a £20 note?

3 (a) Find the total bill for three pairs of jeans each costing £11.95.
  (b) How much change would you have from two £20 notes?

Calculators do not *always* do what you might expect them to do.
Work these questions out and then compare your results with a friend who has a *different* calculator.

4 (a) Find: $5 - 7$
  (b) Is the result displayed as:

  | -2 | or | $2-$ | or | $-$ | 2 |

  (c) Where does the minus sign appear on your calculator?

5 (a) Does $3 + 4 \times 5$ equal 35 or 23?
  (b) Is your calculator designed to carry out:
    (i) the operations in the order in which they are given
    (ii) the multiplication before the addition?

6 (a) Does $8 + 6 \div 2$ equal 7 or 11?
  (b) Explain how you might get each answer.

7 (a) Find the total bill for three articles costing £1.62, £2.53 and £3.45.
  (b) Is the correct result £7.60 or £7.06? What does your calculator show?

8 (a) Find the cost of printing four sets of films at £3.35 each.
  (b) Is the correct cost £13.40 or £13.04? What does your calculator show?

You must take care when reading answers from your calculator.
On many the minus sign is not directly to the left of the number and is easily missed.
If there is a zero at the end of the answer, as in questions **7** and **8**, it will not be shown on the calculator.
£7.60 and £13.40 will be shown as £7.6 and £13.4.

9 (a) Find: $2 \div 3$
  (b) Is the result displayed as:

  | 0.6666666 | or | 0.6666667 |

  (c) Does your calculator chop off the extra figures or round up the last figure?

Often it is quite unnecessary to use all the figures that the calculator displays. Three or four figure answers are usually quite sufficient. Both of the above results are 0.667 when corrected to three decimal places.

10 (a) Find: $1 \div 3 \times 3$
  (b) Is the result   1   or   0.9999999?

11 (a) Find and write down the answer for:
    $2 \div 3$
  (b) Now find:
    0.6666667×3 and $2 \div 3 \times 3$
  (c) Compare your answers for (b).

Your calculator may give any one of

| 2.0000001 | 2 | 1.9999998 | 1.9999997 |

as possible answers for question **11** (b).

12 (a) Without using your calculator, write down the correct answer for -2×-3.

(b) Now find the answer with your calculator.

(c) Is your calculator answer -5?

If your calculator answer for question **12** was -5 then when you press the keys:

$$\boxed{-} \quad \boxed{2} \quad \boxed{\times} \quad \boxed{-} \quad \boxed{3} \quad \boxed{=}$$

the calculator has ignored the $\times$ sign.

---

13 (a) Without using your calculator, write down the correct answer for $3^2 + 4^2$ or $3 \times 3 + 4 \times 4$.

(b) Now find the answer with your calculator.

(c) Is your calculator answer 52?

If your calculator answer for question **13** was 52 then when you press the keys:

$$\boxed{3} \quad \boxed{\times} \quad \boxed{3} \quad \boxed{+} \quad \boxed{4} \quad \boxed{\times} \quad \boxed{4} \quad \boxed{=}$$

the calculator has used them in the order you have pressed them.
3×3 is 9, $9 + 4$ is 13 and 13×4 is 52

---

14 (a) Without using your calculator, write down the correct answer for $\dfrac{8 + 7}{2 + 3}$.

(b) Now find the answer with your calculator.

(c) Is your calculator answer 10.5?

If your calculator answer for question **14** was 10.5 then when you press the keys:

$$\boxed{8} \quad \boxed{+} \quad \boxed{7} \quad \boxed{\div} \quad \boxed{2} \quad \boxed{+} \quad \boxed{3} \quad \boxed{=}$$

$8 + 7$ is 15, $15 \div 2$ is 7.5 and $7.5 + 3$ is 10.5

---

15 For question **13**, some calculators give 169 as the answer.
For question **14**, some calculators give 14.5. Can you find out why they might do this?

16 (a) Check and complete the working shown:

$$\begin{array}{r} 731 \\ - \ 137 \\ \hline 594 \end{array} \quad \text{reversing the digits}$$

$$\begin{array}{r} 594 \\ + \ 495 \\ \hline \end{array} \quad \text{reversing the digits}$$

(b) What is the final answer?

17 (a) Repeat question **16** for the number 654.

(b) Do you get the same result? Investigate.

18 (a) Check and complete the working shown:
$$731\,731 \div 7 = 104\,533$$
$$104\,533 \div 11 = 9503$$
$$9503 \div 13 = \quad ?$$

(b) What is the final answer?

(c) Now try dividing 654 654 first by 7 then by 11 and finally by 13.

(d) How is this connected with the original number?

(e) Does this always happen? Investigate.

19 (a) Find: (i) $7 \times 11 \times 13$ (ii) 1001×654

(b) How does this help you to explain what happens in question **18**?

20 (a) Find: (i) $12\,345\,679 \times 27$
(ii) $12\,345\,679 \times 36$

(b) What two numbers could you multiply together to get 22 222 222?

(c) Do you think that this will work for other multiples of 9? Investigate.

(d) What about $12\,345\,679 \times 81$?

21 (a) Find and write down the answers for:
(i) $100 \times 10\,000$
(ii) $100 \times 1\,000\,000$
(iii) $100 \times 10\,000\,000$
(iv) $10\,000 \times 1\,000\,000$

(b) Can you explain these answers?

(c) How do you think the answer would be shown for $1\,000\,000 \times 1\,000\,000$?

Scientific notation is often used to show very large numbers. $\boxed{1 \quad 08}$ means $1 \times 100\,000\,000$

1.11 *Estimation*

1 Andy and Sue went to the Reading pop festival. They wanted to give their friends an estimate of how many people were there.
   (a) Andy said that, to the nearest 10000, there were about 40000 present.
       What do you think he meant by this figure?
   (b) Sue said that, to the nearest 1000, there were about 43000 present.
       What do you think she meant by this figure?

In question **1** Andy's estimate of 40000 indicated that there were somewhere between 35000 and 45000 people at the pop festival.

Sue's estimate of 43000 indicated that there were between 42 500 and 43 500 people present.

---

2 The actual number of people present at the pop festival was 44 381.
   (a) Write this number correct to the nearest:
       (i) 10000 (ii) 1000 (iii) 100 (iv) 10
   (b) Were either Andy or Sue correct in their estimates?

When you write numbers correct to the nearest 1000 or 100 you must remember to **round up** if necessary.
   44 381 is 44000 correct to the nearest 1000
              (the next figure is a 3)
   44 381 is 44 400 correct to the nearest 100
              (the next figure is an 8)
If the next figure is *five or more* then you must **round up**.

---

3 There are 1625 pupils at Andy and Sue's school.
   (a) Write this figure correct to the nearest:
       (i) 1000 (ii) 100 (iii) 10
   (b) How many more pupils would be needed if the size of the school, correct to the nearest 100, was to reach 1700?

4 There are 17 528 people in Andy and Sue's home town.
   (a) Write this figure correct to the nearest:
       (i) 10 000 (ii) 1000 (iii) 100 (iv) 10
   (b) How many people could the town lose if its population, correct to the nearest 1000, were to remain at 18000?

5 (a) Write each number to the nearest 10:
       (i) 76 (ii) 52 (iii) 88 (iv) 276
       (v) 998
   (b) Write each number to the nearest 100:
       (i) 126 (ii) 588 (iii) 971 (iv) 5748

It is important, *before* using your calculator, to have an **estimate** of the likely result.
In this way you have a first check on your calculator answer.

Find: 73×57 Estimate: $70 \times 60 = 4200$
                    Calculator
                    answer: $73 \times 57 = 4307$
This will tell you whether your calculator answer is about the right size.

---

6 (a) By first writing each number correct to the nearest 10, find estimates for:
       (i) 76×52 (ii) 52×78 (iii) 47×31
       (iv) 19×69
   (b) Use your calculator to find the answers for each part of (a).
   (c) Now check each calculator answer with your estimate.

7 (a) By first writing each number correct to the nearest 100, find estimates for:
       (i) 126×432
       (ii) 588×777
       (iii) 982×527
   (b) Use your calculator to find the answers for each part of (a).
   (c) Now check each calculator answer with your estimate.

8 (a) By first writing each number correct to the nearest 100, find an estimate for:
   (i) $363 \div 121$ (ii) $936 \div 234$
   (iii) $615 \div 123$ (iv) $867 \div 289$
   (v) $792 \div 198$ (vi) $964 \div 241$
 (b) Use your calculator to find the answers for each part of (a), and then check them.

9 (a) By first writing each price correct to the nearest £1, find estimates for:
   (i) £2.95 + £5.87 + £7.23 + £2.17
   (ii) five packets of tights at £1.15 and two lipsticks at £1.85.
 (b) Use your calculator to find the exact answers for each part of (a).
 (c) Check each answer with your estimate.

10 (a) In each part first write the numbers correct to the nearest 10 and then find an estimate for the total number of:
   (i) eggs, in 48 trays each holding one dozen eggs
   (ii) bottles of milk, in 72 crates each holding 24 bottles
   (iii) tins, in 52 boxes of baked beans each containing 48 tins
   (iv) biscuits, in 68 packets each containing 32 biscuits.
 (b) Use your calculator to obtain an answer for each part of (a).
 (c) Check the accuracy of these answers by comparing them with your own estimates.

11 A shirt costs £7.99 and a matching tie costs £1.99.
 (a) Estimate the cost of buying a dozen shirts and ties.
 (b) Use your calculator to find the exact cost, and check this with your estimate.

12 A five litre tin of paint will cover approximately 27 square metres.
 (a) Estimate the number of square metres that can be covered with 64 tins of paint.
 (b) Use your calculator to find the exact area that can be covered, and check this with your estimate.

13 (a) If, in question **12**, you require to cover 1188 square metres, estimate the number of tins of paint that you need.
 (b) Use your calculator to find the exact number of tins that will be needed, and check this with your estimate.

14 (a) By first writing each number correct to the nearest 10, 100 or 1000, find estimates for:
   (i) $19 + 31 + 28 + 57 + 62$
   (ii) $623 + 786 - 216 - 498$
   (iii) $1930 + 8198 + 6299 + 4835$
   (iv) $4086 - 2911 + 9801 - 7348$
 (b) Use your calculator to find the exact answer for each part and then check them.

When doing calculations involving **brackets** it is *first* necessary to work out each bracket. An estimate of the answer is even more important here as a check on your calculator work.

Find: $(48 + 23) \times 98$ Estimate: $(50 + 20) \times 100$
$$= 70 \times 100$$
$$= 7000$$
Calculator answer: 6958

Find: $\dfrac{(487+387)}{(24+14)}$ Estimate: $\dfrac{(500+400)}{(20+10)} = \dfrac{900}{30} = 30$

Calculator answer: $\dfrac{874}{38} = 23$

15 Find estimates, and then use your calculator.
 (a) $49 \times 21 + 732$ (b) $(57 + 32) \times 68$
 (c) $77 \times (187 + 412)$ (d) $629 + 72 \times 19$

16 Find estimates, and then use your calculator.
 (a) $(72 + 48) \div 6$ (b) $(271 + 681) \div 4$
 (c) $(856 - 442) \div 23$ (d) $(936 + 862) \div 29$

17 Find estimates, and then use your calculator.
 (a) $637 \div (23 + 42)$ (b) $3218 \div (638 - 216)$
 (c) $\dfrac{(442 + 765)}{(11 + 16)}$ (d) $\dfrac{(761 + 967)}{(42 + 22)}$

18 Find estimates, and then use your calculator.
 (a) 51^2 (b) 29^3 (c) 19^4
 (d) 113^3 (e) 971^2

1.12 *Checking*

One way of checking your calculations is to make a rough estimate for the answer before you start.

A second way is to repeat the calculation on your calculator but in a **different** way.

(a) Find: $317 + 621$ Check: $621 + 317$

(b) Find: 37×53 Check: 53×37

1   Use your calculator to check that each of the alternatives shown above gives the same answer.

2   (a) Find: $61 + 47 + 25$ and
                 $25 + 47 + 61$

   (b) Do your two answers agree?

3   (a) Find: $17 \times 43 \times 62$ and
                 $62 \times 43 \times 17$

   (b) Do your two answers agree?

4   (a) Find: $567 + 732 + 961$ and
                 $961 + 732 + 567$

   (b) Find: 216×149 and
                 149×216

   (c) Do your two answers for each part agree?

5   (a) Find: $1721 + 3654 + 5972 + 4321$

   (b) Check your answer by starting the calculation at the right hand end.

6   (a) Find: $27 \times 33 \times 62 \times 41$

   (b) Check your answer by starting the calculation at the right hand end.

7   (a) Find: £1.75 + £2.34 + £4.67 + £3.89

   (b) Check your answer by doing the calculation in a different order.

8   The attendance figures over the last six weeks at the local youth club have been:
       17, 25, 14, 22, 19 and 24
   (a) Find the total attendance for this period.

   (b) Check your answers by adding these figures in the reverse order.

9   (a) What can you say about the two columns of figures below?

£	p		£	p
3	71		4	65
2	43		9	81
5	72		5	72
9	81		2	43
4	65		3	71

   (b) Find the total for each column of figures. Do your totals agree?

10   To find the volume of a cuboid you multiply the length by the width by the height.
   (a) Find the volume of a cuboid:
       (i)   17 cm by 12 cm by 9 cm
       (ii) 2.3 cm by 3.5 cm by 4.7 cm
       (iii) 1.75 cm by 2.25 cm by 4.80 cm.
   (b) How can you check your answers?

11   (a) Find the number of:
       (i)   seconds in twelve hours
       (ii) minutes in one week
       (iii) hours in five years.
   (b) Describe how you could check your answers in a different way.

12   The booking office for a pop concert at the NEC had 12 000 tickets available for sale. The numbers of tickets sold on the first three days of booking were:
       6521, 2653 and 1125
   (a) Find how many tickets were left.
   (b) Describe how you could check your answer in a different way.

13   A national supplier distributed 36 288 tins of paint equally amongst 21 wholesalers, each of whom supplied 12 shops.
   (a) How many tins did each shop receive?
   (b) Describe how you could check your answer in a different way.

Another way of checking your answers is to work backwards from the result to see if you get the original figures.

(a) Find: 672 − 345   Check: 327 + 345
                = 327   = ?

(b) Find: 901 ÷ 53   Check: 17 × 53
                = 17   = ?

14 (a) Carry out the check in each of the examples shown above.

  (b) Did you get back to the 672 and the 901?

15 (a) Find: 478 − 232

  (b) Now add 232 to your answer.

  (c) Is the result 478?

16 (a) Find: 1693 − 476

  (b) Check your answer as above.

17 (a) Find: 6993 ÷ 37

  (b) Now multiply your answer by 37.

  (c) Is the result 6993?

18 (a) Find: 4094 ÷ 46

  (b) Check your answer as above.

19 (a) Find: 531 + 222

  (b) Now subtract 222 from your answer to check.

20 (a) Find: 4736 + 2315

  (b) Now subtract 2315 from your answer.

  (c) How could you check your answer in a different way?

21 (a) Find: 82 × 57

  (b) Now divide your answer by 57 to check.

22 (a) Find: 716 × 491

  (b) Now divide your answer by 491.

  (c) How could you check your answer in a different way?

23 (a) Find: (i) $\dfrac{72 \times 54}{81 \times 6}$ (ii) $\dfrac{165 \times 423}{33 \times 45}$

  (b) Describe how you could find each of these answers in a different way.

  (c) How many different ways can you find?

The cost of buying three pairs of trainers at £7.99 each, and three pairs of jeans at £12.99 each can be found in two different ways as shown below.

(a) 3 × £7.99 + 3 × £12.99

(b) 3 × £(7.99 + 12.99)

24 (a) Using each of the above methods, find the cost of buying the three pairs of trainers and jeans.

  (b) Do your answers agree?

25 (a) Find the cost of buying five pairs of tights at £1.99, and five pairs of socks at £3.99.

  (b) Now use the other method to check your answer.

26 Jane says it is cheaper to buy 16 boxes of 48 Mars bars at £5.76 per box.
Bill reckons that it is cheaper to buy the 768 bars at an individual price of 12p.
Who is correct?

27 The population of a particular species of insect doubles itself every day for a week, before starting to die off.

  (a) If there are 213 insects at the beginning of the week, how many are there at the end of the week?

  (b) Describe how you could use your answer to check your working.

28 (a) Which of these calculations are incorrect?

  (i)   1172 + 7538 + 4695 = 14 305

  (ii) 265 + 47 × 13 = 876

  (iii) (61 + 27) × 36 = 1033

  (iv) $\dfrac{6678}{53 \times 42} = 5292$

  (b) Explain how each of the mistakes has happened, and give the correct answers.

29 (a) Explain how the mistake has been made in these calculations.

  (i) 23 × 34 = 1008

  (ii) 1324 × 56 = 69 104

  (b) What should the correct answers be?

1.13 *Approximations*

In many situations you are asked to give your answers **correct** to a specified number of **decimal places.**

7.3④	= 7.3	correct to **1** decimal place
7.3⑥	= 7.4	correct to **1** decimal place
7.3④6	= 7.3	correct to **1** decimal place
7.34⑥	= 7.35	correct to **2** decimal places

Note: If the next figure is 5 or above you must round the last figure **up**.

1 Write these numbers correct to 1 decimal place:
(a) 2.31 (b) 2.35 (c) 2.37 (d) 8.36 (e) 8.30

2 Write these correct to 1 decimal place:
(a) 2.312 (b) 2.354 (c) 2.377 (d) 8.348

3 Write these numbers correct to 2 decimal places:
(a) 5.239 (b) 5.364 (c) 5.407 (d) 7.097

4 Write these correct to 2 decimal places:
(a) 0.4623 (b) 0.4672 (c) 0.4648 (d) 0.0451

5 Write these correct to 3 decimal places:
(a) 0.4623 (b) 0.4672 (c) 0.4648 (d) 0.0497

6 (a) Use your calculator to find:
(i) 3.1×2.4 (ii) 9.7×4.8
(iii) 0.6×7.3 (iv) 16.5×14.9
(b) Now give your answers to 1 decimal place.

7 (a) Use your calculator to find:
(i) 1.6×2.31 (ii) 7.5×4.93
(iii) 6.8×1.44 (iv) 0.15×0.63
(b) Now give your answers to 2 decimal places.

8 (a) Use your calculator to find the area of a rectangle with sides of:
(i) 2.3 cm and 4.5 cm
(ii) 7.5 cm and 8.7 cm
(iii) 4.3 cm and 9.6 cm.
(b) Now give your answers to 1 decimal place.

Daley Thompson was timed at 46.253 seconds for the 400 metres.
This is:
46.3 seconds correct to 1 decimal place or
46.25 seconds correct to 2 decimal places.

9 A 100 metres run was timed at 9.915 seconds. Write this time correct to:
(a) 1 decimal place (b) 2 decimal places.

10 The diameter of a spindle is measured to be 0.7365 mm.
Write this measurement correct to:
(a) 1 decimal place (b) 2 decimal places
(c) 3 decimal places.

11 A penny has a depth of 1.009 mm. Write this depth correct to:
(a) 1 decimal place (b) 2 decimal places.

12 Twenty-five identical pennies, like the one in question **11** are stacked in a pile.
(a) Find the total height.
(b) Write this height correct to:
(i) 1 decimal place (ii) 2 decimal places.

13 A piece of dress material is 1.5 m by 2.25 m.
(a) Find the area of the material.
(b) Write this area correct to:
(i) 1 decimal place (ii) 2 decimal places.

14 The material in question **13** costs £4.89 per square metre.
(a) Find the total cost.
(b) Write this cost correct to:
(i) 1 decimal place (ii) 2 decimal places
(iii) 3 decimal places.
(c) Which of these is a sensible way of giving the cost?

15 A seven metre length of dowelling is to be divided into 24 equal lengths. What is the length of each piece correct to:
(a) 1 decimal place (b) 2 decimal places?

You may also be asked to give your answers correct to a specified number of **significant figures**.

$7③4 = 700$ correct to 1 significant figure
$73④ = 730$ correct to 2 significant figures
$73⑥ = 740$ correct to 2 significant figures
$0.06⑦ = 0.07$ correct to 1 significant figure
Note: The first significant figure is the **first non-zero figure**.

16 Write these numbers correct to 1 significant figure:
(a) 61   (b) 67   (c) 528
(d) 582   (e) 3726

17 Write these numbers correct to 1 significant figure:
(a) 2.19   (b) 3.91   (c) 0.17
(d) 0.053   (e) 0.035

18 Write these numbers correct to 2 significant figures:
(a) 762   (b) 726   (c) 3551
(d) 7073   (e) 7047

19 Write these numbers correct to 2 significant figures:
(a) 0.462   (b) 0.426
(c) 0.0935   (d) 0.0996

20 (a) Use your calculator to find:
　　(i) 67×49 (ii) 7.3×5.2
　　(iii) 116×235 (iv) 0.7×0.7
(b) Give your answers to 1 significant figure.

21 (a) Use your calculator to find the volume of a cuboid with sides of:
　　(i)   2.3 cm, 4.5 cm and 7.2 cm
　　(ii) 7.5 cm, 8.7 cm and 9.1 cm
　　(iii) 4.3 cm, 9.6 cm and 12.5 cm.
(b) Give your answers to 1 significant figure.

22 Find the total number of seconds in a year of 365 days. Give your answer to:
(a) 1 significant figure
(b) 2 significant figures
(c) 3 significant figures
(d) 4 significant figures.

46 235 people attended a football match.

This is
50 000 correct to 1 significant figure or

46 000 correct to 2 significant figures or

46 200 correct to 3 significant figures.

23 19 175 people visited Leeds Castle in 1986. Write this number correct to:
(a) 1 significant figure
(b) 2 significant figures
(c) 3 significant figures.

24 What difference would it have made to your answers for question **23** if the number visiting had been 19 715?

25 9617 people went to a pop festival. Write this number correct to:
(a) 1 significant figure
(b) 2 significant figures
(c) 3 significant figures.

26 What difference would it have made to your answers for question **25** if the number visiting had been 9671?

27 A salesman records the following mileages in the first three months of the year:
　　4653, 5217 and 4971
(a) Write these figures correct to:
　　(i)   1 significant figure
　　(ii) 2 significant figures
　　(iii) 3 significant figures.
(b) Find the total mileage done in the first three months.
　　Give your answer correct to 2 significant figures.

28 A cross channel ferry made 733 crossings last year, with an average of 375 vehicles on each.
(a) Find the total number of vehicles carried.
　　Give your answer correct to 3 significant figures.
(b) If the average charge per vehicle was £53 find the annual revenue.
　　Give your answer correct to 3 significant figures.

1.14 *Accuracy and sensible answers*

When you use your calculator you are often likely to finish with eight figures on the display.

You should be sensible when giving your answers and only include the number of figures which is reasonable.

£8.54 ÷ 3 = £2.8466667 on the calculator.

A sensible answer here would be £2.85 since we don't usually have tenths of pennies.

1   Anne, Gill and Tina treated themselves to a meal in a restaurant. They agreed to share the cost equally.
    If the total cost of the meal was £13.25, how much should they pay each?

2   James bought himself a packet of six biros for £1.60.
    How much did each biro cost?

3   Mr Dane bought a case of twelve tins of baked beans in the cash and carry store for £3.80.
    How much did each tin cost?

4   Anne buys her tights in packets of three which cost £1.49.
    How much does each pair cost?

5   A certain wine costs £2.20 a bottle. However, Mr Sharp discovered that if he bought a box of a dozen bottles he could save £1.
    By how much does this reduce the cost of each bottle?

6   Tina works as a waitress at the Tulip Cafe. She gets paid £10 for working $3\frac{1}{2}$ hours.
    What is the equivalent hourly rate?

7   Farmer Giles sells his corn-on-the-cob by the sack. Each sack contains 24 cobs and costs £5.
    How much is each cob?

8   It took James 3 minutes and 25 seconds to swim three lengths of his school pool.
    How long did each length take?

9   When times are given in athletics or in swimming, they are usually recorded to the nearest hundredth of a second.
    What is the average time taken for each 100m if:
    (a) the   200m took 20.21 seconds
    (b) the   400m took 46.25 seconds
    (c) the 1500m took 3 minutes 41.07 seconds?

10   The length of an olympic-sized swimming pool is 50m.
    What is the average time taken for each length if:
    (a) the   100m took 58.23 seconds
    (b) the   200m took 2 minutes 4.05 seconds
    (c) the   400m took 4 minutes 9.43 seconds
    (d) the 1500m took 17 minutes 2.58 seconds?

11   What difference would it make in question 10 if the length of the pool was only $33\frac{1}{3}$m?

Divide 3 metres of cloth into 7 equal pieces.

$\frac{3}{7}$ = 0.4285714 on the calculator.

A reasonable answer for this would be:
    0.43m   or   0.429m
i.e.   43cm   i.e.   429mm (42cm 9mm)

12   A length of wood is cut into three equal pieces.
    What is the length of each piece, if the length of the original piece was:
    (a) 4m   (b) 5m   (c) 2.45m
    (d) 3.65m   (e) 4.25m?

13   What would have been the original lengths of wood, in question 12, if the length of each piece was actually the size of your answers?

When a measured length is written as 32 cm, this is not an exact measurement.

It usually means that it is closer to 32 cm than to 31 cm or 33 cm.

The actual length will lie between 31.5 cm and 32.5 cm.

When a measured length is written as 32.7 cm, then the actual length lies between 32.65 cm and 32.75 cm.

---

14 A running track 'straight' is measured to be 100 metres, correct to the nearest metre. What could be the largest and smallest actual lengths?

15 If the measurements of the straight in question **14** had been given correct to the nearest tenth of a metre, what could be the largest and smallest actual lengths?

On a rather crude ruler each inch actually measures 2.5 cm instead of 2.54 cm.

If the ruler is twelve inches long, the actual length will be: 12×2.5 cm $= 30$ cm
  instead of: 12×2.54 cm $= 30.48$ cm

So this ruler will be 0.48 cm too short.

---

16 The height of each of the rises on a set of stairs is measured as 30 cm, correct to the nearest cm.
If there are 13 rises, what could be:
(a) the largest possible total rise
(b) the smallest possible total rise?

17 A full swing of the pendulum on a clock is timed as one second, correct to the nearest tenth of a second.
(a) For a full swing, what could be:
  (i) the largest possible time
  (ii) the smallest possible time?
(b) What is the greatest number of seconds by which the clock could be wrong after:
  (i) a minute (ii) an hour (iii) a day
  (iv) a 30 day month (v) a year?

Find the largest possible and smallest possible area of a rectangle when the sides are measured as 3.5 cm and 4.5 cm.

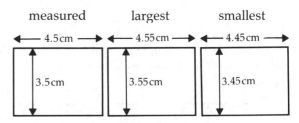

measured	largest	smallest
← 4.5 cm →	← 4.55 cm →	← 4.45 cm →
3.5 cm	3.55 cm	3.45 cm

area 15.75 cm^2 area 16.15 cm^2 area 15.35 cm^2
      (16.1525 cm^2) (15.3525 cm^2)
The actual area of the rectangle could be 0.4 cm^2 larger or smaller than the measured area.

---

18 Find the largest and smallest possible areas of a rectangle whose sides are measured as:
(a) 6 cm and 7 cm, correct to the nearest cm
(b) 6 cm and 7 cm, correct to the nearest mm
(c) 6 cm and 7 cm, correct to the nearest 0.1 mm.

19 (a) Write your answers to question **18** correct to the nearest:
  (i) 1 cm^2 (ii) 0.1 cm^2 (iii) 0.01 cm^2
(b) For each of your answers to (a) say how much too large or small the area could be.

20 The edges of a cuboid are measured as 2.3 cm, 3.6 cm and 4.9 cm, correct to the nearest mm.
Write down:
(a) the volume of a cuboid with these lengths as actual measurements
(b) the largest possible lengths of the three edges
(c) the largest possible volume of this cuboid, correct to 2 decimal places
(d) the smallest possible lengths of the three edges
(e) the smallest possible volume of this cuboid, correct to 2 decimal places
(f) the maximum possible error in the volume, correct to 2 decimal places, when using the lengths as measured.

1.15 *Squares and square roots*

Any number which can be shown as a pattern of dots forming a square is called a square number.

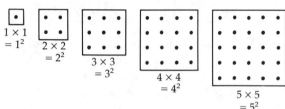

$1 \times 1 = 1^2$ $2 \times 2 = 2^2$ $3 \times 3 = 3^2$ $4 \times 4 = 4^2$ $5 \times 5 = 5^2$

1, 4, 9, 16 and 25 are **square numbers**.
Any whole number multiplied by itself will produce a square number:
$6 \times 6 = 36$ and $7 \times 7 = 49$.
7×7 can also be written as 7^2.
7^2 is read as *seven squared*.

1 Copy and complete:
   (a) $8^2 = 8 \times 8 =$
   (b) $9^2 = 9 \times 9 =$
   (c) $10^2 = 10 \times 10 =$

2 Look at these patterns.

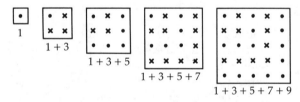

1   $1 + 3$ $1 + 3 + 5$ $1 + 3 + 5 + 7$ $1 + 3 + 5 + 7 + 9$

   (a) Draw the next pattern.
   (b) What is $1+3+5+7+9+11$?
   (c) What is the sum of the first:
        (i) 7 odd numbers   (ii) 10 odd numbers?

3 Write down the area of each large square.

(a)
2 cm

(b)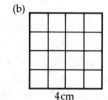
4 cm

(c)
3 cm

4 Find:
   (a) 12^2 (b) 25^2 (c) $(1.2)^2$ (d) $(2.5)^2$

5 Find the area of a square with side:
   (a) 12 cm (b) 25 cm (c) 1.2 cm (d) 2.5 cm

6 Use your calculator to find:
   (a) 17^2 (b) 179^2 (c) 1.7^2 (d) 17.9^2

7 Use your calculator to find the area of a square with side:
   (a) 29 cm (b) 89 cm (c) 2.9 cm (d) 8.9 cm

8 Find:
   (a) 10^2 (b) 100^2 (c) 1000^2 (d) $(0.1)^2$

From $7^2 = 49$ we can find 70^2 and $(0.7)^2$ etc.
$$70^2 = (7 \times 10)^2 = 49 \times 100 = 4900$$
and $(0.7)^2 = (7 \div 10)^2 = 49 \div 100 = 0.49$

9 $8^2 = 64$.
   Write down: (a) 80^2 (b) $(0.8)^2$

10 $9^2 = 81$.
   Write down: (a) 90^2 (b) $(0.9)^2$

11 $15^2 = 225$.
   Write down: (a) $(1.5)^2$ (b) 150^2

12 Copy and complete:
$$11^2 = \qquad 1.1^2 = \qquad 16^2 = \qquad 1.6^2 =$$
$$12^2 = \qquad 1.2^2 = \qquad 17^2 = \qquad 1.7^2 =$$
$$13^2 = \qquad 1.3^2 = \qquad 18^2 = \qquad 1.8^2 =$$
$$14^2 = \qquad 1.4^2 = \qquad 19^2 = \qquad 1.9^2 =$$
$$15^2 = \qquad 1.5^2 = \qquad 20^2 = \qquad 2.0^2 =$$

13 Find 16^2 and 23^2 and write down:
   (a) $(1.6)^2$ (b) $(2.3)^2$
   (c) 160^2 (d) 230^2

14 Find 31^2 and 57^2, and write down:
   (a) $(3.1)^2$ (b) 310^2
   (c) $(5.7)^2$ (d) 570^2

15 Look again at the squares in question **3**. The first square is made up of five squares (four small squares and one large square). How many squares can you find altogether in the second square?

16 Mrs Preedy wishes to buy a carpet for her lounge which is 4 metres by 4 metres.
    (a) How many square metres of carpet will she need?
    (b) If the carpet costs £5 a square metre, what will be the total cost?

17 Mr Price is laying a patio using concrete slabs which are two feet square.
If the patio is to be 30 feet square, how many slabs will he need?

18 Mr Waters is going to replace the tiles in his bathroom.
If the tiles are six inches square and the area to be covered is 25 square feet, how many tiles will he need?

19 The Freedom camping site wishes to provide pitches which are eight metres square, in a field which is 100 metres square.
Assuming that no allowances are made for roadways or facilities, what is the maximum number of pitches that can be provided on the Freedom camping site?

---

49 is the **square** of 7, since $7 \times 7 = 49$.
7 is called the **square root** of 49.
$8^2 = 64$, so 8 is the square root of 64.

The symbol $\sqrt{}$ is used for square root.
So $\sqrt{49} = 7$ and $\sqrt{64} = 8$.

---

20 What number when multiplied by itself is:
    (a) 25   (b) 36   (c) 81
    (d) 144 (e) 225?

21 Write down the square root of:
    (a) 25   (b) 36   (c) 81
    (d) 144 (e) 225

22 Write down:
    (a) $\sqrt{25}$ (b) $\sqrt{81}$ (c) $\sqrt{100}$
    (d) $\sqrt{121}$ (e) $\sqrt{169}$

23 Use your calculator to find:
    (a) $\sqrt{16}$ (b) $\sqrt{400}$ (c) $\sqrt{196}$
    (d) $\sqrt{256}$ (e) $\sqrt{324}$

24 Use your calculator to find:
    (a) $\sqrt{1.21}$ (b) $\sqrt{1.69}$ (c) $\sqrt{1.96}$
    (b) $\sqrt{2.56}$

From $\sqrt{9}\quad = 3$ we can find $\sqrt{900}$
$$\sqrt{900} = \sqrt{9 \times 100} = 3 \times 10 = 30$$
From $\quad\sqrt{169} = 13$ we can find $\sqrt{1.69}$.
$$\sqrt{1.69} = \sqrt{169 \div 100} = 13 \div 10 = 1.3$$

---

25 Find $\sqrt{289}$ and $\sqrt{40}$, and write down:
    (a) $\sqrt{2.89}$ (b) $\sqrt{28\,900}$ (c) $\sqrt{4000}$ (d) $\sqrt{0.4}$

The area of a square is $20\,\text{cm}^2$.
Find the length of its side.

Now $l^2 = 20$
    so $l = \sqrt{20} = 4.472\,136$
    so $l = 4.47$ (2 dec. pl.)
The length of the side is $4.47\,\text{cm}$.

---

26 Find the length of the side of the square.

(a) area $16\,\text{cm}^2$ (b) area $144\,\text{cm}^2$ (c) area $300\,\text{cm}^2$

27 Find the length of the side of a square of area:
    (a) $361\,\text{cm}^2$ (b) $4.84\,\text{cm}^2$ (c) $68.89\,\text{cm}^2$

28 (a) If $l^2 = 484$, what is l?
    (b) If $\pi r^2 = 628$ and π is 3.14, what is r?

29 12.25 square metres of carpet is advertised in the local paper.
What is the size of a square floor it would cover completely?

30 Mr Green has a square lawn of area 300 square metres.
    (a) What is the length of each edge of the lawn?
    (b) If he intends to put 1 metre edging stones all round his lawn, how many will he require?
    (c) A £7 bag of fertiliser covers 25 square metres of lawn. How much will it cost him to do the whole lawn?

1.16 *Directed numbers*

Temperature in centigrade

Temperatures above freezing are positive

Temperatures below freezing are negative

100°C — boiling
90°C
80°C
70°C
60°C
50°C — warm
40°C
30°C
20°C
10°C
0°C — freezing
−10°C
−20°C — below
−30°C — freezing
−40°C

1 By how much has the temperature risen?
(a) 0°C to 20°C   (b) 50°C to 100°C
(c) ⁻20°C to ⁻10°C   (d) ⁻10°C to 20°C

2 By how much has the temperature fallen?
(a) 100°C to 75°C   (b) 20°C to 3°C
(c) ⁻10°C to ⁻20°C   (d) 3°C to ⁻10°C

3 In which is the rise in temperature greatest?
(a) from 17°C to 35°C
(b) from 32°C to 71°C
(c) from ⁻37°C to ⁻15°C
(d) from ⁻17°C to 8°C

4 In which is the fall in temperature greatest?
(a) from 82°C to 56°C
(b) from 32°C to 7°C
(c) from 27°C to ⁻14°C
(d) from ⁻5°C to ⁻43°C

5 Early in the morning two climbers find that the temperature is ⁻11°C.
By mid-day it has risen by 35°C, but it falls again to ⁻17°C by the evening.
(a) What was the temperature at mid-day?
(b) By how much had the temperature fallen between mid-day and the evening?

6 A mine shaft is 310 feet above sea level. The shaft is 400 feet deep.
At what level is the bottom of the shaft?

7 In a laboratory a substance is heated by 45°C until its temperature reaches 23°C. It is then cooled until it reaches ⁻38°C.
(a) What was the original temperature of the substance?
(b) By how much did the temperature rise, or fall, altogether?

8 Copy and complete these tables:

(a)
$$2 + {}^-2 =$$
$$1 + {}^-2 =$$
$$0 + {}^-2 =$$
$${}^-1 + {}^-2 =$$
$${}^-2 + {}^-2 =$$

(b)
$$2 - 2 =$$
$$2 - 1 =$$
$$2 - 0 =$$
$$2 - {}^-1 =$$
$$2 - {}^-2 =$$

(c)
$${}^-2 - 2 =$$
$${}^-2 - 1 =$$
$${}^-2 - 0 =$$
$${}^-2 - {}^-1 =$$
$${}^-2 - {}^-2 =$$

You should know that adding ⁻2 has the same effect as subtracting 2:
$$7 + {}^-2 = 7 - 2 = 5$$
and that subtracting ⁻2 has the same effect as adding 2:
$$7 - {}^-2 = 7 + 2 = 9$$

9 Find:
(a) $5 - 9$ (b) $^-2 + 5$ (c) $^-7 + 3$
(d) $^-7 - 3$

10 Find:
(a) $5 + {}^-2$ (b) $4 + {}^-7$ (c) $^-2 + {}^-4$
(d) $^-8 + {}^-3$

11 Find:
(a) $5 - {}^-2$ (b) $3 - {}^-8$ (c) $^-5 - {}^-2$
(d) $^-3 - {}^-8$

12 Find:
(a) $^-2 + {}^-2 + {}^-2$ (b) $^-3 + {}^-3 + {}^-3 + {}^-3$
(c) $^-7 + {}^-7 + {}^-7$ (d) $^-5 + {}^-5 + {}^-5 + {}^-5$

13 Do you agree that:
(a) $3 \times (^-2) = {}^-6$ (b) $4 \times (^-3) = {}^-12$
(c) $3 \times (^-7) = {}^-21$ (d) $4 \times (^-5) = {}^-20$?

14 Find:
(a) $5 \times (^-3)$ (b) $3 \times (^-6)$ (c) $7 \times (^-8)$
(d) $(^-5) \times 3$ (e) $(^-3) \times 6$ (f) $(^-7) \times 8$

15 Find $(p - q + r)$ when:
(a) $p = 5, q = 7$ and $r = ^-2$
(b) $p = ^-4, q = ^-3$ and $r = 7$

16 (a) John has an overdraft at the bank of £50. If his mother gives him £75, how much does he now have in the bank?
(b) If the overdraft was three times as large, and his mother gave him £100, how much would he now have in the bank?
(c) How much more would he have in the bank in (a) than in (b)?

17 The temperature in a butcher's cold store is three times as far below freezing at the start of the day as it is at the end of the day.
(a) Write down the temperature at the start of the day, if the temperature at the end of the day is:
(i) $^-5°C$ (ii) $^-12°C$ (iii) $^-22°C$.
(b) Write down the temperature at the end of the day, if the temperature at the start of the day is:
(i) $^-30°C$ (ii) $^-48°C$ (iii) $^-24°C$.
(c) If the difference between the two temperatures is 22°C, what are they?

18 Copy and complete these tables:
(a) $2 \times 1 =$ (b) $2 \times ^-2 =$ (c) $^-3 \times 1 =$
$2 \times 0 =$ $1 \times ^-2 =$ $^-3 \times 0 =$
$2 \times ^-1 =$ $0 \times ^-2 =$ $^-3 \times ^-1 =$
$2 \times ^-2 =$ $^-1 \times ^-2 =$ $^-3 \times ^-2 =$
$2 \times ^-3 =$ $^-2 \times ^-2 =$ $^-3 \times ^-3 =$

You should know that when you **multiply** a *positive* and a *negative* number the result is *negative* $2 \times ^-3 = ^-6$
and that when you **multiply** two *negative numbers* the result is *positive* $^-3 \times ^-3 = 9$

---

19 Find:
(a) $8 \times (^-4)$ (b) $(^-5) \times 3$
(c) $(^-3) \times (^-6)$ (d) $(^-7) \times (^-8)$
(e) $(^-8) \times (^-5)$ (f) $(^-12) \times (^-7)$

20 Find $(p \times q) \times r$ when:
(a) $p = 5, q = ^-7$ and $r = 2$
(b) $p = ^-4, q = ^-3$ and $r = 7$
(c) $p = ^-2, q = 5$ and $r = ^-4$
(d) $p = ^-3, q = ^-2$ and $r = ^-4$

21 Copy and complete these tables:
(a) $4 \div 2 =$ (b) $8 \div 4 =$ (c) $^-9 \div 9 =$
$2 \div 2 =$ $8 \div 2 =$ $^-9 \div 3 =$
$0 \div 2 =$ $8 \div ^-2 =$ $^-9 \div ^-3 =$
$^-2 \div 2 =$ $8 \div ^-4 =$ $^-9 \div ^-9 =$
$^-4 \div 2 =$ $8 \div ^-8 =$ $^-9 \div ^-2 =$

You should know that when you **divide** a *positive* and a *negative* number the result is *negative* $8 \div ^-4 = ^-2$
and that when you **divide** two *negative numbers* the result is *positive* $^-9 \div ^-3 = 3$

---

22 Find:
(a) $(^-6) \div 2$ (b) $(^-8) \div 4$
(c) $8 \div (^-2)$ (d) $15 \div (^-3)$
(e) $(^-30) \div (^-6)$ (f) $(^-72) \div (^-6)$

23 Find $(p \div q) \times r$ when:
(a) $p = 8, q = ^-2$ and $r = 3$
(b) $p = ^-16, q = ^-4$ and $r = ^-7$.

24 Which of these statements are *not* true?
(a) $3 \times ^-4 = 4 \times ^-3$
(b) $^-4 \times 3 = ^-7$
(c) $^-5 \times ^-6 = ^-6 \times ^-5$
(d) $^-7 \times ^-6 = 42$
(e) $^-8 \div 4 = 8 \div ^-4$
(f) $^-6 \div ^-3 = 2$

25 Find:
(a) $^-29 + ^-21 - 15 - 32$
(b) $(^-17 + 23) \times ^-5$
(c) $(13 - 45) \div ^-8$
(d) $(2 - 7) \times (4 - 9)$

26 Find $(p - q) \div r$ when:
(a) $p = 17, q = 5$ and $r = ^-4$
(b) $p = 16, q = ^-5$ and $r = ^-7$
(c) $p = 6, q = ^-18$ and $r = ^-3$.

27 Find $(p \times q) + (r \div s)$ when:
(a) $p = 7, q = ^-5, r = ^-4$ and $s = 2$
(b) $p = ^-6, q = ^-3, r = ^-8$ and $s = ^-4$.

1.17 *Indices*

Product	Value	Index	In words
10×10	100	10^2	ten squared
$2 \times 2 \times 2$	8	2^3	two cubed
$5 \times 5 \times 5 \times 5$	625	5^4	five to the power four
$6 \times 6 \times 6 \times 6 \times 6$	7776	6^5	six to the power five

The $^2, ^3, ^4$ and ^{5} are called **indices** or **powers**.
So $2^4 = 2 \times 2 \times 2 \times 2 = 16$ and
  $5^3 = 5 \times 5 \times 5 = 125$

1 Copy and complete:
  (a) $2^5 = 2 \times 2 \times 2 \times 2 \times 2 =$
  (b) $3^4 = 3 \times$ $=$
  (c) $4^3 = 4 \times$ $=$
  (d) $5^2 = 5 \times$ $=$

2 Copy and complete:
  (a) $9 = 3 \times 3 =$ (b) $16 = 2 \times$ $=$
  (c) $27 = 3 \times$ $=$ (d) $125 = 5 \times$ $=$

3 Write using numbers:
  (a) three squared
  (b) four cubed
  (c) two to the power five
  (d) six to the power four.

4 Write using words:
  (a) 5^2 (b) 8^3 (c) 2^4 (d) 3^5 (e) 6^7

5 Find:
  (a) 2^5 (b) 3^4 (c) 4^3 (d) 5^4 (e) 6^3

6 Use your calculator to find:
  (a) 28^2 (b) 43^2 (c) 17^3 (d) 12^3 (e) 5^6

7 Find:
  (a) the area of a square of side 14.3 cm
  (b) the volume of a cube of edge 7.6 cm.

8 Find:
  (a) 10^2 (b) 10^3 (c) 10^4 (d) 100^2
  (e) 100^3

Look at the diagrams below. There are ten millimetres in a centimetre. (10 mm = 1 cm)

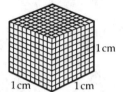

9 (a) How many square millimetres are there in a square centimetre (cm²)?
  (b) How many cubic millimetres are there in a cubic centimetre (cm³)?
  (c) Write each answer using an index.

10 There are one hundred centimetres in one metre. (100 cm = 1 m)
  (a) How many square centimetres are there in one square metre (1 m²)?
  (b) How many cubic centimetres are there in one cubic metre (1 m³)?
  (c) Write each answer using an index.

11 There are one thousand millimetres in one metre, and one thousand metres in one kilometre. (1000 mm = 1 m, 1000 m = 1 km)
  (a) How many square metres are there in one square kilometre (km²)?
  (b) How many square millimetres in:
    (i) 1 m² (ii) 1 km²?
  (c) How many cubic millimetres in:
    (i) 1 m³ (ii) 1 km³?
  (d) Write each answer using an index.

12 Copy and complete:
  (a) $10^2 \times 10^3 = (10 \times 10) \times (10 \times 10 \times 10) = 10^5$
  (b) $10^4 \times 10^3 =$ $=$
  (c) $10^2 \times 10^5 =$ $=$
  (d) Can you find a quick way of writing down the results in (a), (b) and (c)?

13 First write as a product, and then use indices:
  (a) $3^4 \times 3^5$ $= (3 \times 3 \times 3 \times 3) \times ($ $) =$
  (b) $7^2 \times 7^3 \times 7^4 = ($ $) \times ($ $) \times ($ $) =$

$2^3 \times 2^4 = (2 \times 2 \times 2) \times (2 \times 2 \times 2 \times 2) = 2^7$

$(3 + 4 = 7)$

$3^2 \times 3^5 = (3 \times 3) \times (3 \times 3 \times 3 \times 3 \times 3) = 3^7$

$(2 + 5 = 7)$

When **multiplying** you **add** the indices.

14 Which of the following are true?
(a) $2^2 \times 2^3 = 2^6$ (b) $3^2 \times 3^3 \times 3^4 = 3^9$
(c) $4^4 \times 4^1 = 4^5$ (d) $5^1 \times 5^2 \times 5^3 = 5^6$

15 Write using a single power:
(a) $5^4 \times 5^6$ (b) $4^2 \times 4^7$ (c) $6^2 \times 6^3 \times 6^5$
(d) $3^1 \times 3^8$ (e) $7^8 \times 7^8$ (f) $8^3 \times 8^5 \times 8^7$

16 Copy and complete:
(a) $10^3 \div 10^2 = \dfrac{(10 \times 10 \times 10)}{(10 \times 10)} =$

(b) $10^5 \div 10^3 = \qquad =$

(c) $10^4 \div 10^2 = \qquad =$

(d) Can you find a quick way of writing down the results in (a), (b) and (c)?

17 First write as a product, and then use an index to give your answer.
(a) $3^6 \div 3^4 = \dfrac{(3 \times 3 \times 3 \times 3 \times 3 \times 3)}{(3 \times 3 \times 3 \times 3)} =$

(b) $7^5 \div 7^3 = \dfrac{(\qquad\qquad)}{(\qquad\quad)} =$

$2^5 \div 2^3 = \dfrac{(2 \times 2 \times 2 \times 2 \times 2)}{(2 \times 2 \times 2)} = 2^2 \quad (5 - 3 = 2)$

$3^6 \div 3^4 = \dfrac{(3 \times 3 \times 3 \times 3 \times 3 \times 3)}{(3 \times 3 \times 3 \times 3)} = 3^2 \quad (6 - 4 = 2)$

When **dividing** you **subtract** the indices.

18 Which of the following are true?
(a) $2^5 \div 2^4 = 2^1$ (b) $3^4 \times 3^5 \div 3^2 = 3^7$
(c) $4^6 \div 4^2 = 4^3$ (d) $5^7 \times 5^2 \div 5^4 = 5^5$

19 Write using a single power:
(a) $5^4 \div 5^2$ (b) $4^5 \div 4^3$ (c) $6^7 \times 6^3 \div 6^4$
(d) $3^6 \div 3^2$ (e) $7^4 \div 7^3$ (f) $8^4 \times 8^5 \div 8^3$

20 (a) Find:
(i) $5^2 \div 5^2$ (ii) $3^4 \div 3^4$ (iii) $2^5 \div 2^5$
(b) What is:
(i) $25 \div 25$ (ii) $81 \div 81$ (iii) $32 \div 32$?
(c) Do you agree that:
$5^0 = 1, 3^0 = 1$ and $2^0 = 1$?

We can summarise the results as follows:
Multiplying: $2^3 \times 2^4 = 2^7$ (**add** indices)
Dividing: $2^5 \div 2^3 = 2^2$ (**subtract** indices)

From the second result it follows that:

$2^4 \div 2^4 = 2^0$ but $2^4 \div 2^4 = \dfrac{2^4}{2^4} = 1$ so $2^0 = 1$

Also:

$2^3 \div 2^4 = 2^{-1}$ but $2^3 \div 2^4 = \dfrac{2 \times 2 \times 2}{2 \times 2 \times 2 \times 2} = \dfrac{1}{2}$

so $2^{-1} = \dfrac{1}{2}$

$3^2 \div 3^4 = 3^{-2}$ but $3^2 \div 3^4 = \dfrac{3 \times 3}{3 \times 3 \times 3 \times 3} = \dfrac{1}{9}$

so $3^{-2} = \dfrac{1}{9}$

$2^0 = 1 \quad 2^{-1} = \dfrac{1}{2} \quad 2^{-3} = \dfrac{1}{8} = \dfrac{1}{2^3} \quad 2^{-5} = \dfrac{1}{32} = \dfrac{1}{2^5}$

21 Write out as above:
(a) $5^{-2} = \dfrac{1}{5 \times 5} = \dfrac{1}{-}$ (b) $7^{-2} = \dfrac{1}{7 \times} = \dfrac{1}{-}$
(c) $3^{-4} = \dfrac{1}{3 \times} = \dfrac{1}{-}$ (d) $2^{-3} = \dfrac{1}{2 \times} = \dfrac{1}{-}$

22 Which of the following are true?
(a) $3^{-2} = \dfrac{1}{9}$ (b) $2^{-4} = \dfrac{1}{8}$
(c) $2^{-4} = \dfrac{1}{16}$ (d) $5^{-2} = \dfrac{1}{25}$

23 Find:
(a) 2^{-6} (b) 3^{-5} (c) 2^{-7} (d) 3^{-4}
(e) 4^{-2} (f) 5^{-3}

24 Write using a single power:
(a) $2^4 \times 2^3$ (b) $3^4 \times 3^{-2}$ (c) $3^5 \times 3^{-1}$
(d) $5^{-2} \times 5^{-1}$ (e) $2^2 \times 2^5$ (f) $3^3 \times 3^{-4}$
(g) $4^{-3} \times 4^{-5}$ (h) $7^6 \times 7^{-6}$

25 (a) Copy and complete the table to show the powers of 2 and their differences.

Powers	2^0	2^1	2^2	2^3	2^4	2^5	2^6	2^7
of 2	1	2	4					
Differences		1	2					

(b) What do you notice about the differences?
(c) Repeat (a) and (b) for the powers of 3.
(d) What about powers of 4? Investigate.

1.18 *Problems and investigations*

1 Look, at the two multiplications below. Each uses the figures 1, 2, 3, 4, 5 and 6 once.

$123 \times 456 = 56\,088$	$1234 \times 56 = 69\,104$

(a) Can you make an answer larger than 69 104?

(b) Can you make an answer smaller than 56 088?

(c) What is the largest answer you can get?

(d) What is the smallest answer you can get?

(e) Investigate answers you can get using:

  (i) 1, 2, 3, 4 (ii) 1, 2, 3, 4, 5, 6, 7, 8.

2 In each of the calculations below the * stands for one of +, −, × or ÷. Which one?

(a) $16 * 1.2 = 19.2$

(b) $69 * 0.3 = 230$

(c) $132 * 246 * 571 = 949$

(d) $3 * 4.6 * 2.5 = 34.5$

(e) $1.1 * 1.1 * 1.1 * 1.1 = 1.4641$

3 Use your calculator to help you to find the missing numbers.

(a) $*** + 715 = 904$

(b) $8013 - **** = 3875$

(c) $47 \times ** = 2491$

(d) $3888 \div ** = 72$

(e) $8**3 - *41* = 7239$

(f) $93 \times 8* = 7**8$

(g) $83* \times *9 = 41\,013$

(h) $**6 \times 84* = 232\,668$

(i) $3** \times *7 = 14\,171$

(j) $3**4 \div 8* = 48$

4 Each letter stands for one of the figures 1, 2, 3, 4, 5, 6, 7, 8 and 9.

$$
\begin{array}{r}
  \text{B I G} \\
+ \text{C A T} \\
\hline
  \text{L E O}
\end{array}
$$

(a) Can you find a solution where each of the figures is used exactly once?

(b) Can you find more than one solution?

5 In each of the calculations below the * stands for one, or more, of +, −, × or ÷. Find the missing operations.

(a) $(37 * 21) * 223 = 1000$

(b) $(756 * 18) * 29 = 1218$

(c) $27 * (36 * 18) = 675$

(d) $619 * 316 * 425 * 196 = 924$

6 Find the missing numbers.

(a) $2 \times *** + 3 = 561$

(b) $3 \times *** - 2 = 1366$

(c) $19 \times (** + 75) = 2014$

(d) $27 \times (68 - **) = 1053$

(e) $***** \div 17 - 684 = 248$

(f) $(**** + 781) \div 42 = 84$

7 In the diagram below the number in the □ is the sum of the two numbers in the ○'s on either side of it. Find the missing numbers.

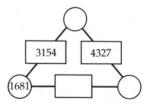

8 Find numbers to put in the ○'s so that their sum is the number in the □ in between them.

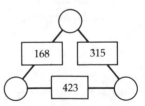

9 (a) What numbers can you make with two 4's using one of +, −, × and ÷?

(b) $4 - (4 \div 4) = 3$ and $(4 + 4) \times 4 = 32$

What other numbers can you make with three 4's using:

  (i) one of +, −, × or ÷

  (ii) two of +, −, × and ÷

(c) $(4 + 4) \div (4 + 4) = 1$,
$(4 \div 4) + (4 \div 4) = 2$
What other numbers between 1 and 100 can you make using four 4's? Can you make some in more than one way?

10 (a) Copy and complete the pattern below.

$$1 \qquad\qquad =$$
$$1 + 2 \qquad\qquad =$$
$$1 + 2 + 4 \qquad\qquad =$$
$$1 + 2 + 4 + 8 \qquad =$$
$$1 + 2 + 4 + 8 + 16 =$$

(b) What will be the next two results?

(c) How are the results connected to the powers of two?

11 In China there is an ancient puzzle known as 'the Towers of Hannoi'. In it you have three posts and a pile of discs.

You have to move the discs, one at a time, from one post to another, so that eventually they finish in the same order.
You may not at any stage place a larger disc on a smaller one.

(a) How many moves do you need to move a pile of three discs?

(b) Investigate the number of moves needed for different piles of discs.

12 Another popular puzzle concerns two teams of frogs sitting on stones. The two teams want to change places.

Each frog may move to the next stone if it is vacant, or jump over one frog from the other team onto a vacant stone. Frogs may not jump over members of their own team, or more than one of the other team.

(a) How many moves are needed for two teams of three frogs to change places?

(b) Investigate the number of moves needed for teams of other sizes.

13 (a) (i) Copy and complete the pattern below.

$$1 \qquad\qquad =$$
$$1 + 2 \qquad\qquad =$$
$$1 + 2 + 3 \qquad\qquad =$$
$$1 + 2 + 3 + 4 \qquad =$$
$$1 + 2 + 3 + 4 + 5 =$$

(ii) What will be the next two results?

(b) (i) Now copy and complete the pattern.
$1 \times 2 =$, $2 \times 3 =$, $3 \times 4 =$,
$4 \times 5 =$, $5 \times 6 =$

(ii) What will be the next two results?

(c) How are the results in (a) and (b) connected?

(d) Can you say what the 10th result will be in each part?

14 (a) Draw two straight lines. In how many points do they meet?

(b) Three straight lines are drawn below. In how many points do they meet?

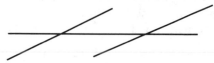

(c) (i) Can you draw three lines meeting in a different number of points?

(ii) What numbers of points could they meet in?

(iii) What is the greatest number of points they could meet in?

(d) Investigate the numbers of points that different numbers of lines could meet in.

15 (a) The three points on the circle below have been joined.
How many different regions has this divided the circle into?

(b) Copy and complete the table below.

Number of points		1	2	3	4	5
Greatest number of regions	1	2	4			
Differences			1	2		

(c) Investigate the number of regions for other numbers of points on the circle.

1.19 *Oral test*

1   What is the next number in the sequence?
    1, 4, 9, 16, 25

2   What is the next number in this sequence?
    1, 4, 7, 10, 13

3   What are the first three multiples of seven?

4   What are the first three multiples of twelve?

5   What are the factors of fifteen?

6   What are the factors of twelve?

7   Saleem sells thirteen shirts on Monday and twenty-five on Tuesday.
    How many does he sell over the two days?

8   Mr Ching had fifty-seven blank tapes in stock, but found that thirteen were faulty.
    How many good ones did he have left?

9   Winston found three boxes each containing twenty-one scout badges.
    How many badges were there altogether?

10   Olga made a tray of twenty-four rice cakes. She invited six friends to tea.
    How many cakes could each person have?

11   Olga's friends only ate one quarter of her twenty-four cakes.
    How many did she have left?

12   Olga's brother eats two-thirds of a tin of baked beans every day.
    How many tins will he finish in three weeks?

13   Pato completed half of his wood carving on Saturday and another one third on Sunday.
    What fraction did he still have left to complete?

14   Mrs Gooding gave half of a bag of potatoes to her daughter, who then gave a friend half of these.
    What fraction of the bag of potatoes did the friend have?

15   How many tenths are there in 1.6?

16   How many hundredths are there in 0.23?

17   What is 2.47 correct to one decimal place?

18   What is 2.748 correct to one decimal place?

19   What is 763 correct to one significant figure?

20   What is 196 correct to two significant figures?

21   The wholesaler supplies boxes of crisps which contain forty-eight large bags. Each of these contains nine small packets.
    Give an estimate for the total number of small packets in the box.

22   What is the square of seven?

23   What is the square of seventy?

24   What is the square root of 81?

25   What is the square root of 900?

26   By how much did the temperature rise from $^-8°C$ to 15°C?

27   By how much does the temperature fall from $^-5°C$ to $^-18°C$?

28   What is three cubed?

29   What is two to the power of four?

30   How would you write five times five using an index?

1.20 *Fact sheet I*

Types of numbers

Natural numbers are the counting numbers:
  1, 2, 3, 4, 5, 6, 7, . . .

The **integers** are directed numbers:
  $^-3, \, ^-2, \, ^-1, 0, 1, 2, 3, \ldots$

Rational numbers are common or decimal fractions:
  $\frac{1}{2}, \frac{3}{4}, \frac{5}{8}, 1\frac{1}{4}, \frac{25}{16}, \ldots$
or 0.5, 0.2, 0.75, 3.125, . . .

Irrational numbers are numbers which cannot be written as fractions:
  $\sqrt{2}, \sqrt{3}, \pi$, etc.

Common sets of numbers

Odd numbers:	1, 3, 5, 7, 9, 11, . . .
Even numbers:	2, 4, 6, 8, 10, 12, . . .
Multiples of 3:	3, 6, 9, 12, 15, 18, . . .
Prime numbers:	2, 3, 5, 7, 11, 13, . . .
Square numbers:	1, 4, 9, 16, 25, 36, . . .
Triangle numbers:	1, 3, 6, 10, 15, 21, . . .
Cube numbers:	1, 8, 27, 64, 125, . . .
Powers of two:	2, 4, 8, 16, 32, 64, . . .

Place value

The 3 in 2 ③ 78 means **three hundreds**
The 3 in 15.2 ③ means **three hundredths**

Decimal places

2.3461 is
  2.346 correct to **three** decimal places
  2.35 correct to **two** decimal places
  2.3   correct to **one** decimal place

Significant figures

1674 is
  1670 to **three** significant figures
  1700 to **two** significant figures
  2000 to **one** significant figure

Squares and square roots

$4^2 = 4 \times 4 = 16$
$\sqrt{16} = 4$
$\sqrt{1600} = 40$
$7^2 = 7 \times 7 = 49$
$\sqrt{49} = 7$
$\sqrt{0.49} = 0.7$

Directed numbers

$^-5 + {}^-7 = {}^-12$
$^-5 - {}^-7 = {}^+2$
$^-5 \times 7 = {}^-35$
$^-5 \times {}^-7 = {}^+35$
$\dfrac{16}{^-2} = {}^-8$
$\dfrac{^-8}{^-2} = {}^+4$

Indices

$3^2 = 3 \times 3$
$5^3 = 5 \times 5 \times 5$
$3^2 \times 3^3 = (3 \times 3) \times (3 \times 3 \times 3) = 3^5$
$\dfrac{7^5}{7^3} = \dfrac{7 \times 7 \times 7 \times 7 \times 7}{7 \times 7 \times 7} = 7 \times 7 = 7^2$

$2^{-3} = \dfrac{1}{2^3}$

$10^4 = 10\,000$
$10^{-3} = 0.001$
$3610 = 3.61 \times 10^3$
$0.0361 = 3.61 \times 10^{-2}$

SECTION 2 Everyday arithmetic

2.1 *Time and timetables*

The two clocks below both show the same time.

22.15

10.15 p.m.

1 If the time showing on the clock on the left was 10.15 a.m., what time would the clock on the right show?

2 Write these times using the 24-hour clock.
(a) 11.35 a.m. (b) 7.35 a.m.
(c) 7.30 a.m. (d) 2.35 p.m.
(e) 8.30 p.m. (f) 11.30 p.m.

3 Write these times using the 12-hour clock.
(a) 10 35 (b) 05 31
(c) 09 00 (d) 00 41
(e) 15 35 (f) 12 01
(g) 20 30 (h) 23 59

4 Write down the time which is 37 minutes before the given time.
(a) 11.40 a.m. (b) 5.15 p.m.
(c) 1.07 p.m. (d) 10 43
(e) 09 05 (f) 00 25

5 Write down the time which is 1 hour 35 minutes after the given time.
(a) 10.40 a.m. (b) 5.55 p.m.
(c) 11.07 p.m. (d) 10 13
(e) 09 28 (f) 22 35

6 How much time has elapsed between:
(a) 10.15 a.m. and 11.45 a.m.
(b) 11.53 a.m. and 1.15 p.m.
(c) 3.37 p.m. and 1.15 a.m?

7 How long is it between:
(a) 11 25 this morning and 20 15 this evening
(b) 21 53 today and 01 47 tomorrow?

8 John leaves for school at 8.15 a.m. and arrives 37 minutes later.
(a) What time does he get to school?
(b) If John arrives at school seven minutes after Winston, at what time did Winston arrive?

9 Jane leaves school at 16 45 and arrives home at 17 23.
(a) How long does her journey take?
(b) If Jane then has to wait 40 minutes for her tea when does she have it?

10 Look at these bus times.

Coventry	depart 09 45
Banbury	arrive 10 23
Banbury	depart 10 40
Oxford	arrive 11 17

(a) How long was the journey from:
(i) Coventry to Banbury
(ii) Banbury to Oxford?
(b) How long did the bus wait in Banbury?

11 The bus in question **10** leaves Oxford on its return journey to Coventry at mid-day. It takes the same total length of time for the journey.
When does the bus arrive back in Coventry?

12 Look at the cinema programme.
(a) How long does the main film last?
(b) How long does the full programme last?
(c) How long is it from the end of the cartoon to the beginning of the next showing of the cartoon?

Cartoon	17 15
Intermission	17 45
Main film	18 05
Intermission	19 40
Cartoon	20 10
Intermission	20 40
Main film	21 00
Doors close	22 45

13 The timetable given below shows the programme for the school industrial conference day.

Morning		Afternoon	
Registration	9.15	Plenary session	1.50
Plenary session	9.30	Working groups	2.35
Coffee	10.45	Tea	3.10
Working groups	11.05	Plenary session	3.25
Lunch	12.30	Depart	3.45

(a) What was the length of the industrial conference day?

(b) How long was allowed for the lunch break?

(c) How much time was allocated to:
(i) plenary sessions
(ii) working groups?

(d) How much time was there between the end of the plenary session in the morning and the beginning of the second plenary session in the afternoon?

14 Part of a bus timetable is shown in the figure below.

Service	X20	S20	217	X50	X20
Henley	07 50	08 20	09 25	09 49	10 49
Wootton	07 54	08 25	09 30	09 54	10 54
Bearley(PO)	–	–	09 35	–	–
Bearley(ST)	07 58	–	09 37	09 57	10 57
Stratford	08 10	08 55	09 48	10 10	11 10

(a) Which service would you catch to go from Henley to Bearley(PO)?

(b) Which service can't you catch to go from Henley to Bearley(ST)?

(c) Which service takes the least time to get from Henley to Stratford?

(d) Which service takes the most time to get from Wootton to Stratford?

(e) If you miss the first X20 service from Henley, how long do you have to wait for the next X20?

(f) If Jo catches the S20 to Stratford, how long will she have to wait for Anne who comes on the X50?

15 Part of a train timetable is shown below.

		SX	FO	SO	
Bradford	06 04	06 43	14 24	19 03	20 44
Harrogate		06 30	14 20	18 50	21 13
Leeds	06 32	07 15	15 05	18 16	20 17
Wakefield	06 47	07 31	15 21	17 58	19 58
London	08 57	09 30	17 38	15 50	17 55

(a) What do you think the codes SX, FO and SO tell you about the trains?

(b) How long does the 06 04 from Bradford take to get to London?

(c) Where does the train, which arrives at Leeds at 18 16, start from?

(d) How long does the 17 55 from London take to get to Bradford?

(e) Which train takes the longest time between Leeds and Wakefield? How long?

16 Jason lives in Leeds and wants to spend a day in London on a Saturday. He has to come back on the early train.
(a) How long can he spend in London?
(b) How much time will have elapsed before he arrives back in Leeds?

17 Sally lives in Harrogate. She has to collect her birthday present from a friend in London on a Friday.
(a) What is the latest time she can leave?
(b) How long will she have at the station in London?
(c) What time will she get back home to Harrogate?

18 A charter plane flies to Munich, leaving Manchester at 7.50 a.m. The plane spends one hour at Munich airport loading and unloading and gets back to Manchester at 12.50 p.m.
(a) If the return flight to Manchester takes the same time as the outward flight:
(i) how long is each flight
(ii) what time does the plane land at Munich?
(b) If the return flight takes twenty minutes longer, how long is each flight?

2.2 *Reading dials, charts and tables*

1 A modern gas meter shows how many cubic feet have been used on a simple display like the one below. The last figure is ignored.

Cubic feet	1001	2

(a) How many cubic feet of gas have been used altogether?

(b) If gas costs 35p per cubic foot, how much would the gas have cost?

2 The diagram below shows the dials on an old electricity meter.

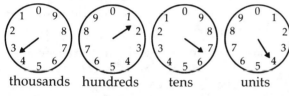

thousands   hundreds   tens   units

(a) What is the reading on the meter?

(b) If the reading last time was 2434, how many units of electricity have been used since then?

(c) Show the meter when another 1134 units have been used.

3 When you use Economy 7 electricity you have a new double meter like the one below. The last figure on each display is ignored.

kWh	Low	0112	7

kWh	Normal	0299	2

(a) How many kWh are shown at the low rate?

(b) How much more electricity has been used at the normal tariff than at the low rate?

(c) If electricity cost 5.7p per kWh at the normal tariff and 1.9p at the low rate, what is the total cost of the electricity shown on the meter?

4 The table shows the cost of a day excursion. What is the cost for:

(a) a 12-year-old

(b) two adults and two children of 6 and 2

(c) three adults and three children of 9?

Day excursion	Rates
Adults	£2.50
Children (11–16)	£1.50
Children (7–11)	£0.75
Children (3–7)	£0.50
Children under 3	free
OAPs	£1.50

5 The cost of seats at the Royal Shakespeare Swan Theatre is shown in the table below.

Price zone	Low (£)	Mid (£)	Peak (£)
A	11.00	12.00	13.00
B	8.50	9.50	10.50
C	6.50	7.50	8.50
D	3.00	3.00	3.00

(a) How much do seats cost in the B price zone at peak times?

(b) What is the cheapest C price zone ticket?

(c) How much will five seats in the B price zone cost in mid season?

(d) A school takes a party of 40 at the cheapest time. They have 10 tickets in each of zones A and B and 20 tickets in zone C.
What will be the total cost for the party?

6 Alec needs a new 155 × 13 tyre for his car. He sees this advertisement in the paper. What will be the cost of the tyre?

145 × 10	13.75 + 2.07 VAT
145 × 13	15.95 + 2.39 VAT
155 × 13	16.95 + 2.54 VAT
165 × 13	18.95 + 2.84 VAT

7   The charges per night at the camping site at Croyde Bay are shown in the table below.

Dates	Caravans	Tents	
22/5 – 29/5	£7.00 (up to 6)	£4.00	(1–2)
		£5.00	(3–4)
		£7.00	(5–6)
30/5 – 16/7	£5.50 (up to 6)	£3.00	(1–3)
		+ 50p for each extra person	
17/7 – 28/8	£8.50 (up to 6)	£4.00	(1–2)
		£5.00	(3–4)
		£7.00	(5–6)

(a) How much will it cost a family of four to spend a week at Croyde Bay in a caravan during June?

(b) How much will it cost James and David to spend a week in a tent during August?

(c) How much will it cost a family of six to have a two week holiday in a caravan starting on 22 May?

(d) How much will it cost a family of five to spend six days in a tent in June?

8   The chart below shows the distances by road, between towns in England, in both miles and kilometres.

kilometres

London	185	114	277	304	92	311
115	**Bristol**	299	122	312	119	340
71	186	**Dover**	399	418	206	425
172	76	248	**Exeter**	435	229	462
189	194	260	270	**Leeds**	270	39
57	74	128	142	168	**Oxford**	291
193	211	264	287	24	181	**York**

miles

(a) How far is it from London to York:

   (i) in miles (ii) in kilometres?

(b) How far is it from Bristol to Leeds:

   (i) in miles (ii) in kilometres?

(c) How far is it, in miles, from Leeds to Bristol:

   (i) via London (ii) via Oxford?

9   The table below shows the number of calories in various foods.

Food	Quantity	Calories
Apple	1 large	115
Orange	1 small	50
Bread	1 slice	130
Cheese	30 g	105
Butter	1 pat	50
Milk	1 glass	165

(a) Mandy has a large apple, 30 g of cheese and a glass of milk for her lunch. How many calories is this?

(b) Jo has three slices of bread, 60 g of cheese, a pat of butter and a small orange. How many more calories is this?

10 David and his parents want to go skiing. They have a brochure for La Plagne in France. This is shown below.

Dates	Apartment		Studio		Hotel(hb)	
	7	14	7	14	7	14
Dec 20	185	295	175	279	249	353
Dec 27	211	294	200	273	279	399
Jan 3	187	282	180	261	218	329
Jan 10	159	267	149	247	199	319

(a) David wants to go for the two weeks immediately after Christmas and fancies staying in a studio. How much will it cost each of them?

(b) David's father says he can only get time off work for one week from 27 December and would prefer an apartment. How much will it cost the three of them?

(c) David's mother wants meals provided. What will be the cost for the family of the cheapest hotel week?

(d) In the end David and his mother decide to have two weeks in a studio, one of which includes 27 December and David's father joins them for just one week. What is the cheapest way of arranging this package?

2.3 Changing units of measure

In the metric system the same sort of words are used for all the measurements.

kilo means 1000　　so 1 **kilometre** = 1000 metres
centi means $\frac{1}{100}$　　so 1 **centilitre** = $\frac{1}{100}$ litre
milli means $\frac{1}{1000}$　　so 1 **milligram** = $\frac{1}{1000}$ gram

1　John's ruler is 30 cm long.
　How long is the ruler in:
　(a) millimetres　(b) metres?

2　Mrs Ashad bought a 250 g packet of butter.
　What is the mass of the butter, in kilograms?

3　Andy's medicine came in a 20 cl bottle.
　What is its capacity in:
　(a) litres　(b) millilitres?

4　It is 250 metres from Saleem's house to the Post Office.
　How far is this in:
　(a) kilometres　(b) centimetres?

5　Saleem's airmail letter weighed 800 milligrams.
　What is this in:
　(a) grams　(b) kilograms?

6　(a) What is the perimeter of the rectangle, in centimetres?

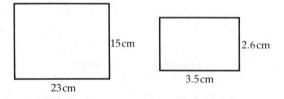

　(b) Write each of your answers in:
　　(i) millimetres　(ii) metres.

7　(a) Find the total capacity in litres of three jugs which hold:
　　1.5 l, 0.75 l and 0.33 l.
　(b) Write your answer in:
　　(i) centilitres　(ii) millilitres.

8　Mrs Green found four part-used bags of sugar which contained:
　　250 g, 178 g, 475 g and 748 g.
　(a) How much sugar was there altogether?
　(b) Write your answer in:
　　(i) kilograms
　　(ii) milligrams.

---

1 **square centimetre** is a square 10 mm by 10 mm.
　So 1 cm² = 100 mm²
1 **square metre** is a square 100 cm by 100 cm.
　So 1 m² = 10 000 cm²

9　(a) What is the area of the rectangle, in square centimetres?

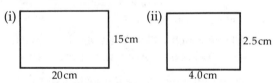

　(b) Write each of your answers in:
　　(i) square millimetres
　　(ii) square metres.

---

1 **cubic centimetre** is a cube 10 mm by 10 mm by 10 mm.
　So 1 cm³ = 1000 mm³
1 **cubic metre** is a cube 100 cm by 100 cm by 100 cm.
　So 1 m³ = 1 000 000 cm³

10　(a) What is the volume of the cuboid, in cm³?

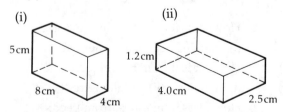

　(b) Write each of your answers in:
　　(i) cubic millimetres　(ii) cubic metres.

1 inch is approximately 2.54 cm.
_____

5 cm is approximately 1.97 inches.
_____

11 (a) John's 30 cm ruler also has inches marked on it.
What length, in inches, is 30 cm?

(b) Sally has a 12 inch ruler with centimetres marked on it.
What length, in centimetres, is 12 inches?

12 (a) Ian has a metre rule.
What length, in inches, is this?

(b) Jane's father has an old 36 inch rule.
What length, in centimetres, is 36 inches?

_____

8 kilometres is approximately 5 miles.
So 1 km is approximately 0.625 miles
1 mile is approximately 1.6 km.
_____

13 When Jason's father was in Cyprus the sign posts were marked in both kilometres and miles.

(a) How far in miles is:

(i) 16 km (ii) 72 km (iii) 120 km?

(b) How far in km is:

(i) 25 miles (ii) 60 miles (iii) 95 miles?

_____

50 litres is approximately 11 gallons.
1 litre is approximately 1¾ pints.
_____

14 In each of the following give your answer correct to 2 decimal places.

(a) How many gallons are there in one litre?

(b) How many litres are there in one gallon?

(c) How much more than 1 gallon is five litres?

(d) Would you have more petrol in a 50 gallon drum or in a 225 litre tank?

(e) How many pints of lager will you get from a 50 litre barrel?

(f) How many litres are there in eight pints?

1 kilogram is approximately 2.2 pounds (lb).
There are 16 ounces in 1 pound (lb).
_____

15 Jackie has kitchen scales which only show weights in kilograms. She has a recipe book which only shows quantities in pounds and ounces.
She decides to draw a graph to convert the weights from kilograms to pounds.

(a) Copy and complete the table:

Kilograms	0.5	1	1.5	2	2.5	3
Pounds	1.1	2.2				

(b) Use the values in your table to copy and complete the graph below:

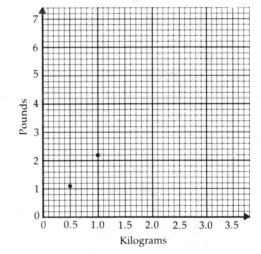

Use your completed graph to find:

(c) the number of pounds (to 0.1 lb) in:

(i) 0.75 kg (ii) 1.25 kg (iii) 2.25 kg

(d) the number of kilograms (to 0.1 kg) in:

(i) 1 lb (ii) 4 lb (iii) 2.5 lb

16 Draw an enlarged version of the bottom left-hand corner of your graph in question **15**.
This should show kilograms and tenths of kilograms up to 1 kilogram on one axis, and pounds and ounces up to 2 lb on the other.
Use this graph to find the number of:

(a) ounces in 0.2 kg

(b) pounds and ounces in 0.6 kg

(c) kg in 1 lb 4 oz.

2.4 *Measures in the home*

Mass   A bag of sugar has a mass of about 1 kilogram.

$1000\,g = 1\,kg$

A small packet of sweets has a mass of about 100 grams.

---

1 Janine is weighing out the ingredients for a recipe. She needs 200 g of butter, 450 g of flour and 550 g of raisins.
What is the mass of her ingredients:
(a) in grams   (b) in kilograms?

2 The cost of first class post is 18p for letters up to 60 g, 26p for letters between 60 g and 100 g.
What is the cost of posting:
(a) one letter of 75 g
(b) two letters of 55 g
(c) one letter of 9 g and two letters of 80 g?

3 Sam's father orders twelve bags of coal each weighing 25 kg.
(a) What is the total weight ordered?
(b) When the coal is delivered he discovers that one bag is half empty.
What is the weight of this bag?
(c) If the coal cost £4.50 a bag:
(i) how much was the bill
(ii) how much should he pay?

Length

The length of your stride is about 1 metre.

$100 \text{ cm} = 1 \text{ m}$

The width of your little finger is about 1 centimetre.

$10\,mm = 1\,cm$

The height of ten sheets of paper is about 1 millimetre.

---

4 The length of the walls of Jason's room are:
320 cm, 323 cm, 324 cm and 325 cm.
What is the total length of the four walls:
(a) in centimetres   (b) in metres?

5 Ian is building a model aeroplane.
From a 45 cm length of balsa wood he cuts smaller lengths of:
7.6 cm, 4.2 cm and 5.8 cm.
(a) What total length does he cut off?
(b) How much is left of the original piece?
(c) Now write both your answers in mm.

6 Pato is helping his father lay edging stones around their lawn.

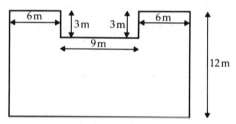

A sketch of the lawn is shown above.
(a) What is the perimeter of the lawn?
(b) If each edging stone is 1.5 m long, how many will they need altogether?

(c) How many more will they need if there is also a square flower bed of side 2.25 m in the centre of the lawn?

Area   The area of a football pitch is about half a hectare.

$1\,ha = 100\,m \times 100\,m$

The area of your front door is about 1.5 square metres.

$1\,m^2 = 100\,cm \times 100\,cm$

The area of this page is about 500 square centimetres.

---

7 Find the area of the lawn shown in the sketch in question **6**, giving your answer in:
(a) square metres
(b) hectares.

8 (a) A standard door measures 190 cm by 80 cm.
        Find the area of the door in:
        (i) cm² (ii) m².

   (b) Measure the height and width of your classroom door correct to the nearest cm.

   (c) Find its area in:
        (i) cm² (ii) m².

9 (a) A folded newspaper is 30 cm by 44 cm. Find its area when unfolded in:
        (i) cm² (ii) m².

   (b) When decorating Jason's father covers the floor with newspaper.
        How many unfolded newspapers will he need to cover the floor of a room 6 metres long and 2.2 metres wide?

Volume The size of a telephone box is about 2 cubic metres.

$1 m^3 = 100 cm \times 100 cm \times 100 cm$
$ = 1\,000\,000 cm^3$

A small sugar cube is about 1 cubic centimetre.

$1 cm^3 = 10 mm \times 10 mm \times 10 mm$
$ = 1000 mm^3$

---

10 A telephone box is 1.2 m wide, 0.8 m deep and 2.3 m tall.
    Find the volume of air in this box.
    Give your answer in cubic metres.

11 Jason's bedroom is 300 cm wide, 320 cm long and 250 cm high.
    Find the volume of air in this room, giving your answers in:
    (a) cubic centimetres (b) cubic metres.

12 (a) Measure the length, width and height of your classroom correct to the nearest cm.

    (b) Find the volume of air in the room. Give your answer in m³, correct to three significant figures.

    (c) According to the health regulations every person requires 5 m³ of air space. How many people will this allow the room to hold?

13 (a) Measure the length, width and depth of this book correct to the nearest mm.

    (b) Find its volume in:
        (i) mm³ (ii) cm³.

14 Fiona is about to move house. She has packed a lot of her things into cake tins which measure 22 cm by 22 cm by 10 cm. How many of these tins can she pack into a cardboard box which measures 45 cm by 45 cm by 30 cm?

Capacity A large bottle of orange squash holds about 1 litre.

$1 l = 10 cm \times 10 cm \times 10 cm$
$ = 1000 cm^3$

A small teaspoon holds about 5 millilitres.

$1000 ml = 1 l$

A pint of milk is just over 0.5 l.

---

15 Andy has collected a 0.25 l bottle of cough mixture from the chemist.
    (a) How many 5 ml doses will this give?
    (b) If he has to take 5 ml six times a day, how many days will the medicine last him?

16 You can get six full glasses of wine out of a 0.75 l bottle.
    What fraction of a litre does each glass hold?

17 You can get thirty-two tots of spirits out of a 0.66 l bottle.
    Approximately how many millilitres is each tot?

18 Jenny has collected a number of bottles of different sizes. They hold:
    1.5 l, 1 l, 0.75 l, 0.67 l, 0.5 l and 0.33 l.
    (a) What is the total capacity of these six bottles?

    (b) In how many different ways could you store 1.5 l of vinegar?

    (c) By using more than one bottle, what different amounts of home-made wine could you store?

    (d) How could you obtain exactly 0.25 l of water by using these bottles?

    (e) What other quantities could you obtain?

2.5 *Bills and money calculations*

1   John bought a bar of chocolate for 42p.
How much change did he get from £1?

2   Sally bought note paper costing 75p and
envelopes costing 57p.
(a) How much did she spend?
(b) What change did she get from £5?

3   Ian and Gerry spend a wet Saturday
afternoon in the Clara's Coffee Club. They
had two coffees costing 35p each, two Cokes
costing 36p each and two cakes costing 27p
each.
(a) What was their total bill?
(b) How much change did they get from £5?

4   Raj took his two younger brothers into town
on the bus. The fare for each of them was
45p.
How much did he have to pay altogether?

5   Pato bought four bars of chocolate each
costing 37p.
How much did they cost him altogether?

6   Jenny bought seven Christmas cards which
each cost 12p and seven stamps for them.
How much did she have to pay altogether if
she then sent these cards by:
(a) first class post
(b) second class post?

7   How many 13p stamps can you buy for 65p?

8   How many 18p stamps can you buy for £3?
How much change will you get?

9   If Sue spends 80p on stamps how many 13p
and how many 18p stamps does she buy
altogether?

10   Mrs Green's family likes baked beans which
cost 32p a tin.
(a) How much will five tins cost?
(b) How many tins can she buy for £3.84?
(c) How many tins can she buy for £5 and
how much change will she get?

11   John, Alison, David and Sue decided to go
out to a restaurant for a celebration. John's
meal cost £4.95, Alison's cost £3.70, David's
cost £4.35 and Sue's cost £5.20.
(a) What did the total bill come to?
(b) If John paid the bill with a £20 note, how
much change did he get?
(c) If they decided to share the cost equally,
how much did each have to pay?

12   The menu in the Cosy Corner Cafe is:

Bacon, egg and chips	£1.25
Sausage, beans and chips	£1.10
Mixed grill	£1.75
Toast and butter	£0.40
Pot of tea	£0.60
Cup of coffee	£0.35

What is the cost of:
(a) a mixed grill and a pot of tea
(b) bacon, egg and chips, toast and butter
and a cup of coffee
(c) two sausage, beans and chips, one bacon,
egg and chips, two toast and butters, two
cups of coffee and a pot of tea?

13   Copy and complete Mrs Bell's greengrocery bill
shown below.

5 lb potatoes at 7p per lb.
3 lb carrots at 12p per lb.
1 lb sprouts at 15p per lb.

```
BILL
Potatoes
Carrots
Sprouts

————
————
```

14   The cost of tickets in the Regal cinema are:
adult £2.15, child £1.25, OAP £0.95.
If Mrs Sasha wants to take her family, how
much will it cost for:
(a) herself and her three children
(b) herself, her husband and one child
(c) the whole family including her elderly
father?

In the next questions use your calculator.

15 Curtain material costs £2.72 per metre. Find the cost of:
(a) 4 metres (b) 2.5 metres
(c) $1\frac{3}{4}$ metres.

16 Dress material costs £4.38 per metre. Find the cost of:
(a) 3 metres (b) 1.5 metres
(c) $2\frac{2}{3}$ metres.

17 Dishwasher powder is sold in three sizes. The 4 kg size costs £4.28, the 2 kg size costs £2.48 and the 500 g size costs £0.75.
(a) What is the cost of 1 kg of powder in:
  (i) the largest size
  (ii) the middle size
  (iii) the smallest size?
(b) Which size gives you the best value?
(c) What is the cost of 300 g of the smallest size?

18 Coffee is sold in various sizes. A 750 g tin costs £11.65 in one store whilst a 500 g tin of the same make costs £7.73 in a second store, and a 150 g jar costs £2.30 in a third store.
(a) What is the cost of 100 g of coffee in:
  (i) the first store
  (ii) the second store
  (iii) the third store?
(b) Which store gives you the best value for your money?
(c) How much would you save by buying 750 g of coffee in small size jars rather than in the largest tin?
(d) How much would you save by buying three 500 g tins of coffee rather than two 750 g tins?

19 (a) Which is the better buy, a 450 g economy size packet of Sugar Pops at 89p or a smaller 250 g packet of Sugar Pops at 49p?
(b) Is the new 550 g family size packet at £1.07 a better buy or not? Explain.

20 A domestic electricity bill is made up of charges for two items:
(i) standing charge
(ii) units used.
The standing charge is £6.40 and the charge for each unit is 5.52p.
Find the total bill when the number of units used is:
(a) 100 (b) 500
(c) 750 (d) 1200.

21 Due to a reduction in the price of fuel the electricity board decides to make a refund of 0.20p per unit used.
By how much would each of the bills in question **20** be reduced?

22 A domestic gas bill is made up of charges for two items:
(i) standing charge
(ii) therms used.
The standing charge for a 13 week quarter is £7.60 and each therm used costs 38p.
Find the total bill when the number of therms used is:
(a) 100 (b) 500 (c) 750 (d) 1200.

23 By how much would each of the bills in question **22** be changed if the period covered was:
(a) only 11 weeks instead of 13 weeks
(b) 14 weeks instead of 13 weeks?

24 A domestic telephone bill is made up of charges for three items:
(i) system (ii) apparatus
(iii) units used.
The charges for the system and the apparatus are £13.95 and £3.10 and 4.40p for each unit.
Find the total bill when the number of units used is:
(a) 100 (b) 750
(c) 960 (d) 1257.

25 The telephone company decides to change their charging structure. Instead of charging 4.40p for each unit used, they are going to charge the first 100 units used at 5.00p and the remainder at 4.33p.
Which of the bills in question **24** are going to be cheaper and by how much?

2.6 *Ready reckoners and conversion graphs*

1   Copy and complete the table below which shows the number of gallons for different numbers of litres.

Litres	5	10	15	20	25	30	35	40	45
Gallons	1.1	2.2	3.3						

Approximately how many litres are there in:
(a) 5 gallons   (b) 7 gallons   (c) 9 gallons?

2   Copy and complete the table started below.

×	9	19	29	39	49	59	69	79	89	99
1	9	19	29							
2	18	38								
3	27									
4										
5										
6										
7										
8										
9										
10										
20										

3   What is the cost of buying:
(a)   7 badges at 89p   (b)   17 badges at 89p
(c)   3 books at £2.59   (d)   13 books at £2.59?

4   An exercise book costs 29p. Use your table from question **2** to find the cost of:
(a) 4 books   (b) 7 books
(c) 9 books   (d) 15 books.

5   Use your completed table from question **2** to write down the cost of buying:
(a)   5 cans of Coke at 29p each
(b) 24 bottles of squash at 59p each
(c)   8 packets of sweets at:
(i)   19p each (ii) 39p each (iii) 79p each.

6   The graph below shows the number of litres in different numbers of gallons.

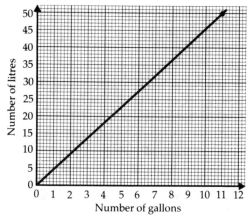

Use the graph to write down:
(a) the number of litres in:
(i) 3 gallons   (ii) 7.5 gallons
(b) the number of gallons in:
(i) 20 litres   (ii) 35 litres.

7   Petrol costs 36.4p per litre.
(a) What is the cost of buying:
(i) 20 litres   (ii) 35 litres?
(b) Use the graph above to find the cost of:
(i) 4 gallons   (ii) 8.5 gallons.

8   Jane works as a petrol pump attendant, where the petrol costs 35.7p per litre. She makes up a number of ready reckoners. Use the graph in question **6** to help you to copy and complete her tables:

(a)

Number of gallons	1	2	3	4	5
Number of litres					

(b)

Number of litres	5	10	15	20	25
Cost (£)		3.57			

(c)

Number of gallons	1	2	3	4	5
Cost (£)					

9 Mr Rawlings wants to convert all the maths examination marks, which were out of 72, to percentages. He does this using a graph like this.

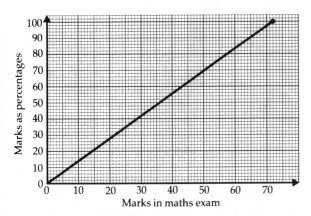

Use the graph to write down the percentage score for a mark of:
(a) 36   (b) 18   (c) 54
(d) 48   (e) 60

10 Jani wants to draw a conversion graph for changing Centigrade temperatures to Fahrenheit.
She knows that at freezing point
    0°C = 32°F
and that at boiling point
    100°C = 212°F.

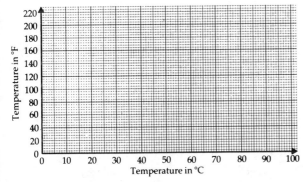

(a) Copy the graph above onto cm/mm graph paper and mark the points (0,32) and (100, 212).
(b) Join your points with a straight line.
(c) Use your graph to find:
    (i) 35°C in °F
    (ii) 50°F in °C.

11 Susan sells potatoes and other vegetables in a farm shop. She wants to draw a conversion graph for pounds and kilograms.
She knows that 1 kg is about 2.2 lb, and that 10 kg is 22 lb.

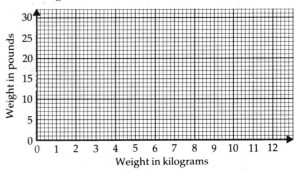

(a) Copy the graph above onto cm/mm graph paper and show the points (1, 2.2) and (10, 22).
(b) Join your points with a straight line.
(c) Use your graph to find:
    (i) 5.2 kg in lb   (ii) 18 lb in kg.

12 (a) Potatoes are 9p per lb. Find the cost of:
        (i) 5 kg   (ii) 12 kg
        (iii) 6.7 kg   (iv) 3.2 kg.
    (b) Peas are 25p per kg. Find the cost of:
        (i) 5 lb   (ii) 15 lb
        (iii) 7.5 lb   (iv) 22.5 lb.

13 Jimmy wants to buy a moped. He needs to know what the monthly repayments will be if he borrows some money.
He has obtained the following information.

Loan	12 months	24 months	36 months
£200	£18.34	£10.00	£7.24
£500	£45.86	£25.00	£18.10

(a) Draw three lines to show the monthly repayments on loans up to £500 for each of the schemes.
(b) Use your graph to write down the monthly repayments on a loan of £230 over:
    (i) 24 months
    (ii) 12 months
    (iii) 36 months.

2.7 *Percentages, decimals and fraction*

1 What fraction of each square is shaded?

(a)   (b)   (c)

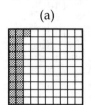 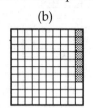 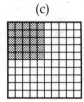

Per cent means *out of 100.*
Percentages can be written as fractions.
$21\% = \frac{21}{100}$ $7\% = \frac{7}{100}$ $109\% = \frac{109}{100}$

---

2 Write each of your answers to question **1** as a percentage.

3 Make four copies of the square on the right and shade in:
(a)   8%   (b)   12%
(c)   40%   (d)   97%

4 Write each percentage as a fraction and simplify where possible:
(a)   50%   (b)   25%   (c)   75%   (d)   20%
(e)   80%   (f)   15%   (g)   44%   (h)   65%
(i)   5%   (j)   150%

5 Write each percentage in question **3** as a fraction, and simplify where possible.

Fractions can be written as percentages.
$$\frac{7}{20} = \frac{7 \times 5}{20 \times 5} = \frac{35}{100} = 35\%$$

$$\frac{4}{5} = \frac{4 \times 20}{5 \times 20} = \frac{80}{100} = 80\%$$

---

6 Write each fraction as a percentage:
(a) $\frac{3}{10}$ (b) $\frac{3}{20}$ (c) $\frac{2}{5}$ (d) $\frac{3}{4}$ (e) $\frac{3}{50}$
(f) $\frac{1}{2}$ (g) $\frac{1}{4}$ (h) $\frac{1}{8}$ (i) $\frac{3}{8}$ (j) $\frac{1}{200}$

7 What fraction of each square is shaded?

(a)   (b)   (c)

8 Write each of your answers to question **7** as a percentage.

Percentages can be written as decimals.
$17\% = \frac{17}{100} = 0.17$ $3\% = \frac{3}{100} = 0.03$
$115\% = \frac{115}{100} = 1.15$

---

9 Write each percentage as a decimal.
(a)   19%   (b)   47%   (c)   85%   (d)   7%
(e)   109%

Decimals can be written as percentages.
$0.53 = \frac{53}{100} = 53\%$ $0.06 = \frac{6}{100} = 6\%$
$1.03 = \frac{103}{100} = 103\%$

---

10 Write each decimal as a percentage.
(a)   0.13   (b)   0.65   (c)   0.09   (d)   0.7   (e)   0.125

Find: 25% of £3.20

$25\% = \frac{1}{4}$ so   25% of £3.20 $= \frac{1}{4}$ of £3.20 $=$ 80p

---

11 Find:
(a)   50% of £16   (b)   25% of £48
(c)   75% of £80   (d)   10% of £15
(e)   10% of £150   (f)   5% of £80.

Find: 7% of £4.00

7% of £4.00 $= \frac{7}{100}$ of £4.00   (find $\frac{1}{100}$ then $\frac{7}{100}$)
$\frac{1}{100}$ of £4.00 $=$ 4p   so $\frac{7}{100}$ of £4.00 $= 7 \times$ 4p $=$ 28p

---

12 Find:
(a)   1% of £7   (b)   3% of £5   (c)   12% of £4
(d)   8% of £6   (e)   9% of £3   (f)   15% of £8.

Use your calculator to find 17% of £43.

17% of £43 = 0.17 × £43 = £7.31

---

13 Using your calculator find:
(a) 17% of £52 (b) 17% of £76
(c) 23% of £91 (d) 27% of £46
(e) 45% of £83 (f) 73% of £12.

14 Using your calculator find:
(a) 17% of 43 metres (b) 23% of 72 kg
(c) 27% of 95 litres (d) 45% of 63 km
(e) 73% of 125 gallons (f) 108% of £2000.

15 Use whichever method you prefer to find:
(a) 50% of 8 metres (b) 25% of 14 km
(c) 75% of 44 litres (d) 20% of 5 kg
(e) 4% of 82 grams (f) 7% of 40 cm
(g) 29% of 451 tons (h) 93% of 75 cl
(i) 12% of £350 (j) 17% of £7425.

16 Saleem spends 80% of his pocket money and saves the rest.
(a) What percentage does he save?
(b) If Saleem has £2.50 each week how much does he spend?
(c) If Saleem has £3.20 each week how much does he save?

17 The town of Tidmarsh has a population of 2400. 53% of these are male.
How many of the population of Tidmarsh are female?

18 Denise spends 30% of her wages on rent, 25% on food, 15% on travel, 10% on entertainment and saves the rest.
(a) What percentage of her wages does she spend altogether?
(b) What percentage does she save?
(c) If Denise earns £84 a week, how much does she save?

19 Saleem invests £200 in a building society which pays him 8% of this in interest in a year.
(a) How much interest does he receive on his money?
(b) If the interest rate were increased to 9%, how much more would Saleem get?

20 VAT is usually added to the price of a meal. This is 15% of the cost of the meal.
(a) How much VAT is added to the cost of a £3.40 meal?
(b) What is the final cost of the meal?

21 Mr Price managed to persuade the garage to give him a 12% discount off the price of a new car.
(a) If the car costs £6200, what is the discount?
(b) How much does the car actually cost him?

22 27% income tax is payable on a person's net income.
(a) How much income tax is payable on a net annual income of £4200?
(b) What percentage of this income is left after the tax has been paid?

23 Many people have to pay 6% of their income into their pension fund.
(a) How much would you have to pay on an annual income of £5450?
(b) How much of this income is left after paying the pension contribution?

24 Andy wants to buy a hi-fi.
In one shop he finds that the price is quoted as £180, but VAT at 15% has to be added to this price.
In another shop the price is quoted as £218, but they give a 5% discount on this price for cash.
(a) How much is the VAT charged in the first shop?
(b) How much is the discount in the second shop?
(c) In which of the two shops will Andy get a better deal?
(d) How much will Andy save if he chooses the cheaper shop?

25 Andy's father gets a 15% discount off his car insurance for being the only driver and is allowed a 30% discount for his no claims bonus.
Does it matter in what order these discounts are calculated?
Explain your answer.

2.8 *Finding percentages and percentage change*

In the 5th year at the Dragon School there are 23 boys and 27 girls. What percentage are boys?

Boys: 23 out of 50 $= \frac{23}{50} = \frac{46}{100} = 46\%$

Girls: 27 out of 50 $= \frac{27}{50} = \frac{54}{100} = 54\%$

Note: $46\% + 54\% = 100\%$

1   Of the fifty pupils in the 5th year at the Dragon School seven go home to lunch.
   (a) What percentage of the pupils go home to lunch?
   (b) What percentage stay at school?

2   Forty-three of the fifty pupils take computer studies as an option course.
   (a) What percentage is this?
   (b) What percentage of the pupils take other options?

3   Express as a percentage:
   (a) 36 out of 100   (b)   4 out of 50
   (c)   2 out of 25   (d) 13 out of 25
   (e)   9 out of 20   (f) 19 out of 20
   (g)   1 out of 10   (h)   9 out of 10.

Express 24 out of 40 as a percentage.

   24 out of 40 $= \frac{24}{40} = \frac{12}{20}$
   $\qquad\qquad = \frac{12\times5}{20\times5} = \frac{60}{100}$
   $\qquad\qquad = 60\%$

   or $\frac{24}{40} = \frac{3}{5}$
   $\qquad\quad = \frac{3\times20}{5\times20} = \frac{60}{100}$
   $\qquad\quad = 60\%$

4   Express as a percentage:
   (a) 50 out of 100   (b) 25 out of 50
   (c) 20 out of 40   (d) 13 out of 26
   (e) 30 out of 40   (f) 15 out of 20
   (g)   6 out of 30   (h) 21 out of 30.

5   James scored 26 out of 40 in a maths test. What is this score as a percentage?

6   Write each of the marks which James scored in other tests as percentages:
   (a) 15 out of 20   (b) 36 out of 40
   (c) 45 out of 60   (d) 52 out of 80.

7   The Elfin stamp company sells packets of stamps which contain different numbers of foreign stamps.
   Write the number of foreign stamps in each packet as a percentage of the total number of stamps:
   (a) 20 foreign, 20 British
   (b) 30 foreign, 10 British
   (c) 31 foreign, 19 British
   (d)   8 foreign, 24 British
   (e) 16 foreign, 24 British
   (f) 52 foreign, 28 British.

8   For each part of question 7, write the number of British stamps in each packet as a percentage of the total number of stamps.

Use your calculator to express as a percentage:
(i)   52 out of 80   (ii) 7 out of 16.

(i)   52 out of 80 $= \frac{52}{80} = 0.65 = \frac{65}{100} = 65\%$

(ii)   7 out of 16 $= \frac{7}{16} = 0.4375 = \frac{43.75}{100} = 43.75\%$

9   Express as a percentage:
   (a) 44 out of 80   (b) 25 out of 40
   (c) 20 out of 32   (d) 13 out of 52
   (e) 11 out of 16   (f) 15 out of 75
   (g)   6 out of 48   (h) 21 out of 56.

10   The following marks were scored in a maths test which was out of 80.
      James 76, Allison 74, Sarah 71, Andy 43
   (a) Write each of these marks as a percentage.
   (b) Write the total of the four marks as a percentage of 320.
   (c) What is the average percentage score?

11 Last year John had £2 per week pocket money. This year it has been increased to £2.50.
   (a) By how much has John's pocket money increased?
   (b) Express this increase as a percentage of last year's pocket money.

12 Last week Raj earned £25 from his part-time job.
   This week he only earned £20.
   (a) By how much have Raj's wages decreased?
   (b) Express this decrease as a percentage of last week's wages.

The temperature has risen from 60°F to 75°F. What percentage change has taken place?

   The change is 75°F − 60°F = 15°F
   The percentage change is $\frac{15}{60} = \frac{1}{4}$
   $$= 25\%$$

The **change** is always found as a percentage of the **original** value.
In this case it is 60°F.

---

13 During the day, the temperature rises from 75°F to 90°F.
   (a) What change has taken place?
   (b) What is the percentage change?

14 During the evening, the temperature falls from 60°F to 45°F.
   (a) What change has taken place?
   (b) What is the percentage change?

15 In both questions **13** and **14** the temperature has changed by 15°F.
   Explain why the percentage change is larger in question **14** than in question **13**.

16 Yesterday Pato delivered 50 papers. Today he delivered eight more.
   What percentage increase in the number of papers is this?

17 Today Pasha eats 25 sweets. Tomorrow he will eat 4 fewer sweets.
   What percentage decrease is this?

18 What is 50 increased by 8%?

David's father's wages have increased by 5%.
They were £120 per week.
What are they now?

The increase is 5% of £120.
$$5\% \text{ of } £120 = \tfrac{5}{100} \times £120$$
$$= \tfrac{1}{20} \times £120$$
$$= £6$$

So the new wages are £120 + £6 = £126.

---

19 Yesterday Pato delivered 80 papers. Today he increased this number by 25%.
   (a) How many extra papers did Pato deliver today?
   (b) How many papers did he deliver altogether?
   (c) If he had only managed to deliver 90 papers today, what would have been the percentage increase?

20 Today Pasha ate 50 sweets. He decides to reduce this number by 20% tomorrow.
   (a) How many fewer sweets will he eat tomorrow?
   (b) How many sweets will he eat altogether?
   (c) If in fact he only eats 38 sweets tomorrow, what is the percentage decrease?

21 Last year a Ford car dealer sold 400 cars. This year he hopes to increase his sales by 18%.
   (a) How many cars does he hope to sell this year?
   (b) He actually manages to sell 480 cars. How much better is his percentage increase than he expected?

22 Raj wants to buy a hi-fi and discovers that he can insure it for five years by paying an extra 6% on the purchase price.
   (a) If the price of the hi-fi is £250, how much will it cost him altogether if he wants to insure it as well?
   (b) In another shop he finds that the price of the hi-fi is £264 including the five years insurance.
   What percentage increase has this shop charged for the insurance?

2.9 *VAT, discount, profit and loss*

On most things you buy a **selling tax** is added. This is called **value added tax**, or VAT for short. This is normally 15% of the original cost of the article.

The original cost of a TV is £200, and VAT at 15% is then added.
The VAT is 15% of £200 = £30.
So the price of the TV is £200 + £30 = £230.

1   A list of goods in Jo's electrical shop is given below, together with their cost before VAT is added.

A stereo radio cassette recorder	£100
A portable black and white TV	£50
A washing machine	£200
A microwave cooker	£300
A refrigerator	£150
A large deep freezer	£250

   For each article find:
   (a) the amount of VAT at 15%
   (b) the final price of the article.

2   What is the amount of VAT at 15% on an article whose original cost is:
   (a) £1   (b) £2   (c) £5
   (d) £1.20   (e) £4.40?

3   Jane and her four friends are having a meal in the How Chi Chinese restaurant. The bill comes to a total of £20 to which VAT is then added.
   (a) What is the amount of VAT at 15%?
   (b) What is the final bill?
   (c) If they share the bill equally among them how much does each one actually have to pay?

4   Jason buys a book whose price before VAT is added is £3.60.
   (a) How much is the VAT?
   (b) What does he have to pay for the book?

5   Jason's father has to have a tree cut down in his garden.
   He is quoted a price of £40 excluding VAT.
   What does he actually have to pay?

Sometimes shops try to encourage customers by giving a **discount** if they pay cash rather than using a cheque or credit card.
The discount is usually quoted as a percentage of the price of the article.

6   Jane wants to buy herself a sewing machine. She finds one whose price is quoted as £120 with a 10% discount for cash.
   (a) How much is the discount?
   (b) What is the actual cost if she pays cash?

7   Jane's father is buying a new car which is listed at £5000. If he has no car to trade in the garage will give him a 12% discount.
   By how much does this reduce the price of the car?

8   Sarah is having a moped for her sixteenth birthday. She finds two she likes.
   In one shop the price is quoted as £375, whilst in a second shop the price is quoted as £400 with a 6% discount for cash.
   (a) How much will the moped cost her in the second shop if she pays cash?
   (b) Which shop has the better deal?

9   Mr Keen needs some paving slabs for his patio. His local shop charges £1 for each slab but allows an 8% discount on a quantity of fifty or more.
   (a) How much will it cost him to buy:
       (i) 47 slabs   (ii) 50 slabs   (iii) 51 slabs?
   (b) If he only needs 48 slabs what is the cheapest way of buying these?
   (c) How would your answer to (b) differ if the discount were only 3%?

10 Which is the better buy, £12 with a 10% discount or £11.50 with a 6% discount?

When goods are sold the seller wants to make a **profit**. The profit is usually given as a percentage of the price the seller had to pay for the goods in the first place.

11 Raj bought a radio for £20. When he sells it he wants to make a 7% profit.
   (a) What is the amount of this profit?
   (b) What must he sell the radio for to make this profit?
   (c) If he actually sold the radio for £21, find the profit as a percentage.

12 Mr Sharp bought an old desk for £50. He first renovated it and then he sold it making a 12% profit.
   (a) What did he sell the desk for?
   (b) If he had sold the desk for £57, what would his percentage profit have been then?

13 Jason bought an old motor bike for £250 and a year later sold it for £280.
   (a) What was his percentage profit when he sold the bike?
   (b) If he had actually made a 13% profit what amount would he have sold the bike for?

Often, instead of making a profit, shopkeepers have to sell less popular items at a **loss**.
Again the loss is given as a percentage of the price the shopkeeper originally paid for the article.

14 Mr Starr the newsagent paid £2.40 for some large Easter eggs. In order to sell what were left after Easter, he sold them at a 25% loss.
   (a) How much did he lose on each Easter egg?
   (b) What was the selling price of the eggs after Easter?

15 Natasha thought she had found a bargain when she bought a sailboard for £160. However, it was too heavy and she sold it again for £120.
   (a) What was her percentage loss?
   (b) If it had originally cost £150 what would have been her percentage loss?

16 Mr Hogg bought a car for £2000. He wanted to make a 20% profit so he advertised the car at this inflated price.
   In fact he had to give a 10% discount on the advertised price when he sold the car.
   (a) At what price did he advertise the car?
   (b) What did he eventually sell the car for?
   (c) How much profit did he actually make?
   (d) What was his percentage profit on the original cost of the car?

17 Soni makes artificial flower arrangements for the local craft stall. The materials for each arrangement cost her £3.20. She wants to make a 10% profit when she sells them.
   (a) What should each arrangement be priced at to make this profit?
   (b) If they are actually priced at £3.60 and she sold them at this price, what would her percentage profit be?
   (c) At the end of each week she needs to sell what are left to buy her next materials. If she is prepared to sell them at a 5% loss how much does she lose on each one?

18 Tina can buy hairdryers priced at £7.50 and get a 10% discount. She sells them to her friends for £7.02.
   (a) How much does each hairdryer cost Tina?
   (b) How much profit does Tina make on each?
   (c) What is Tina's percentage profit?
   (d) What percentage discount is each of her friends getting off the original price of the hairdryer?

19 The basic premium on Sarah's moped insurance is £120. She is allowed various discounts: 10% for being the only rider, 5% for having passed her test and 20% for having a year without any claims.
   (a) Find what premium Sarah actually pays by first taking off the 10% discount, then taking the 5% off this result and finally taking the 20% off the second result.
   (b) What difference would it make if you took the 20% off first, and then the 10% and 5%?

2.10 *Interest, loans and hire purchase*

When people want to save they often put their money into National Savings or a building society. They are then paid **interest** on the money they **invest**.

1 Jane is given £200 and decides to put the money into a National Savings account which is paying 11% interest each year.
   (a) How much interest does she receive at the end of the first year?
   (b) How much interest will she receive over three years?

2 Jane adds the interest to her account at the end of the first year.
   (a) How much will she now have in her account?
   (b) How much interest will she receive at the end of the second year?

3 Jane's father has saved £1000 in a National Savings income bond, which pays 12% interest each year.
   (a) How much interest does Jane's father receive at the end of the first year?
   (b) How much interest will he receive over five years?

4 Jane's father can receive the interest from his income bond each month.
   (a) How much interest will he receive at the end of the first month?
   (b) How much interest will he receive over eighteen months?

5 Raj discovers that building societies pay different amounts of interest.
   On their ordinary accounts, some are paying 6% per year, some 6.25% and some 6.5%.
   (a) If he were to invest £400, how much interest would he receive from each?
   (b) How much more interest would he receive from the best than from the worst, over a period of five years?

6 A building society is paying 8% interest per year. Josie invests £300 for six months. How much interest will she receive?

7 Mr Gray has just sold his house for £30 000. He invests this money for three months with a local authority which is paying 12% interest per year.
   How much interest will he receive?

8 Mrs Ahmed invests £7300 in a bank which pays interest on a daily basis. Their rate is $\frac{1}{365}$ of the yearly interest rate of 9%.
   How much interest will she receive if she leaves her money in for:
   (a) one week   (b) thirty days
   (c) six months?

If you **re-invest** your interest your savings will grow more quickly.
The working below shows what £100 invested at 8% per year will become after each full year.
Year 1:   £100 becomes £100 + 8% of £100
              = £108
   Note: Each £1 invested becomes £1.08.
Year 2:   £108 becomes £108 × 1.08 = £116.64
Year 3:   £116.64 becomes £116.64 × 1.08
              = £125.97

9 What will:
   (a) £100 invested for one year at 7% become
   (b) £1 invested for one year at 7% become
   (c) £107 invested for one year at 7% become
   (d) £114.49 invested for one year at 7% become?

10 If the interest is re-invested, what will £100 invested at 6% per year become after:
   (a) 1 year   (b) 2 years   (c) 3 years?

11 If the interest is re-invested, what will £200 invested at 9% per year become after:
   (a) 1 year   (b) 2 years   (c) 3 years?

When you want to **borrow** money you have to pay the lender interest.

The amount of interest will depend on the length of the loan and the amount you want to borrow.

12 Pato wants to borrow £200 for twelve months to buy a hi-fi.
His father says he will lend him the money if he is willing to pay 12% interest.
(a) How much will the interest be?
(b) How much money will he have to pay back altogether in twelve months time?

13 After twelve months Pato is only able to pay his father the interest and £100, so he now wants to borrow the remaining £100 for another twelve months.
(a) How much will the interest be?
(b) How much altogether will he have paid back to borrow the £200?

14 James wants to borrow £200 to buy a motorbike. The shop says they will lend him the money at 8% interest providing he pays it back in twelve equal monthly instalments.
(a) How much will the interest be?
(b) How much will James have to pay back altogether?
(c) How much will he have to pay back each month?

15 Sue wants to borrow £300 to buy a knitting machine. The shop says they will lend her the money at 6% interest per year. She wants to pay the loan back over a period of two years.
(a) How much will the interest be for the two years?
(b) How much will Sue have to pay back altogether?
(c) How much will she have to pay back each month?

16 A bank charges 18% interest per year on its loans.
How much interest will be charged on a loan of £5000 for three years?

At the end of the first year in question **15**, Sue had paid back £168, which was half of her loan. For the second year she has to pay another £168 even though she is now only borrowing £150.
This means she is actually paying 12% interest for the second year instead of 6%.

When you borrow money on **hire purchase** two rates of interest are quoted. A **flat rate** which enables you to work out the cost of your loan, and the **annual percentage rate (APR)** which is the real cost of borrowing the money.

17 David's father wants to buy a new Ford car.
He needs to borrow £1800.
There is a special offer on at present.

> Interest only 5% (APR 9.7%)
> thirty-six weeks to pay!

(a) How much interest at 5% will he be charged for the three years?
(b) How much will he have to pay back altogether?
(c) How much will he have to pay back each month?

18 David's father has £1800 in the building society which is paying 7.5% interest per year.
If he withdraws his £1800 to pay for his new car, how much interest will he lose over the three years?

19 Saleem wants to buy the latest computer on hire purchase. The shop is quoting a flat rate of 12% (APR 23.3%). He needs to borrow £420 over two years.
(a) How much interest at 12% will he be charged for the two years?
(b) How much will he have to pay back altogether?
(c) How much will he have to pay back each month?

20 How much better off would Saleem be, if he withdrew the £420 from a bank, where he was getting 8% interest per year, and paid cash?

2.11 *Wages and taxes*

1 Mr Green earns £480 per month.
How much does he earn in one year?

2 Mrs Green earns £84 per week.
How much does she earn in one year?

3 Julie Green has a part-time job and is paid
£2.40 per hour. She works 15 hours per
week.
(a) How much does she earn per week?
(b) How much does she earn per year?

4 Sammy Green does two daily paper rounds.
He is paid £2.10 for the morning one, and
£2.60 for the evening one.
(a) How much does he earn per day?
(b) How much does he earn for a six day
week?
(c) If he doesn't do the evening round on
Sundays, how much does he earn each
week?
(d) How much did he earn in February 1987?

5 Mrs Green has to pay 7% of her weekly
wages of £84 in NI contributions.
(a) How much is this per week?
(b) How much is this per year?

6 Mr Green has to pay 9% of his monthly
salary of £480 in NI contributions.
(b) How much is this per month?
(b) How much is this per year?

7 If Mr Green was only paying the 'contracted
out' rate of NI of 6.85%, how much less
would he have to pay each month?

8 (a) Jason earns £4992 per year.
(i) How much is this per month?
(ii) How much is this per week?
(b) If Jason pays 7% of his wages in National
Insurance contributions, how much does
he pay each year?
(c) If Jason only pays the lower rate of 4.85%
NI contributions, how much less would
he pay each year?

When people work overtime, they are usually
paid at a higher rate of pay.
Time-and-a-half means that they are paid at
one and a half times their normal hourly rate of
pay.
Double time means that they are paid at twice
their normal hourly rate of pay.

9 Mr Ray earns £2.60 per hour for a 40-hour
week.
(a) What is his basic weekly wage?

On Saturday he works 4 hours extra on time-
and-a-half.
(b) What is his hourly rate of pay for this
extra work?
(c) What is his total overtime pay?
(d) What is his total pay for the week?

10 Miss Ray earns £3.65 per hour and normally
works a 42-hour week. Last week she
worked for an extra 8 hours on double time.
(a) What was her basic wage for the week?
(b) What was her total overtime pay?
(c) What was her total pay for the week?

11 Mrs McKay works for 38 hours most weeks
and is paid £2.42 per hour. Next week she is
offered the possibility of an extra 6 hours at
time-and-a-half.
How much extra pay would she receive?

12 Mr Macdonald works a 5-day week from 8
a.m. to 5 p.m. with a one hour break for
lunch (unpaid). His basic rate of pay is £4.50
per hour.
(a) How many hours does he work per
week?
(b) What is his basic wage for the week?

Last week he worked overtime from 5 p.m.
to 7 p.m. on Thursday and Friday at double
time.
(c) What was his total overtime pay?

Most people who have an income have to pay **taxes** to the government. The amount of tax depends on your **net income** after certain allowances have been subtracted.

13 A single person can earn up to £2425 in any one year before having to pay any income tax.
(a) What is the equivalent weekly wage?
(b) What is the equivalent monthly salary?

14 A married person can earn up to £3795 in any one year before having to pay any income tax.
(a) What is the equivalent weekly wage?
(b) What is the equivalent monthly salary?

15 A single person pays tax on any net income above £2425.
What is the taxable income for a single person who earns:
(a) £5000 in one year
(b) £100 each week for a year
(c) £500 each month for a year?

16 A married person pays tax on any net income above £3795.
What is the taxable income for a married person who earns:
(a) £5000 in one year
(b) £100 each week for a year
(c) £500 each month for a year?

17 Providing a person's taxable income is less than £17900, the rate of income tax charged is 27%.
How much tax do you have to pay each year on a taxable income of:
(a) £1200 (b) £1400 (c) £2000 (d) £2200?

18 How much tax at 27% do you have to pay on a taxable income of:
(a) £6500 in one year
(b) £600 each month
(c) £150 each week?

19 James earns £4425 per year. He is entitled to the single person's allowance of £2425 before tax is deducted.
How much tax will he have to pay at a rate of 27%?

20 Jane's father earns £9795 per year before the married person's allowance of £3795 is deducted.
How much tax will he have to pay at 27%?

21 Alison, who is not married, earns £4800 each year as a secretary. She is entitled to the single person's allowance of £2425.
(a) What is her taxable income?
(b) How much tax at 27% does she have to pay:
   (i) each year (ii) each week?

22 Alison has to pay 7% of her salary of £4800 in National Insurance contributions.
(a) How much is this contribution:
   (i) each year (ii) each week?
(b) How much does she have left to spend after these contributions and tax at 27% have been deducted:
   (i) each year (ii) each week?

23 John Smith earns £8500 a year. He is married and pays £1700 interest on his mortgage. His taxable income is found by subtracting the married person's allowance of £3795, and the mortgage interest from his earnings.
(a) What is his taxable income?
(b) How much tax does he have to pay at 27%?

24 People who have a taxable income of more than £17900 have to pay tax at higher rates. These are shown in the table below.

Taxable income	Rate
up to 17900	27%
next 2500	40%
next 5000	45%
next 7900	50% etc.

Mr Ahmed's taxable income is £25000 a year. How much tax does he have to pay on:
(a) the first £17900 of his taxable income
(b) the next £2500
(c) the remaining £4600?

25 Using the rates in question **24**, find how much tax you would have to pay on a taxable income of:
(a) £10000 (b) £20000 (c) £30000.

2.12 *Household finance*

1 Saleem has a £10 note to do some shopping. If there is any change he can go to the cinema which costs £1.25.
   He has to buy 2 lb of steak at £1.95 a lb, five packets of butter each costing 52p and a packet of washing powder which costs £2.25.
   (a) Has he enough money to go to the cinema?
   (b) Can he afford an ice cream in the interval?

2 Jane has £70 saved for her summer holiday. She wants to spend seven days youth hostelling in Scotland.
   (a) If the return bus fare is £19.60 and each night at the youth hostel will cost her £2.75, how much will she have to spend on food and sight-seeing?
   (b) If she allocates £3.50 per day for food how much will she have left each day for other things?

3 Jane's mother is trying to plan her weekly budget. Her housekeeping is £80 per week. She usually spends £35 at the supermarket on food, £15 at the greengrocer's on fruit and vegetables. She has four pints of milk every day. Each pint costs 25p. Her daughter goes to playschool three mornings a week at £1.25 a morning, and she gives her son 70p each day for his lunch at school and another 50p for the bus fare.
   (a) How much does she spend each week?
   (b) If she allows herself £3.25 a week for make-up and tights how much can she save for clothes and outings?

4 Jane's father brings home £154 a week and gives his wife £80 for housekeeping. The rent and rates come to £45 a week, and he spends £6 each week on bus fares.
   (a) If he allows £2 a day for electricity and gas, how much does he spend weekly?
   (b) How much does he have left over each year to allocate for other things?

5 The Jones family can't decide whether to go to the Longleat Safari Park or to the Alton Towers Adventure Park.
   At Longleat the entrance fee is £2.50 per person and £5 for the car.
   At Alton Towers an adult ticket costs £4.50 and a child's ticket costs £2.75.
   Which of the two is cheaper for two adults and three children?

6 Sandra wants a new carpet for her bedroom which measures 4 metres by 3 metres.
   (a) What area of carpet will she need?
   (a) If the carpet costs £3.75 a square metre how much will it cost altogether?

7 David wants some new wallpaper for his room which is also 3 metres by 4 metres.
   (a) What is the perimeter of his room?
   (b) If each piece of wallpaper is half a metre wide, how many widths will he need, ignoring doors and windows?
   (c) If he can get four lengths out of each roll, how many rolls will he need?
   (d) If the wallpaper costs £2.75 a roll and the paste he needs costs another £1.50, how much will it cost to paper the room?

8 Rick's father wants to use weedkiller on his lawn which is 25 metres square.
   (a) What is the area of the lawn?
   (b) How many cans will he need, if a can of weedkiller will treat $125\,m^2$ of lawn?
   (c) If each can is £2.69, how much will it cost him altogether?

9 Rick's mother is buying bedding plants for the flower borders around their lawn.
   (a) What is the perimeter of the lawn in question 8?
   (b) How many plants will she need, to put a plant every third of a metre?
   (c) Plants come in trays of 25; each tray costs 45p. How much will the plants cost her?

10 Raj wants to hire a motorbike for a day.
He is quoted two different rates:
(i) 25p per mile or (ii) £15 plus 5p per mile.

(a) Which is the better bargain if he intends to do 100 miles?

(b) Which is the better bargain if he is only going to do 72 miles?

(c) For what mileage would the two costs be the same?

11 Sam is a keen squash player. He likes to play for an hour, four times a week. He wants to join the local club where the annual subscription is £75 and the cost of the courts is 10p for twenty minutes.

(a) How many hours will he play in a whole year?

(b) How much will it cost him for one full year?

(c) How much would he save by playing at squash courts where the charge is 60p per hour?

12 Jason is trying to decide whether he can afford to leave home and live in a flat. He earns £450 a month but he wants to save at least £50 of this towards a car.
The rent of the flat excluding the rates is £35 per week. The rates come to £364 a year. His rail season ticket to work costs £468 a year. He needs to allow at least £40 per week for food, heating and lighting.

(a) Can he afford the flat?

(b) How much will this leave him each week for other things?

13 Jenny is about to buy a car and she wants to know how much it will cost her to run.
Each year the tax is £100, the insurance is £180 and a service and MOT will be £90. A gallon of petrol costs £1.75 and the car does 30 miles per gallon.
She expects to do about 6000 miles each year.

(a) How many gallons of petrol will she use in one year?

(b) How much will this petrol cost her altogether?

(c) What will be the total cost of using the car for one year?

(d) How much will it cost per mile?

14 Mr and Mrs Wardle like camping holidays. Each year they take their three sons and their car abroad on the car ferry to France. The ferry charges for single journeys are:
    cars £19.50, adults £11.50, children £5.75.
The camp site charges, in francs, are usually:
    cars 15 F, adults 20 F, children 10 F.
The rate of exchange last year was 10 F = £1.
Find out the cost for the Wardle family of:

(a) a single crossing on the ferry

(b) one night at a French camp site, in pounds

(c) a two week holiday for the ferry and the camp site charges.

15 During their holiday the Wardles usually travel about 1760 miles in France.
Their car does 32 miles per gallon. Petrol in France costs 4.65 F per litre. 50 litres is about 11 gallons.

(a) How many gallons do they use on holiday?

(b) How many litres is this?

(c) What is the total cost of the petrol they use?

16 (a) What is the total cost of the Wardle family's holiday, in questions **14** and **15**?

(b) Would it have been cheaper for them to go on a two week package holiday by coach costing £70 per person?

(c) How much would it cost the Wardles to have a three week camping holiday in France, assuming that they did the same mileage?

17 Alison is moving to a new house. Her old house is sold for £44 000 through an estate agent who charges a commission of 2% on the first £20 000 and 1.5% on the remainder, plus £250 for the solicitor's charges.
How much did it cost Alison to sell the house?

18 Alison's father has a mortage of £40 000 and is charged 11% interest.

(a) How much is the interest each month?

(b) If he gets tax relief of 27% on the interest paid on the first £30 000, by how much does this reduce the monthly figure?

2.13 *Ratio and proportion*

A **ratio** is a comparison between two similar quantities.
In the bag there are three white beads and seven red beads.

The ratio of white beads to red beads is 3 to 7.
This is written as 3 : 7.

1 Write down the ratio of white squares to black squares.

(a)   (b)   (c)

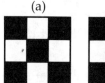

2 Sally is seven years old.
Her younger brother is three years old.
Write down the ratio of their ages.

3 Anne is 15 years old and 160 cm tall.
David is 16 years old and 171 cm tall.
(a) Write down the ratio of their ages.

(b) Write down the ratio of their heights.

4 In a class of 29 pupils there are 15 boys.
Write down the ratio of boys to girls.

5 Thirteen people are standing in the queue for the cinema. Three of these are adults.
Write down the ratio of:
(a) adults to children   (b) children to adults.

6 Two squares have sides of 4 cm and 5 cm.
Write down the ratio of:
(a) the lengths of their sides

(b) their areas.

7 Two cubes have sides of 2 cm and 3 cm.
What is the ratio of their volumes?

8 The ratio of John's marks in two maths tests is 1 : 2.
(a) If he got 20 marks in the first test how many did he get in the second test?
(b) If he got 30 marks in the second test how many did he get in the first test?

9 A rectangle has sides in the ratio 1 : 1.
What is special about rectangles like this?

10 A rectangle has sides in the ratio 1 : 3.
(a) Write down the length of the longer side if the length of the shorter side is:
    (i) 1 cm   (ii) 2 cm   (iii) 5 cm.
(b) Write down the length of the shorter side if the length of the longer side is:
    (i) 3 cm   (ii) 6 cm   (iii) 12 cm.

11 The ratio of boys to girls at a party is 3 : 4.
(a) How many girls are there at the party if there are:
    (i) 3 boys   (ii) 6 boys   (iii) 15 boys?
(b) How many boys are there if there are:
    (i) 4 girls   (ii) 8 girls   (iii) 12 girls?

Each strip below is divided into two parts.
The ratio of the two parts is written alongside.

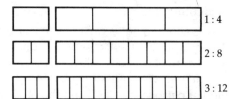

The lengths of the two parts of the strip are the same in each case, so the ratios are the same.

12 Copy and complete:
(a) $1 : 3 = 2 : \square$ (b) $1 : 5 = 2 : \square$
(c) $2 : 3 = 4 : \square$ (d) $2 : 5 = 4 : \square$

13 Write each of the ratios in its lowest form:
(a) 2 : 6   (b) 12 : 3
(c) 10 : 15   (d) 12 : 8

14 Copy and complete:
(a) $1 : 6 = 2 : \square$ (b) $1 : 7 = \square : 21$
(c) $2 : \square = 4 : 10$ (d) $\square : 4 = 9 : 12$

15 A 5 cm tape is divided in the ratio $1 : 4$.
What is the length of each part?

16 A 10 cm tape is divided in the ratio $1 : 4$.
What is the length of each part?

Divide £20 between Jo and Di in the ratio $1 : 3$.

Jo is to have one part and Di is to have three parts, so there are four parts altogether.
Jo has $\frac{1}{4}$ of £20 = £5 and Di has $\frac{3}{4}$ of £20 = £15

---

17 £20 is to be divided between Ian and Anne in the ratio $1 : 4$.
(a) What fraction will each receive?
(b) How much will each receive?

18 £60 is to be divided between Ian and Anne in the ratio $1 : 4$.
(a) What fraction will each receive?
(b) How much will each receive?

A 25 cm tape is to be cut in the ratio $2 : 3$.
What is the length of each piece?

The ratio $2 : 3$ means there are 5 parts.

One piece will be $\frac{2}{5}$ of 25 cm = 10 cm.

The other piece will be $\frac{3}{5}$ of 25 cm = 15 cm.

---

19 A 250 g packet of butter is to be divided in the ratio $2 : 3$.
What will be the weight of each piece?

20 An 80 cm tape is to be divided into two pieces in the ratio $3 : 5$.
What will be the length of each piece?

21 A one litre bottle of squash is to be divided into two bottles in the ratio $1 : 2$.
What smaller size bottles are needed?

Proportions are used for comparisons between three or more quantities.
The strip is divided in the proportions $1 : 3 : 4$.

22 Write down the proportions of white, dotted and black squares.

(a)   (b)   (c)

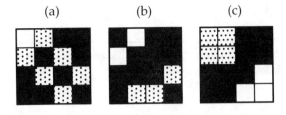

23 The sides of a cuboid are in the proportions $1 : 2 : 7$.
(a) If the length of the shortest side is 10 cm, what are the lengths of the other two sides?
(b) If the length of the longest side is 21 cm, what are the lengths of the other two sides?

Divide £24 among Ian, Mark and Paul in the proportions $1 : 2 : 5$.

Ian is to have 1 part, Mark 2 parts and Paul 5 parts, so there are 8 parts altogether.
Ian has $\frac{1}{8}$ of £24 = £3,
Mark has $\frac{2}{8}$ of £24 = £6
Paul has $\frac{5}{8}$ of £24 = £15
(Note: £3 + £6 + £15 = £24)

---

24 £60 is to be divided among Ann, Jo and Pat in the proportions $1 : 2 : 3$.
(a) What fraction will each receive?
(b) How much will each receive?

25 In a concrete mix cement, sand and gravel are mixed in the proportions $1 : 3 : 6$.
(a) If a 12 kg bag of cement is used, how much gravel and sand will be needed?
(b) If the total weight of the mix comes to 50 kg, how much gravel was used?

26 A recipe for shortbread uses butter, sugar and flour in the proportions $2 : 1 : 3$.
(a) If 100 g of butter are used, how much sugar and flour are needed?
(b) If 450 g of flour are used, how much butter and sugar are needed?

2.14 *Maps and scale drawing*

Ratios are used to indicate sizes on a scale drawing or map.

The house on the left is a **scale drawing** of the one on the right. The scale used is 1 : 2.
Each length on the left-hand drawing is half the corresponding length on the right.

1   The triangle on the left is a scale drawing of the one on the right.

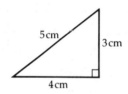

(a) What is the height of the small triangle?

(b) What is the length of the hypotenuse?

(c) What is the scale of this drawing?

2   The rectangle below is drawn to scale.

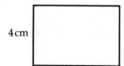

Estimate, in centimetres, the length of the rectangle. What scale has been used?

3   The rectangle below is drawn to scale.

The length of the longer side is 9 cm.
Estimate, in centimetres, the length of the shorter side. What scale has been used?

When a model is made to a scale of 1 : 8, this indicates that 1 unit of length on the model corresponds to 8 units of length on the object.

4   Julie has a china house which is a replica of a country mansion. It is made to a scale of 1 : 12. The doors on the replica are 16 cm high, and 7 cm wide. The chimneys are 4 cm tall.

(a) What is the height of a door on the original house?

(b) What is the width of a door on the house?

(c) What is the height of a chimney on the original house?

(d) If the country mansion has sixteen windows, how many windows will the replica have?

5   A car is 120 cm wide and 400 cm long. James is making a model of the car to a scale of 1 : 8.

(a) What is the length of the model car?

(b) What is the width of the model car?

(c) If the wheels on the model car are 7 cm in diameter, what is the diameter of the wheels on the original car?

6   The drawing below is the plan of John's house, using a scale of 1 : 200.

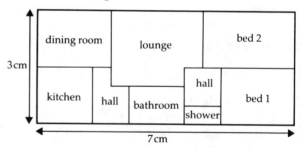

(a) What is the size on this plan of:

(i) the dining room   (ii) the kitchen?

(b) What are the actual dimensions of:

(i) the dining room   (ii) the kitchen?

(c) A snooker table, 200 cm by 100 cm, is to be shown on the plan. How large will it be?

7   You are to make a scale drawing of each rectangle below.

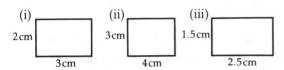

(i) 2 cm, 3 cm   (ii) 3 cm, 4 cm   (iii) 1.5 cm, 2.5 cm

(a) If you use a scale of 4 : 1 what will be the length and breadth of each rectangle?

(b) Make an accurate scale drawing of each.

8   Using a scale of 5 : 1 and centimetre squared paper, make a scale drawing of each shape.

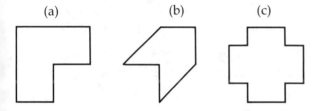

(a)   (b)   (c)

9   The drawing below is the plan of Anne's kitchen.

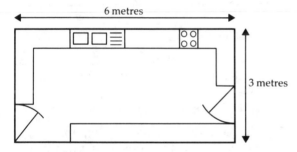

6 metres

3 metres

You are to make a scale drawing of this plan using a scale of 1 : 50.
(a) What will be the length and width of the kitchen?

(b) What will be the size of the cooker on your drawing?

(c) Make an accurate scale drawing of the kitchen.

10   A rectangle 9 cm by 12 cm is to be re-drawn using a scale of 1 : 3.
(a) What will be the width and length of the new rectangle?

(b) What is the area of the first rectangle?

(c) What is the area of the new rectangle?

(d) How are these two areas related?

On **maps** a scale can be given in different ways:
either using a ratio   1 : 25 000
or using a statement   1 cm represents 1 km.
Using the ratio 1 : 25 000 one unit of length on the map represents 25 000 units of the same length on the ground.

---

11   A map has a scale of 1 : 25 000. On the map the distance between a pub and a church is 4 cm.
What is the actual distance between the pub and the church in:
(a) centimetres   (b) metres   (c) kilometres?

12   A map showing two towns which are 6 km apart has a scale of 1:20 000.
(a) How many centimetres are there in 6 km?

(b) What length on the map represents the distance between these two towns?

13   On a map 1 cm represents 2 km.
(a) What distance on the ground is represented on the map by:
(i) 4 cm   (ii) 5 cm   (iii) 12 cm?

(b) What distance on the map represents:
(i) 4 km   (ii) 5 km   (iii) 12 km?

(c) What area on the ground is represented by one square centimetre on the map?

(d) What area on the map is used to represent sixteen square kilometres on the ground?

14   The map below has a scale of 1 : 20 000.

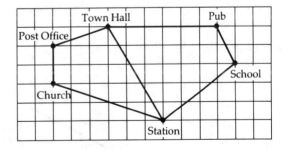

Town Hall   Pub
Post Office
School
Church
Station

(a) Find the actual distance, in km, between:
(i)   the church and the Post Office
(ii)   the Town Hall and the pub
(iii) the station and the school.

(b) What area does each square represent?

2.15 *Similarity and enlargement*

The two shapes below are similar. Their angles are the same size and their corresponding sides are in the same ratio.

1   The two triangles are similar.

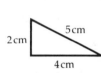 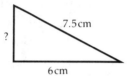

(a) What is the ratio of their corresponding sides?

(b) What is the missing length in the second triangle?

2   The two hexagons are similar.

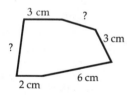 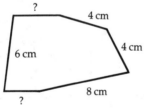

(a) What is the ratio of their corresponding sides?

(b) What are the missing lengths in each hexagon?

3   The two triangles are similar.

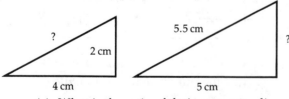

(a) What is the ratio of their corresponding sides?

(b) What are the missing lengths?

4   The two rectangles are similar with their corresponding sides in the ratio 3 : 4.

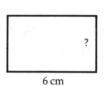 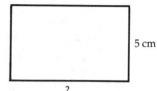

What is:

(a) the length of the larger rectangle

(b) the width of the smaller rectangle?

5   The two triangles are similar with their corresponding sides in the ratio 2 : 3.

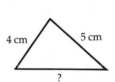 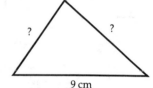

What are the missing lengths in each triangle?

6   A 3 cm by 2 cm photograph is to be enlarged. What is its new size if it is enlarged in the ratio:

(a) 1 : 2   (b) 1 : 3   (c) 2 : 3   (d) 4 : 5?

7   Julie has a 10 cm by 8 cm photograph. She wants a smaller copy for her handbag and an enlargement for her dressing table.

(a) What is the length of the smaller copy if its width is 4 cm?

(b) What is the width of the enlargement if its length is 15 cm?

8   Andrew collects brass rubbings. If they are too large he has to make smaller copies. One copy he has is 25 cm by 30 cm having been reduced in the ratio 5 : 12.

(a) What is the length of the original?

(b) What is the width of the original?

(c) If the length of a sword on the original was 48 cm, how long is it on the copy?

(d) If the length of a dog is 8 cm on the copy, what is its length on the original?

When a shape is enlarged each of the lengths is increased in the same ratio. The enlarged shape is similar to the original shape.

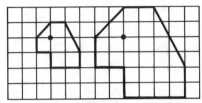

9 (a) What is the ratio of the lengths of the corresponding sides in the shapes above?

(b) Do you agree that the scale factor of the enlargement is two?

(c) On a copy of the above grid draw the shape when it is enlarged using a scale factor of three.

10 In the diagram below the triangle LMN is enlarged to triangle PQR using C as the centre of enlargement.

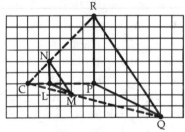

(a) Measure CP and CL. Is CP = 3 × CL?

(b) Measure CQ and CM. Is CQ = 3 × CM?

(c) Measure CR and CN. Is CR = 3 × CN?

(d) What is the ratio of the lengths of the corresponding sides of the two triangles?

(e) Do you agree that the scale factor of the enlargement is three?

11 (a) Copy the grid and the triangle LMN from question **10**.

(b) Draw the line CX so that CX = 4 × CL.

(c) Draw the line CY so that CY = 4 × CM.

(d) Draw the line CZ so that CZ = 4 × CN.

(e) Join XY, YZ and ZX.

(f) Do you agree that △XYZ is similar to △LMN?

(g) What is the enlargement scale factor?

12 (a) Write down the area of a square of side:
(i) 5 cm   (ii) 10 cm

(b) What happens to the area of the square when the lengths of its sides are doubled?

13 (a) Write down the area of a square of side:
(i) 2 cm   (ii) 6 cm

(b) What happens to the area of the square when the lengths of its sides are trebled?

14 (a) Write down the area of a rectangle of sides:
(i) 3 cm and 4 cm   (ii) 6 cm and 8 cm

(b) What happens to the area of the rectangle when the lengths of its sides are doubled?

15 (a) Write down the area of a rectangle of sides:
(i) 5 cm and 6 cm   (ii) 15 cm and 18 cm

(b) What happens to the area of the rectangle when the lengths of its sides are trebled?

16 (a) Copy the grid and the shape below.

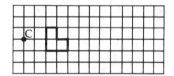

(b) Use the ideas of question **11** to enlarge the shape using a scale factor of two and a scale factor of three.

(c) What is the area of the above shape?

(d) What are the areas of each of your enlarged shapes?

(e) How are the areas of the enlarged shapes related to the scale factors?

17 (a) Write down the volume of a cube of side:
(i) 5 cm   (ii) 10 cm

(b) What happens to the volume of the cube when the lengths of its sides are doubled?

18 (a) Write down the volume of a cube of side:
(i) 2 cm   (ii) 6 cm

(b) What happens to the volume of the cube when the lengths of its sides are trebled?

2.16 *Speed, distance and time*

The average speed of a car, which travels 240 miles in 4 hours, is 60 miles per hour.

$$\text{average speed} = \frac{\textbf{distance travelled}}{\textbf{time taken}}$$

1   Find the average speed of a car which does:
    (a) 80 miles in two hours
    (b) 96 miles in three hours
    (c) 32 miles in half an hour
    (d) 12 miles in quarter of an hour
    (e) 15 miles in twenty minutes
    (f) 69 miles in one and a half hours.

2   A train leaves Leamington at 7.42 a.m. and travels 96 miles to London in two hours.
    (a) At what time will the train arrive?
    (b) What is its average speed?

3   A man leaves London by car at 7.00 a.m. and has travelled 84 miles by 9.00 a.m.
    He arrives at 11.00 a.m. having covered the remaining 136 miles of his journey to Bude. What is his average speed:
    (a) during the first two hours
    (b) during the second two hours
    (c) over the whole of his journey?

4   Anne can cycle 56 miles in four hours. David can cycle 50 miles in three hours twenty minutes.
    (a) What is Anne's average speed?
    (b) What is David's average speed?
    (c) How much faster does David cycle?
    (d) If they both started together and cycled at these speeds, how much further will David have cycled in one hour?

5   A van travels a distance of 168 kilometres along a motorway in three hours.
    (a) Find the average speed of the van.
    (b) How far does it travel in 4 hours if it keeps up this speed?

The distance travelled in 3 hours, by a man on a bicycle doing 12 miles per hour, is 36 miles.

distance travelled = average speed × time taken

6   How far can a man cycle doing:
    (a) 12 miles per hour for 2 hours
    (b) 10 miles per hour for 6 hours
    (c) 14 miles per hour for half an hour
    (d) 16 miles per hour for quarter of an hour
    (e) 18 miles per hour for twenty minutes
    (f) 12 miles per hour for 2½ hours?

7   A coach left Coventry at 9.11 a.m. and travelled to Luton at an average speed of 56 miles per hour. The journey took one hour and a quarter.
    (a) At what time did the coach arrive?
    (b) How far is Luton from Coventry?

8   Sally can cycle at 16 miles per hour. Julie can cycle at 19 miles per hour.
    (a) How far can each cycle in three hours?
    (b) How much further can Julie cycle than Sally in two hours?

9   At noon David sets out from a point S and walks in a straight line at a constant speed of 8 km per hour. At 1.00 p.m. Susan sets out from S and walks in a straight line in the opposite direction at 5 km per hour. How far apart are they at 2.00 p.m?

10 For the first two hours of his journey from York to London Andy drove at an average speed of 50 miles per hour. For the rest of the way he drove at an average speed of 64 miles per hour for one and a half hours.
    (a) How far did Andy travel:
        (i) in the first two hours
        (ii) in the last hour and a half?
    (b) How far is it from York to London?
    (c) What was Andy's average speed for the whole journey?

Sarah cycles the 32 miles to her grandmother's house at an average speed of 16 m.p.h. It takes her two hours.

$$\textbf{time taken} = \frac{\textbf{distance travelled}}{\textbf{average speed}}$$

11 Find the time taken for each journey.
   (a) 24 miles at an average speed of 12 m.p.h.
   (b) 60 miles at an average speed of 15 m.p.h.
   (c) 30 miles at an average speed of 60 m.p.h.
   (d) 12 miles at an average speed of 48 m.p.h.
   (e) 45 miles at an average speed of 30 m.p.h.
   (f) 80 miles at an average speed of 64 m.p.h.

12 A train leaves Exeter at 4.30 p.m. It travels the 174 miles to London at an average speed of 58 m.p.h.
   (a) How long does the journey take?
   (b) At what time does the train arrive?

13 Tom can do a 288 mile trip on his motorbike at an average speed of 48 m.p.h., whilst his father covers the same distance in his car at an average speed of 64 m.p.h.
   (a) How long does Tom take to do this trip?
   (b) How long does his father take?
   (c) If Tom left home an hour before his father, who would arrive first?

14 It takes Raj eight minutes to cycle two miles. If he does not alter his speed, how long will it take him to cycle a further four miles?

15 It takes Pia thirty minutes to walk to see her aunt. She can go three times as fast when she is cycling.
   How long does it take her when she is cycling?

16 It takes Rick twenty minutes to drive sixteen miles on the motorway.
   If he does not alter his speed, how long will it take him to drive forty miles?

17 A train takes two and a half hours to travel between two towns 150 km apart.
   (a) Find how much longer the return journey takes if the speed is reduced by 10 km per hour.
   (b) What is its average speed overall?

18 Saleem leaves home at the same time each day to cycle to school. On most days he cycles at his normal speed, the journey takes him 15 minutes and he arrives at school ten minutes early.
   (a) How long does it take him on a day when he cycles at twice his normal speed?
   (b) How much sooner will he arrive at school?
   (c) How long does it take him on a day when he cycles at only half his normal speed?
   (d) How late will he now be for school?

19 Penny can run the 1500 metres in exactly 4 minutes.
   (a) If she were able to maintain this speed how far could she run in one hour?
   (b) What is her average speed, in km per hour?
   (c) How long does it take her to do the first 800 metres?

20 Carl Lewis runs the 100 metres in 10 seconds.
   (a) If he were able to maintain this speed how far would he run in:
      (i) one minute
      (ii) one hour?
   (b) What is his average speed, in km per hour?

21 Sebastian Coe runs the 400 metres in 45 seconds.
   (a) If he were able to maintain this speed how far would he run in:
      (i) one minute
      (ii) one hour?
   (b) What is his average speed, in km per hour?

22 A motorist travelling at a steady speed of 120 km per hour usually covers a section of motorway in 37.5 minutes.
   One day a speed limit was imposed and he took 12.5 minutes longer to cover the same section.
   What was the speed limit?

2.17 *Direct and inverse proportion*

Two quantities are in **direct proportion** if they increase or decrease at the same rate.
When you buy milk, which is sold by the bottle, you would expect to pay *twice* as much if you buy *twice* the quantity, or *half* as much if you buy *half* the quantity.

1   A litre bottle of Coca-Cola costs 53p.
   (a) What is the cost of buying:
      (i) 5 bottles   (ii) 12 bottles?
   (b) Now write down the cost of:
      (i) 10 bottles   (ii) 6 bottles.
   (c) How many bottles can you buy for £15.90?

2   A 2 lb bag of sugar costs 42p.
   (a) What is the cost of buying:
      (i) 4 bags   (ii) 15 bags?
   (b) Now write down the cost of:
      (i) 12 bags   (ii) 5 bags.
   (c) How much sugar can you buy for £10.08?

3   Mrs Patel can buy 5 lb of carrots for 65p.
   (a) How much will she pay for:
      (i) 15 lb   (ii) 1 lb   (iii) 3 lb?
   (b) How many pounds can she buy for £3.90?

If 12 tickets at the cinema cost £27, use the unitary method to find how much 5 tickets cost.
      12 tickets cost £27
   so 1 ticket costs £27 ÷ 12   = £2.25
   so 5 tickets cost £2.25 × 5 = £11.25

4   7 bus fares cost £2.80.
   (a) What is the cost of 1 fare?
   (b) What is the cost of 5 fares?

5   4 tape cassettes cost £5.
   (a) What is the cost of 1 cassette?
   (b) What is the cost of 5 cassettes?

6   24 exercise books cost £18.
   (a) What is the cost of 15 exercise books?
   (b) What is the cost of 32 exercise books?

7   Mrs Payne bought eight metres of hall carpet for £47.92. Her friend Jean liked the carpet but only needed five metres. What will this cost?

8   The ingredients for a cake which will serve eight people are as follows:

> 80 grams of flour, 64 grams of sugar, 4 eggs, 36 grams of butter, 500 grams of fruit, and 2 tablespoons of milk

   What quantities will be needed in a similar cake for 12 people?

9   A train travels 96 miles in three hours. How far will it travel in five hours?

10   A car travels 160 miles on five gallons of petrol. How much petrol is needed for a journey of 240 miles?

When a car travels at a constant speed the amount of petrol used is **directly related** to the distance travelled.

11   The petrol consumption figures quoted for a popular family car are:

urban	43 miles per gallon
56 m.p.h.	55 miles per gallon
75 m.p.h.	42 miles per gallon

   (a) How many miles would you expect to travel on a nine gallon tank of petrol at a constant:
      (i) 56 m.p.h.   (ii) 75 m.p.h?
   (b) How many gallons would you expect to use doing 100 miles around the town?
   (c) If you have 2 gallons left in the tank how much further can you go at 56 m.p.h. than at 75 m.p.h?

12 The manufacturers say their car will do 63 miles on each gallon of petrol at 56 m.p.h. If the car travels at this speed, how far would you expect to be able to travel on:
(a) 2 gallons (b) 8 gallons (c) 10.5 gallons?

A jet aeroplane can fly 600 miles in one hour. If its speed was *doubled* it would take only *half an hour* to fly the same distance.
On the other hand if the speed were *reduced* to a third it would take *three times as long*.
Quantities like this are in **inverse proportion**.

---

13 For a particular flight, a plane flying at 300 m.p.h. takes 40 minutes.
How long will the plane take for the same flight, when its speed is:
(a) 600 m.p.h. (b) 150 m.p.h. (c) 400 m.p.h?

14 For a particular journey a car travelling at 48 m.p.h. takes 2 hours.
At what average speed would the car be travelling if for the same journey it took:
(a) 1 hour (b) 3 hours
(c) 1 hour 30 minutes?

15 It takes two men three days to build a wall. If the men always work at the same speed, how many days will it take:
(a) one man (b) four men (c) three men?

If it takes 5 men 6 days to paint a house, how long will it take 3 men to paint the same house?
5 men take 6 days
so 1 man will take $5 \times 6 = 30$ days
so 3 men will take $30 \div 3 = 10$ days

---

16 It takes 5 girls 8 hours to pick 240 kg of strawberries.
How long would it have taken 4 girls to pick the same 240 kg?

17 It takes 6 boys 5 hours to pick 300 kg of apples.
How long would it have taken 10 boys to pick the same amount?

18 Three cub scouts take 45 minutes to wash six cars.
How many cars can the three cub scouts wash in one hour?

19 A factory employs 12 men and makes 240 tennis racquets in four weeks.
(a) How many men would be needed if these tennis racquets had to be made in
(i) two weeks (ii) three weeks?
(b) How many tennis racquets could be made in four weeks by:
(i) 9 men (ii) 15 men (iii) 25 men?
(c) How long will it take the 12 men to make:
(i) 300 racquets (ii) 180 racquets?

20 (a) Copy and complete the table for a car travelling at a constant speed.

Number of gallons	0	1	2	3	4	5	6
Miles covered			70				

(b) Draw a graph to show this information.
(c) Join the points with a straight line.
(d) Use your graph to find the number of:
(i) miles covered on 3.5 gallons
(ii) gallons needed for 80 miles.
(e) Check that when you double the miles you also double the gallons used.

21 (a) Copy and complete the table for a plane flying 3000 miles at different speeds.

Time in hours	0	1	2	3	4	5	6
Speed in m.p.h.						600	

(b) Draw a graph to show this information.
(c) Join the points with a smooth curve.
(d) Use your graph to find:
(i) the speed for a time of 2.5 hours
(ii) the time for a speed of 800 m.p.h.
(e) Check on your graph that when you double the speed you halve the time.

22 30 students address 540 letters in 36 minutes.
(a) How long will they take to address 900?
(b) How many letters can they address in 100 minutes?
(c) How many students could address the same number of letters in 30 minutes?
(d) How many letters can 20 students address in one hour?

2.18 *Problems and investigations*

1 Part of a continuous cinema programme is shown below.
Use the given information to fill in the missing times.

Advertisements	1400	1650		2230
Tom and Jerry	1405	1655	1945	
The Outback		1710	2000	2250
Intermission	1500		2040	
Goldfinger	1510	1800		2340

2 A particular type of carpet comes in 4-metre widths, cut to any length, or can be bought in exact sizes.
Mr Payne wants to carpet the following floor areas.

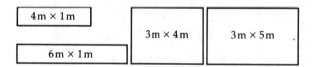

(a) What is the area of floor that is to be covered?

(b) What is the shortest length of carpet he needs to give the minimum amount of waste?

(c) Is it cheaper for Mr Payne to buy a length of carpet which costs £8.95 per square metre, or to pay £9.50 per square metre for what he actually uses?

3 What is the longest pole that can be carried horizontally along the corridor shown below?

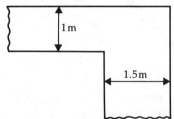

What difference will it make if the pole is 20 cm wide?

4 If you use electricity, you can choose to be charged at the normal tariff, or on the Economy 7 rate. These rates are:

	Standing charge	Cost per unit
Normal tariff	£6.40	5.7p
Economy 7	£9.80	1.9p

(a) Draw a graph to show the cost of using up to 100 units on each tariff.

(b) Use your graph to find when Economy 7 becomes a more economical tariff.

5 In a recent pay negotiation three different offers were made by the management side. These were:

Flat rise:	£10 per week for everyone
Split offer:	5% of wages plus £4 per week
Percentage:	8% of wages

(a) Draw a graph to show the amounts of the pay offers on weekly wages from £100 to £150.

(b) Use your graph to find which weekly wages would benefit most from each offer.

6 Rachel is investigating coloured borders using square floor tiles.
Find the number of tiles needed to form a border around the following types of shape:
(a) a line of tiles:

(b) a rectangle of tiles:

(c) other shapes:

7 (a) Using your calculator find:

(i) $\frac{1}{7}$ (ii) $\frac{2}{7}$ (iii) $\frac{3}{7}$.

(b) Compare your results. What do you notice?

(c) Can you predict how the decimals will start off for:

(i) $\frac{4}{7}$ (ii) $\frac{5}{7}$ (iii) $\frac{6}{7}$?

(d) Check your answers to see if you were right.

8 (a) Using your calculator find:

(i) $\frac{1}{13}$ (ii) $\frac{3}{13}$ (iii) $\frac{4}{13}$.

(b) Compare your results. What do you notice?

(c) Which is the next 'thirteenth' which uses the same figures?

(d) Find $\frac{2}{13}$.

(e) Investigate the patterns of figures in the other 'thirteenths'. What do you notice?

9 Sarah and Pasha both work 'non-standard' weeks in a busy holiday hotel. Sarah has a day off every six days, whilst Pasha only has one off every nine days.

(a) If Pasha's day off is today and Sarah's day off is tomorrow, when will they next have a day off together?

(b) If they both had the same day off one week, how long would it be before they next had a day off together?
What about the one after that?

(c) Investigate other pairs of 'non-standard' weeks and different days off.

10 At the end of a singles tennis match Josie and Tina shake hands.
They each shake hands with the umpire, so there are now three handshakes.

(a) If John and Pato join them for a mixed doubles match and they all shake hands at the end, how many handshakes are possible?

(b) What if they now all shake hands with the umpire as well?

(c) Try to predict the number of possible handshakes with different numbers of people.

11 On a piece of squared paper two squares of different sizes are drawn.

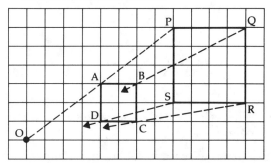

(a) Copy the diagram and continue the lines joining the vertices, as shown above. What do you notice?

(b) Measure OA and OP.
How are these two lengths connected to the sizes of your two squares?

(c) Now repeat this for:

(i) the same squares in different positions

(ii) two other squares of different sizes.

(d) Try to make some general statements about the lines, the lengths and the sizes of the squares.

12 Bob is given £100 by his uncle.
He wants to invest it to earn as much interest as possible.
He discovers that whilst all the building societies are offering 12% interest
    some add the interest monthly
    some quarterly
    some half yearly
and some only once a year.

(a) Use your calculator to find what the £100 will be worth at the end of one year if:

(i) 12% is added on at the end of the year

(ii) 6% is added on at the end of each half year

(iii) 3% is added on at the end of each three months

(iv) 1% is added on at the end of each month.

(b) Compare the best return with the worst return if the £100 is left in for ten years.

2.19 *Oral test*

1   What is the cost of eight jackets at twenty pounds each?

2   What is the cost of eight drinks at ninety-nine pence each?

3   What is the cost of eight note-pads at forty-five pence each?

4   Jason bought twelve cards for a total cost of ninety-six pence.
How much were they each?

5   Joelle bought twenty-five cans of Coke for a total cost of ten pounds.
How much were they each?

6   Tani bought eight cups of coffee for a total cost of two pounds and eighty-eight pence.
How much were they each?

7   The Ahmed family eats two-thirds of a sponge cake every day.
How many cakes will they eat in one week?

8   £18 is to be split in the ratio two to seven.
How much is each part?

9   Saleem bought a cup of tea and a chocolate bar for a total cost of forty-two pence. The cup of tea cost him twenty-five pence.
How much was the chocolate bar?

10 Mr Inan's antique clock was made in eighteen hundred and twenty.
How long ago was that?

11 Pato's uncle is eighty-seven this year.
In which year was he born?

12 Jo arrived at the race twenty minutes before it started. His friend Jim was five minutes late.
How long was Jo there before Jim arrived?

13 Sally arrived at the race twenty-three minutes before it started. Her friend Sue arrived nine minutes before it started.
How long had Sally been there when Sue arrived?

14 When Mrs Green put some vegetables into her deep freeze the temperature was nine degrees Celsius.
When she came to take them out four hours later the temperature was minus five degrees Celsius.
How much did the temperature fall in this time?

15 Winston bought a car radio marked at thirty pounds plus VAT at fifteen per cent.
How much VAT did he pay?

16 In a sale all goods are reduced by 12%.
How much will Jason save on a television which originally cost £200?

17 In a sale all goods are reduced by 12%.
How much will Sandra actually pay for a dress which originally cost £20?

18 In a sale all goods are reduced by 12%.
If David actually paid £44 for a suit how much did it originally cost?

19 It takes Javid eight minutes to cycle two miles.
If he does not alter his speed how much longer will it take him to cycle a further four miles?

20 It takes Sucha thirty minutes to cycle to see his aunt. He can drive there three times as fast.
How long does it take him when he drives?

21 It takes Patrick twenty minutes to drive sixteen miles on a motorway.
If he does not alter his speed how long will it take him to drive forty miles?

2.20 *Fact sheet II*

You should be familiar with the following units of time, mass, length, area, volume and capacity.

Time

12 hour clock	10.30 a.m. 8.15 p.m.
24 hour clock	10 30 20 15

Mass

kilograms (kg)
grams (g) $1000 \text{g} = 1 \text{kg}$
pounds (lb)
ounces (oz) $16 \text{oz} = 1 \text{lb}$

Length

kilometres (km)
metres (m) $1000 \text{m} = 1 \text{km}$
centimetres (cm) $100 \text{cm} = 1 \text{m}$
millimetres (mm) $10 \text{mm} = 1 \text{cm}$
miles (M)
yards (yd) $1760 \text{yd} = 1 \text{M}$
feet (ft) $3 \text{ft} = 1 \text{yd}$
inches (in) $12 \text{in} = 1 \text{ft}$

Area

hectares (ha)
square metres (m²) $10\,000 \text{m}^2 = 1 \text{ha}$
square cm (cm²) $10000 \text{cm}^2 = 1 \text{m}^2$
square mm (mm²) $100 \text{mm}^2 = 1 \text{cm}^2$

Volume

cubic metres (m³)
cubic cm (cm³) $1\,000\,000 \text{cm}^3 = 1 \text{m}^3$
cubic mm (mm³) $1000 \text{mm}^3 = 1 \text{cm}^3$

Capacity

litres (l)
centilitres (cl) $100 \text{cl} = 1 \text{l}$
millilitres (ml) $10 \text{ml} = 1 \text{cl}$
gallons
pints $8 \text{ pints} = 1 \text{ gallon}$

Note:
1 cubic centimetre is the same as 1 millilitre
1000 cubic centimetres is the same as 1 litre
1 cubic centimetre of water has a mass of 1 gram

Some useful conversions

$8 \text{km} = 5 \text{ miles}$	$1 \text{kg} = 2.2 \text{lb}$
$1 \text{m} = 39.25 \text{ in}$	$1 \text{l} = 1.75 \text{ pints}$
$2.54 \text{cm} = 1 \text{ inch}$	$4.55 \text{l} = 1 \text{ gallon}$

Percentages

$50\% = 0.5$ $25\% = 0.25$ $75\% = 0.75$
$20\% = \frac{20}{100}$
$\frac{7}{25} = \frac{28}{100} = 28\%$
$5\% \text{ of } £120 = \frac{5}{100} \times £120 = £6$

Ratio

A ratio of 5 to 3 is $5 : 3$ or $\frac{5}{3}$

$4 : 12 = 2 : 6 = 1 : 3$ or $\frac{1}{3}$

Divide 14 in the ratio $3 : 4$

$3 + 4 = 7$ so $\frac{3}{7}$ of 14 is 6 and $\frac{4}{7}$ of 14 is 8

Divide 132 in the proportion $1 : 3 : 7$
$1 + 3 + 7 = 11$
$\frac{1}{11} \times 132 = 12,$
$\frac{3}{11} \times 132 = 36,$
$\frac{7}{11} \times 132 = 84$

Speed

$$\text{Average speed} = \frac{\text{distance travelled}}{\text{time taken}}$$

Distance travelled = speed × time

3.1 *Lines and angles*

1 Measure the length of the line below with your ruler.

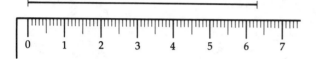

Give your answer correct to:
(a) the nearest centimetre
(b) the nearest millimetre.

2 (a) Estimate the length of each line to the nearest centimetre.

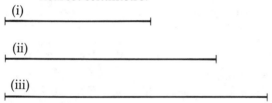

(i)

(ii)

(iii)

(b) Now measure the length of each line and give your answers correct to the nearest millimetre.

3 (a) Find the perimeter of each shape:

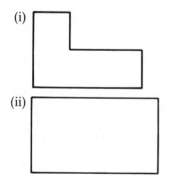

(i)

(ii)

(b) In (ii), how much longer is the rectangle than it is wide?

4 Draw a line of the given length:
(a) 6 cm (b) 4 cm 5 mm (c) 3.4 cm.

5 A ribbon is cut into four pieces of lengths:
2.5 cm, 3.4 cm, 4.7 cm and 5.9 cm.
How long was the ribbon originally?

An angle of 90° is called a **right angle**.

An angle which is less than a right angle is an **acute angle**.

An angle between 90° and 180° is an **obtuse angle**.

An angle which is more than 180° but less than 360° is a **reflex angle**.

6 (a) Say whether the angle is acute, obtuse, reflex or a right angle.

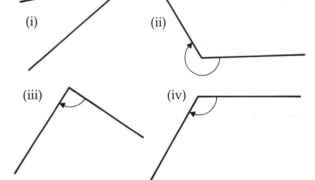

(i)   (ii)

(iii)   (iv)

(b) Now use your protractor to measure each angle.

7 Use your protractor to draw accurately angles of:
(a) 50° (ii) 67° (c) 120° (d) 133° (e) 235°

8 Draw a line AB of length 7.5 cm. Construct angles of 45° and 57° at A and B. Complete the triangle ABC as shown in the sketch below.
Measure AC and BC.

The angle at a point is 360°.

An angle on a straight line is 180°.

These angles are called **corresponding** angles.

These angles are called **alternate** angles.

These angles are called **vertically opposite** angles.

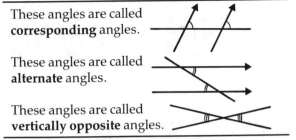

9 Write down the size of the unmarked angle:

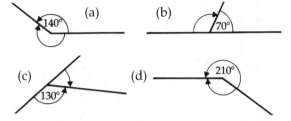

(a) (b)

(c) (d)

The **sum of the angles** of a triangle is 180°.
So $a + b + c = 180°$.

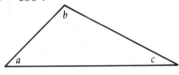

10 Write down the size of the unmarked angle in each triangle.

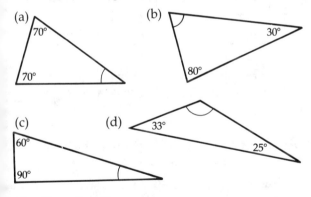

(a) (b)

(c) (d)

11 Copy the diagrams.
Show the sizes of all the unmarked angles.

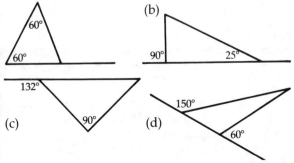

(a) (b)

(c) (d)

12 What is the size of each of the angles marked with a letter?

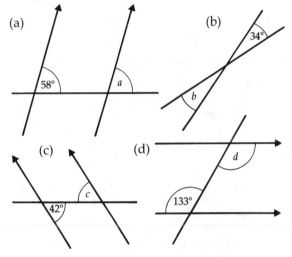

(a) (b)

(c) (d)

13 Write down, giving reasons, the pairs of angles which are the same size.

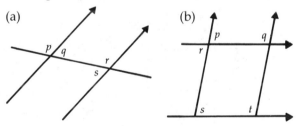

(a) (b)

14 Find, giving reasons, the size of all the unknown angles.

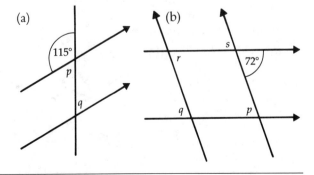

(a) (b)

3.2 *Symmetry and reflection*

The shapes below all have **lines of symmetry**.

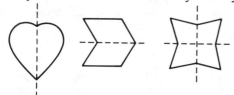

1 Copy the shape.
Mark on it any lines of symmetry.

(a)   (b)

(c)   (d)

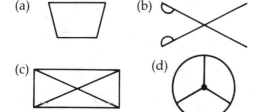

2 Half of a shape is drawn. The dotted line is a line of symmetry of the complete shape.
Draw the completed shape.

(a)   (b)

(c)   (d)

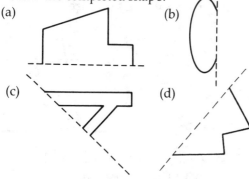

3 Part of a shape is drawn. The dotted lines are lines of symmetry of the complete shape.
Draw the completed shape.

(a)   (b)

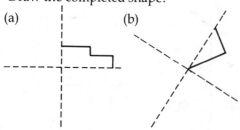

4 The pattern started below has one line of symmetry.
Using squared paper, copy and complete the pattern.

(a)   (b)

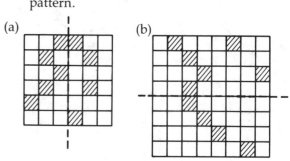

5 The pattern started below has two lines of symmetry.
Using squared paper, copy and complete the pattern.

(a)   (b)

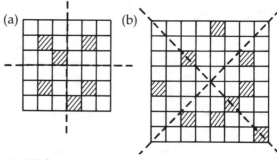

6 Make up a pattern of your own which has:
(a) one line of symmetry

(b) two lines of symmetry

(c) three lines of symmetry.

7 Make a collection of pictures which show objects which have one or more lines of symmetry.

8 (a) If you were to look in a mirror and close your left eye which eye would appear to close in your reflection?

(b) If you were to look in a mirror and rotate your left hand in a clockwise direction in which direction would its reflection appear to move?

9 Copy the diagram and show the reflection of the shape in the mirror line.

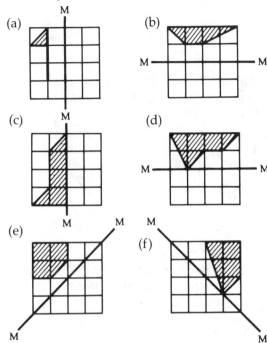

(a)

(b)

(c)

(d)

(e)

(f)

10 Each of the diagrams show an object and its reflection in a mirror line. Copy the diagram and mark in the mirror line.

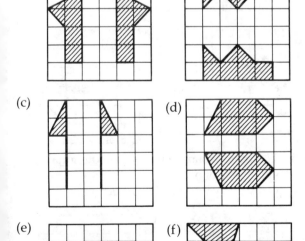

(a)

(b)

(c)

(d)

(e)

(f)

The diagram below shows a flag and its reflections in the x-axis and the line AB.

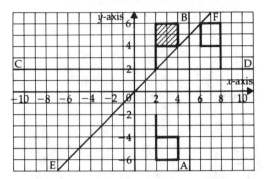

11 Copy the grid above, and show the reflection of the flag in:
   (a) The y-axis
   (b) The line CD
   (c) The line EF.

12 (a) On a grid like the one above, draw the triangle with corners at $(1, 1)$, $(3, 1)$ and $(3, 2)$.
   (b) Draw the reflection of this triangle in:
      (i)   the y-axis
      (ii) the x-axis
      (iii) the line joining $(3, 0)$ to $(3, 5)$
      (iv) the line joining $(0, 0)$ to $(5, 5)$
      (v)   the line joining $(5, 0)$ to $(0, 5)$.

13 Each of the points below is reflected in the line joining $(4, 0)$ to $(4, 6)$.
      $(1, 2)$ $(2, 4)$ $(3, 1)$ $(0, 5)$
   (a) Write down the coordinates of each image.
   (b) Can you say what you have to do to the original coordinates to obtain the coordinates of the image?

14 Repeat question **13** when the points are reflected in the line joining:
   (a) $(0, 6)$ to $(6, 6)$
   (b) $(0, 0)$ to $(6, 6)$.

15 Draw a shape of your own choice on a grid like the one above. Show reflections of your shape in:
   (a) the x-axis   (c) the line $y = n$
   (b) the y-axis   (d) the line $y = x$.

3.3 *Symmetry and rotation*

The shapes below all have **rotational symmetry** of the given order.

order 4   order 3   order 5

1 Copy the shape.
Give the order of rotational symmetry.
Mark on it any lines of symmetry.

(a)   (b)

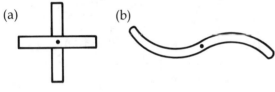

(c)   (d)

2 Half of a shape is drawn below. The shape has rotational symmetry of order 2.
Draw the completed shape.

(a)   (b)

(c)   (d)

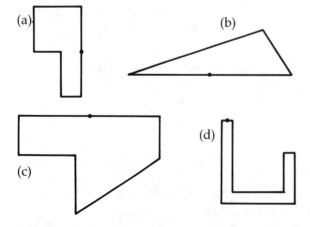

3 Part of a shape is drawn below. The shape has rotational symmetry of order 4.
Draw the completed shape.

(a)   (b)

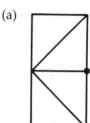
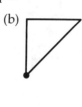

4 Part of a shape is drawn below. The shape has rotational symmetry of order 3.
Draw the completed shape.

(a)   (b)

5 The pattern started below has rotational symmetry of order 2.
Copy and complete the pattern.

(a)   (b)

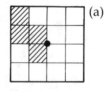
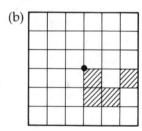

6 The pattern started below has rotational symmetry of order 4.
Copy and complete the pattern.

(a)   (b)

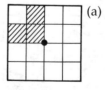
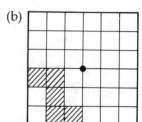

7   Make up a pattern of your own which has rotational symmetry of:
(a) order 2   (b) order 4
(c) order 3.

8   Make a collection of pictures which show objects which have rotational symmetry.

9   Copy the diagram and show the shape when it is given a half-turn rotation about the marked point.

(a) 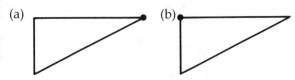 (b)

(c) 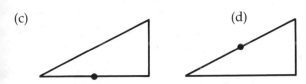 (d)

10   Copy the diagram and show the shape when it is given a quarter-turn rotation, in an anti-clockwise direction, about the marked point.

(a) 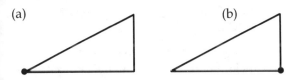 (b)

11   Each of the diagrams shows an object and its image after a half-turn rotation.
Copy the diagram and mark the centre of rotation.

(a)
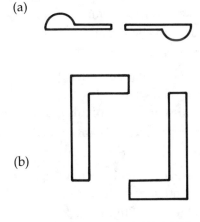
(b)

The diagram below shows a flag and its images after a half-turn rotation about A and after a quarter-turn rotation about B.

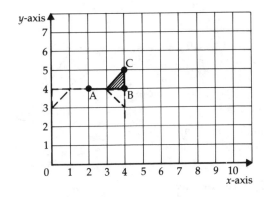

12   Copy the grid above and show the image of the flag after:
(a) a half-turn rotation about B
(b) a half-turn rotation about C
(c) a quarter-turn (clockwise) rotation about A.

13   (a) On a grid like the one above draw the triangle with points at (1, 1), (3, 1) and (3, 2).
(b) Draw the image of this triangle after a rotation of:
(i)   a half-turn about (3, 2)
(ii) a half-turn about (4, 2)
(iii) a half-turn about (2, 2)
(iv) 90° (anti-clockwise) about (1, 1)
(v)   90° (clockwise) about (3, 1)
(vi) 90° (clockwise) about $(1\frac{1}{2}, 1\frac{1}{2})$.

14   Each of the points below is given a half-turn rotation about (0, 0).
    (1, 0)   (0, 4)   (3, 1)   (2, 5)
(a) Write down the coordinates of each image.
(b) Can you say what you have to do to the original coordinates of each point to obtain the coordinates of the image?

15   Repeat question 14 when the points are given an anti-clockwise quarter-turn rotation about (0, 0).

3.4 *Triangles*

An **equilateral** triangle has three angles the same size and three sides the same length.

An **isosceles** triangle has two angles the same size and two sides the same length.

A **scalene** triangle has no angles or sides the same size.

The **sum of the angles** of any triangle is 180°.

1 Find the size of the third angle in each triangle.

(a)

70°   40°

(b)

50°

40°

(c)

30°   30°

(d)

110°   30°

2 For each part of question **1**, say whether the triangle is:
(a) equilateral, isosceles or scalene
(b) right-angled, acute-angled or obtuse-angled.

3 On a copy of each diagram show the size of each unknown angle.

(a)

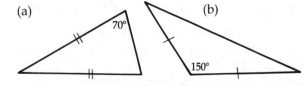

70°

(b)

150°

You can draw any triangle when you are given the lengths of the three sides.
First draw one side, then use your compasses as shown for each of the other two sides.

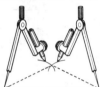

4 (a) What is the size of each angle in an equilateral triangle?

(b) Using your compasses, construct an equilateral triangle with each side 6 cm.

(c) Check with your protractor that the angles are 60°.

(d) Show on your triangle any lines of symmetry.

(e) What order of rotational symmetry has this triangle?

5 (a) Use your set-square to measure the height of the triangle in question **4**.

(b) Now find the area of this triangle.

6 (a) Construct a triangle with two sides of 6 cm and one side of 4 cm.

(b) Measure each angle of this triangle.

(c) Show any lines of symmetry.

(d) Measure the height of this triangle.

(e) Find the area of the triangle.

7 (a) Construct a triangle with sides 4.5 cm, 6 cm and 7.5 cm.

(b) Measure each angle of this triangle.

(c) What type of triangle have you drawn?

(d) What is the area of this triangle?

8 (a) Construct a triangle with two sides of 13 cm and one of 10 cm.

(b) Find the area of this triangle.

You can draw any triangle when you are given the lengths of two sides and the size of the angle between them.
First draw one side, then use your protractor to draw the angle, and then mark off the second side.

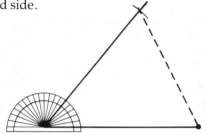

9 (a) One angle of an isosceles triangle is 50°. What is the size of the other two angles if they are equal?

(b) Construct a triangle with two sides of 6 cm and the angle between them 50°.

(c) Measure the other two angles and check your answer to (a).

10 (a) Construct a triangle with sides of 10 cm and 5 cm, and the angle between them 60°.

(b) Measure the length of the third side.

(c) Measure the other two angles.

(d) Find the area of this triangle.

You can draw any triangle when you are given the length of one side and the size of two angles.
First draw the side, then use your protractor to draw the two angles.

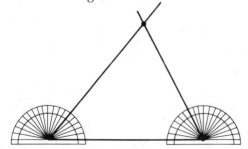

11 (a) Construct a triangle with angles of 35° and 55°, and the side between them of 8 cm.

(b) Measure the third angle, and the lengths of the other two sides.

When the angle between two lines is 90° they are **perpendicular**.
You can construct a line perpendicular to another using your compasses, as shown below.

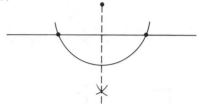

12 Draw a line AB of length 8 cm. Mark a point P above the line. Use your compasses to draw the perpendicular from P to AB.

13 (a) Construct a triangle ABC with AB = 8 cm, BC = 10 cm and CA = 9 cm.

(b) Construct the perpendicular from A to BC.

(c) Measure the length of this line and hence find the area of the triangle.

14 (a) Repeat question **13**, constructing the perpendiculars from C to AB, and B to AC.

(b) Do you get the same area each time?

(c) Did you find that the three perpendiculars all met at the same point?

You can find the **midpoint** of a line AB using the above ideas.
Using your compasses, form two **arcs** as shown. Draw a line joining the two points where these arcs meet.
This line **bisects** AB and is **perpendicular** to AB.

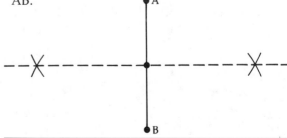

15 (a) Redraw the triangle in question **13**.

(b) Construct the perpendicular bisectors of each side. What do you notice?

3.5 *Quadrilaterals and polygons*

You should recognise each of these shapes and know their **names** and main **properties**.

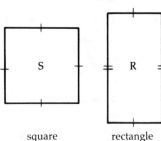 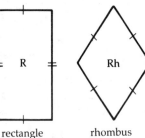 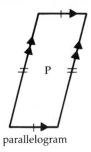 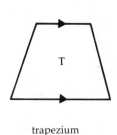 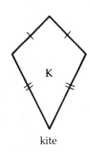

square	rectangle	rhombus	parallelogram	trapezium	kite

1 (a) Make a copy of each of the six quadrilaterals above.

   (a) On each show any lines of symmetry.

   (c) For each, say if the shape has rotational symmetry and, if so, of what order.

2 (a) Make a cut-out copy of each shape which has lines of symmetry.

   (b) Fold the shape in half along its line of symmetry and write down any properties of the shape that this shows.

3 (a) Make a cut-out copy of each shape which has rotational symmetry.

   (b) Draw around the outline of the shape.

   (c) Give the shape a half-turn rotation and write down any properties that this shows.

4 Copy and complete the table below which shows the main properties of each shape.

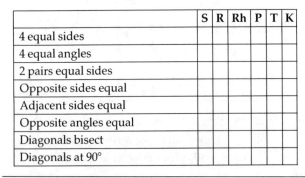

	S	R	Rh	P	T	K
4 equal sides						
4 equal angles						
2 pairs equal sides						
Opposite sides equal						
Adjacent sides equal						
Opposite angles equal						
Diagonals bisect						
Diagonals at 90°						

5 Copy each shape and show the sizes of the unknown sides and angles.

(a)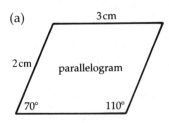

(b)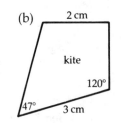

Any quadrilateral can be split into two triangles.

The **sum of the angles** of any quadrilateral is 360°.

So $a + b + c + d = 360°$

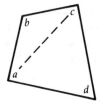

6 Write down the size of the unknown angle in each shape.

(a)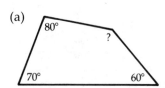

(b)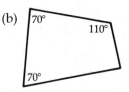

7 Two of the angles in a quadrilateral are each equal in size, and double the size of the other two angles.
Find the size of each of the angles.

You should recognise each of these shapes and know their **names** and main **properties**.

pentagon
(5 sides)

hexagon
(six sides)

octagon
(eight sides)

decagon
(ten sides)

A **regular polygon** has all its sides the same length and all its angles the same size.

8 (a) Copy each of the regular polygons shown above.

(b) Show on your copies any lines of symmetry.

(c) For each polygon say what the order of rotational symmetry is.

9 (a) Copy and complete the table below for regular polygons.

	Number of sides	Lines of symmetry	Order of rot. sym.
Triangle	3		
Square	4	4	4
Pentagon	5		
Hexagon	6		
Octagon	8		
Decagon	10		
n-agon	n		

(b) Look at the results in your table. What can you say about the number of lines and the order of rotational symmetry for any polygon?

10 Look again at the regular polygons.
   (a) How many diagonals does each one have?

   (b) Can you say how many diagonals a polygon will have with:
   (i)   12 sides (ii) 17 sides
   (iii) 51 sides (iv) n sides?

Any polygon can be divided into triangles.

(i)

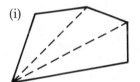

(ii)

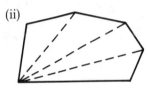

The sum of the angles in a polygon is 180° multiplied by two less than the number of sides.
   $S = (n - 2) \times 180°$

11 Use the above result to complete the table:

Number of sides		3	4	5	6	7	8	10
Sum of angles		180°						

12 Use your results in question **11** to help you to write down the size of each angle in a regular polygon.

13 (a) Use the diagrams below to find the size of each exterior angle of the regular polygon.

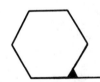

(b) Multiply each of your results in (a) by the number of sides of the polygon. What do you notice? Does this always work?

3.6 *Circles and loci*

You should know the names of these parts of a circle and be able to identify them.

- centre
- radius
- diameter
- circumference
- semicircle
- chord
- arc
- sector
- segment
- tangent

circumference
radius
diameter
sector
tangent
chord
segment
arc

1 (a) Use your compasses to draw a circle with radius 5 cm.

   (b) Draw any diameter AB and mark any two points P and Q on the circumference.

   (c) Join PA and PB. Measure angle APB.

   (d) Join QA and QB. Measure angle AQB.

   (e) Repeat (c) for other points on the circumference. What do you notice?

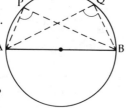

2 (a) Draw a circle with radius 6 cm.

   (b) Draw any chord CD and mark two points R and S on the circumference.

   (c) Join RC and RD. Measure angle CRD.

   (d) Join SC and SD. Measure angle CSD.

3 Repeat question **2** for other points on the circumference. What do you notice?

4 Repeat question **2** for points on the circumference on the opposite side of the chord CD. What happens this time?

5 Add the angles you found in questions **2** and **4**. Is the sum 180°?

6 (a) Repeat questions **2**, **4** and **5** for a different chord CD.

   (b) Is the sum of the angles still 180°?

7 (a) Draw a circle with radius 6 cm and mark its centre C.

   (b) Draw any chord PQ.

   (c) Show on your circle any lines of symmetry.

   (d) Cut out the circle and fold it along the line of symmetry. What do you notice?

8 In question **7**, do you agree that the chord:
   (a) is bisected by the line of symmetry
   (b) is perpendicular to the line of symmetry?

9 (a) Draw a circle with radius 5 cm and mark a point P on the circumference.

   (b) Join point P to the centre of the circle C.

   (c) Draw a line through P which makes an angle of 90° with PC as shown below.

   (d) Do you agree the line in (c) touches the circle at P?

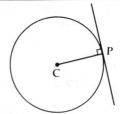

A line which touches a circle is called a **tangent**. The tangent to a circle, at a point P on its circumference, is **perpendicular** to the radius through P.

The diagram below shows all the points which are 2 cm from the point C.
They lie on the circumference of a circle with centre C and radius 2 cm.

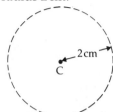

10 (a) Mark two points P and Q which are 8 cm apart.

(b) Show the points which are 5 cm from P.

(c) Show the points which are 6 cm from Q.

(d) How many points are 5 cm from P and 6 cm from Q? Label them X and Y.

11 (a) The diagram below shows two points which are the same distance from A and B. Can you see how they were obtained?

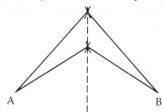

(b) Construct six other points which are the same distance from A and B.

(c) What do you notice about the points in (a) and (b)?

12 (a) Draw two lines XY and XZ which meet at X.

(b) Show all the points which are the same distance from X and Y.

(c) Show all the points which are the same distance from X and Z.

(d) How many points are the same distance from all three of the points X, Y and Z?

13 (a) Draw a line AB which is 5 cm long.

(b) Show all the points which are 3 cm from this line.

(c) How would you mark out a running track?

14 Farmer Hale has a goat which he keeps tethered, using a rope which is 4 metres long.

Draw a diagram to show the grass that the goat can graze if the rope is:
(a) fixed to a pole P

(b) allowed to slide on a horizontal bar which is 10 metres long

(c) fixed to the corner of a rectangular shed which is 8 m long and 6 m wide.

15 For each part of question **14**, find the area of the ground on which the goat can graze.

16 The diagram below shows four houses set back from a road. Two bus stops are to be sited. One is to be the same distance from A and B, the other the same distance from C and D.

Copy the diagram and use a construction to find out where the bus stops should be sited.

17 The Bell Telephone Company wishes to install a new telephone distribution box so that it is the same distance from each of three houses. The distances between the houses are 2, 3 and 4 kilometres respectively.
(a) Make a scale drawing and find where the box should be sited.

(b) Use your drawing to find the length of cable needed to connect the box to each house.

3.7 *Perimeter and area: rectangles*

The **perimeter** of a shape
is the distance
around the outside
of the shape.

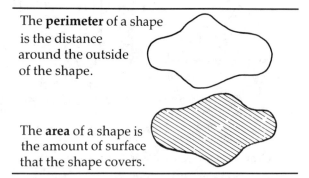

The **area** of a shape is
the amount of surface
that the shape covers.

1 Find the perimeter of each shape.

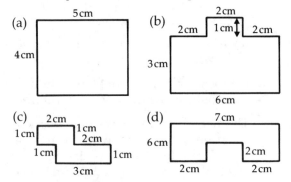

2 Copy and complete these statements.
(a) The perimeter of a square is . . . the
length of one side.

(b) The perimeter of an equilateral triangle is
. . . the length of one side.

(c) The perimeter of a regular hexagon is
. . . the length of one side.

(d) The perimeter of a regular polygon is . . .
the length of one side.

3 Describe in your own words how you would
find the perimeter of a rectangle.

4 A rectangle has sides of 3.5 cm and 5.5 cm.
Write down three different ways of finding
the perimeter.

5 A rectangle has sides of p cm and q cm.
Write down three different ways of finding
the perimeter.

6 Find the area of each rectangle.

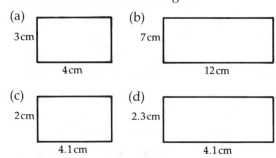

7 Describe how you would find:
(a) the area of a rectangle

(b) the length of one side of a rectangle if
you know the area and the other side.

8 In each diagram you are told the area of the
rectangle and the length of one side. Find the
length of the other side.

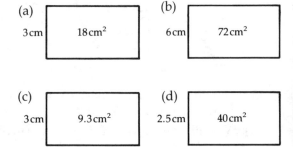

With more complicated shapes you can find
the area if you can divide the shape into
separate rectangles, as shown below.

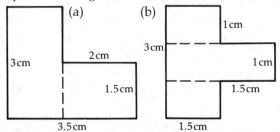

9 For each of the shapes above find:
(a) the area, in two different ways

(b) the perimeter.

10 Find the area of each of the shapes, if they are drawn on a 1 cm grid.

(a) 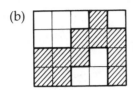 (b)

11 Find the area of each of the shapes in question **10** if they had been drawn on:
(a) a 2 cm grid   (b) a 3 cm grid
(c) a 4 cm grid   (d) a 10 cm grid.

12 Say how the area of each of the shapes in question **10** changes as the size of the grid is changed.

13 Find the area of a square of side:
(a) 6 cm   (b) 12 cm
(c) 2.1 mm   (d) 6.3 cm.

14 Copy and complete each of these statements.
(a) When the side of a square is doubled the area is . . . .
(b) When the side of a square is trebled the area is . . . .

15 Find the length of the side of a square if its area is:
(a) 9cm^2 (b) 64cm^2
(c) 400cm^2 (d) 6.25cm^2.

16 Sally wishes to build a patio using square paving slabs. She is going to cover the area shown in the diagram below.

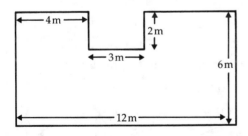

(a) Find how many paving slabs she will need for her patio if each one is 1 m by 1 m.
(b) If the cost of each slab is £1.40 and the labour will cost £25 what will be her total bill?

17 Ali wants to paint the walls of his room and needs to know how much paint he will need to buy.
(a) Find the area of each wall excluding any doors or windows.

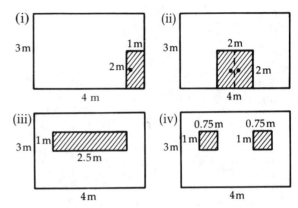

(b) If one litre of paint covers 2m^2, how many litres will he need altogether?
(c) If a 5 litre can of paint costs £6.95, how much will it cost him altogether?

18 Jason has been asked to re-tile a bathroom wall.
(a) Find the area of wall needing tiles.

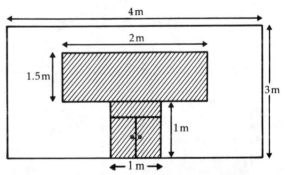

(b) If each tile is 20 cm by 20 cm, how many tiles will be needed altogether, excluding any allowance for breakages?
(c) If a pack of 36 tiles costs £4.75, how many packs will be needed, and what will be the total cost?

19 (a) Use squared paper to make a scale drawing of each of the walls of your bedroom.
(b) Find the total area of the walls excluding any doors or windows.

3.8　*Area: triangles and parallelograms*

1　Find the area of each of the triangles, which are drawn on a 1 cm grid.

(a)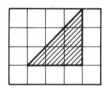

(b)

(c)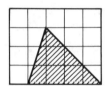

(d)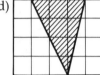

2　(a)　Find the area of each rectangle below.

(i)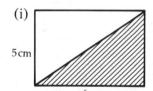

(ii)

(iii)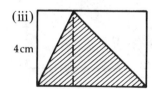

(iv)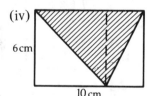

(b)　Using your results for (a) write down the area of each shaded triangle.

3　Find the area of each triangle.

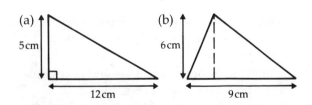

(a)

(b)

The **area of a triangle** is half the area of the surrounding rectangle.

The area of a triangle is $\frac{1}{2} \times$ (base) $\times$ (height)

4　Use the above result to find the area of each triangle.

(a)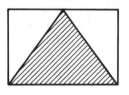

(b)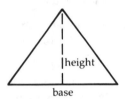

(c)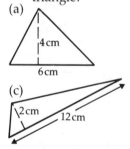

(d)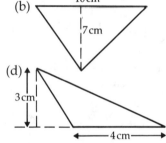

5　Find the area of a triangle with:
(a)　base　4 cm, height　3 cm
(b)　base　12 cm, height　5 cm
(c)　base　5 cm, height　17 cm
(d)　base　3.2 cm, height　4.5 cm
(e)　base　7.3 cm, height　6.8 cm.

6　Find the height of each triangle.

(a)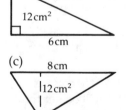

(b)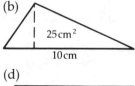

(c)

(d)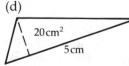

Any parallelogram can be split up into two identical triangles.

The **area of a parallelogram** is (base) × (height)

7 Find the area of the shaded triangle and write down the area of the parallelogram.

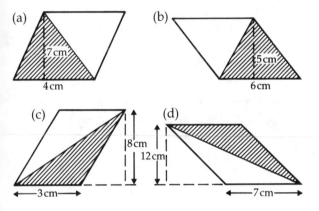

8 Find the area of each parallelogram.

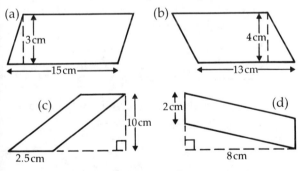

9 Find the height of the parallelogram with:
   (a) area 40 cm², base   5 cm
   (b) area 32 cm², base 16 cm
   (c) area 8.4 cm², base   4 cm
   (d) area 9.6 cm², base 2.4 cm.

10 Find the unknown length in each diagram.

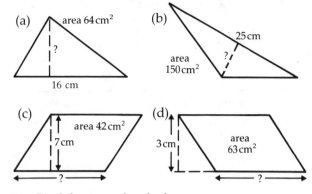

11 Find the area of each shape.

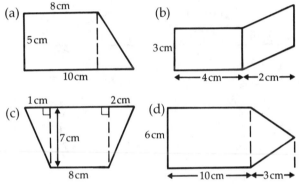

12 The diagram shows a lean-to greenhouse. Find:
   (a) the area of glass in one end
   (b) the total area of glass
   (c) the cost of glazing the greenhouse if glass costs £1.60 per m².

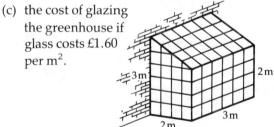

13 The diagram shows one end of a house.

The area of brickwork is 70 m². Find the total height of the roof.

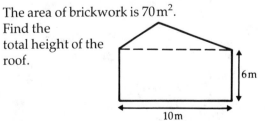

3.9 *Perimeter and area: circles*

The **perimeter of a circle** is the distance around its **circumference**.

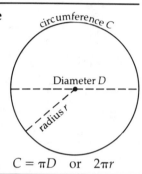

The circumference is calculated from:
either 2 × π × (radius)
or π × (diameter)

π is approximately 3.14. $C = \pi D$ or $2\pi r$

1 Find the circumference of the circle. (Use π = 3.14)

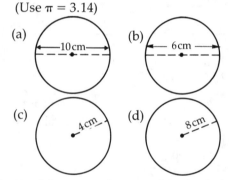

(a) —10cm—   (b) — 6cm —   (c) 4cm   (d) 8cm

2 Find the circumference of the circle for the given diameter. (Use π = 3.14)
(a) 2cm   (b) 5cm   (c) 12cm   (d) 2.5cm

3 Find the circumference of the circle for the given radius. (Use π = 3.14)
(a) 3cm   (b) 8cm   (c) 14cm   (d) 2.5cm

4 Find the diameter of the circle. (Use π = 3.14)

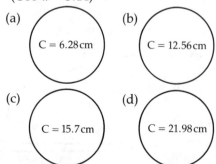

(a) C = 6.28cm   (b) C = 12.56cm   (c) C = 15.7cm   (d) C = 21.98cm

5 (a) Write, in your own words, how you would find the radius of a circle when given the circumference.
(b) Find the radius of a circle with the given circumference. (Use π = 3.14)
(i) 6.28cm   (ii) 12.56cm   (iii) 20.41cm.

6 The diameter of a cotton reel is 4cm.
(a) Find the circumference of the reel.
(b) Find the length of cotton on the reel if it was wound round 500 times.
(c) You are told that there are 100 metres of cotton on the reel. How many times has the cotton been wound round?

7 Mr Cape, the school groundsman, has been asked to mark out the 400 metre track. The diagram below shows what he did.

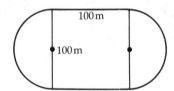

100m   100m

(a) Find the perimeter of this track.
(b) How many metres was it longer than it should have been?

8 What should be the diameter of each semicircle on the running track in question 7 if the perimeter is 400 metres?

9 Kerry is wrapping up a present which is in a cylindrical box as shown below.

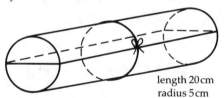

length 20cm
radius 5cm

(a) What is the circumference of the box?
(b) How much string does she need to tie it up, allowing an extra 10cm for the knots?

The area of a circle is calculated from:

$\pi \times (\text{radius})^2$

or

$\pi \times (\text{radius}) \times (\text{radius})$

$A = \pi r^2$

If you are given the area of a circle you can find the radius by working backwards from the result.

radius → | find square | → | × π | → area

radius ← | find square root | ← | ÷ π | ← area

10 Find the area of the circle.
(Use π = 3.14)

(a) 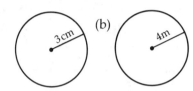 (b)

(c) 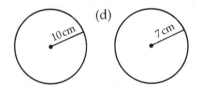 (d)

11 Find the area of a circle with the given radius. (Use π = 3.14)
(a) 2cm (b) 5cm (c) 12cm (d) 2.5cm

12 (a) Write in your own words how you would find the area of a circle when given the diameter.

(b) Find the area of a circle with the given diameter. (Use π = 3.14)
   (i) 4cm   (ii) 10cm
   (iii) 24cm   (iv) 4.2cm

13 Find the area shaded in each diagram.

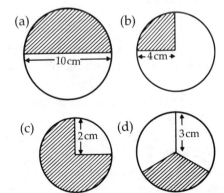

14 Find the radius of the circle for the given area. (Use π = 3.14)
(a) 31.4cm² (b) 12.56cm² (c) 78.5cm²

15 Find the area of a semicircle:
(a) of diameter 6 cm
(b) of radius 8cm.

16 Find the radius of a semicircle of area 14.13cm².

17 Find the area of each shape.

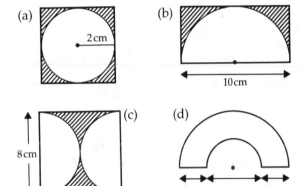

18 Find the area of each shape.

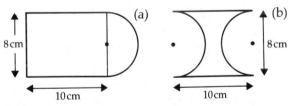

19 A front door is made from a rectangle with a semicircle at the top.
Find the total area of the door if the rectangle is 1 metre wide and 2 metres tall.

20 If the door in question **19** consists of a glass panel surrounded by a frame of timber which is 10cm wide, find the area of glass needed.

3.10 *3-D shapes and their nets*

You should recognise and be familiar with the following common 3-D shapes.

     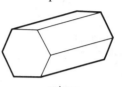

| cube | cuboid | cylinder | cone | sphere | prism | pyramid |

1 Look at the shapes above.
   (a) Copy and complete the table.

	Number of edges	Number of faces	Number of vertices
Cube		6	8
Cuboid			
Prism			
Pyramid			

   (b) For each of the shapes in the table, find the number of faces plus the number of vertices.

   (c) Compare your results for (b) with the number of edges. What do you notice?

2 One of the nets of a cube is like this.

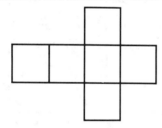

   Which of the following are also nets of cubes?

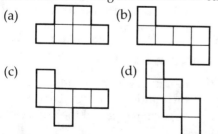

   (a)   (b)

   (c)   (d)

3 One of the nets of a cuboid is like this.

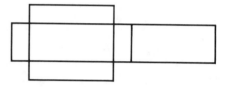

   Draw two other nets for this cuboid.

4 Make a sketch of a net for:
   (a) the cylinder shown at the top of this page
   (b) the pyramid shown at the top of this page.

5 (a) Say what shape you think can be made from the net.

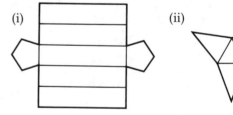

   (i)   (ii)

   (b) Make a sketch of each shape in (a).

6 (a) Make your own collection of as many different cardboard containers as you can find.
   (b) Make a sketch of each one of your containers.
   (c) Draw one of the possible nets for each container.

7 Draw one of the possible nets for a 'Toblerone' box.

8 (a) Find the area of each of the three faces of the cuboid.

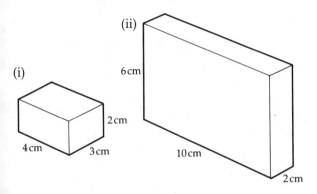

(b) What is the total surface area of each cuboid in (a)?

9 (a) Find the area of each of the rectangular faces of the prism.

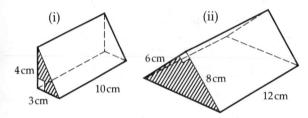

(b) What is the area of the end face of each prism in (a)?

(c) What is the total surface area of each prism in (a)?

10 What is the total surface area of each of the 3-D shapes below?

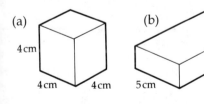

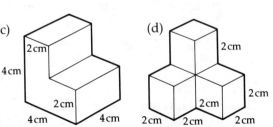

The net of a cylinder is made up of two circles and a rectangle as shown below. The width of the rectangle is the same as the circumference of the cylinder.

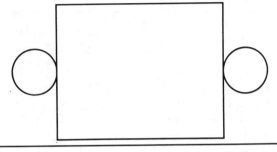

11 (a) Find the circumference of the cylinder. (Use $\pi = 3.14$)

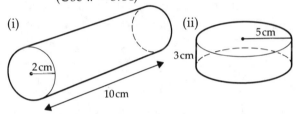

(b) Find the area of the curved surface of each cylinder in (a).

(c) Find the area of the circular end of each cylinder in (a).

(d) Find the total surface area of each cylinder in (a).

12 Mr Packer runs a packaging company. He makes various different shaped cardboard boxes.
Find how much cardboard is needed to make each box. Ignore any flaps or wastage.

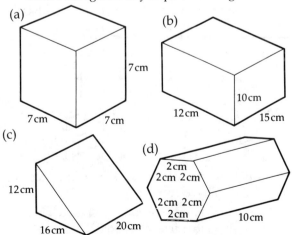

3.11 *Volume: cubes and cuboids*

The **volume** of an object
is the amount of space
that it takes up.
Volume is usually
measured in
cubic metres (m^3),
cubic centimetres (cm^3)
or **cubic millimetres** (mm^3).

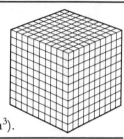

1   Each shape below is made up of 1 cm cubes.
How many 1 cm cubes are there in each one?

(a)   (b) 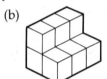   (c)

2   (a) Find the volume of each cube.

(i) 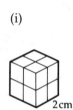   (ii) 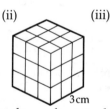   (iii)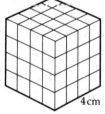

2 cm   3 cm   4 cm

(b) What do you have to do to the length of
the edge of a cube to find its volume?

(c) Find the volume of a cube with edge:

(i) 5 cm   (ii) 6 cm   (iii) 10 cm

3   (a) Write down the number of 1 cm cubes on
the bottom layer of each cuboid.

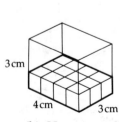   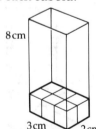

(b) How many layers will there be in each of
the above cuboids?

(c) How many 1 cm cubes will there be
altogether in each cuboid?

4   Repeat question **3** for each of these cuboids.

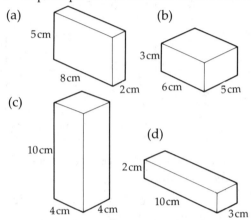

(a)   (b)

(c)   (d)

5   Write in your own words how you would
find the volume of a cuboid.

The volume of a cuboid is obtained by
finding   (length) × (breadth) × (height)

6   Find the volume of each cuboid.

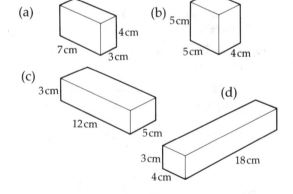

(a)   (b)

(c)   (d)

7   Find the volume of each cuboid with:

	length	breadth	height
(a)	3 cm	4 cm	5 cm
(b)	7 m	8 m	2 m
(c)	6 cm	9 cm	2.5 cm
(d)	12 cm	5 cm	3.6 cm

8   Find the volume of each of the shapes in
question **10**, on page 105.

9 Jane says that a cube of edge 15 cm has a larger volume than a cuboid of length 17 cm, breadth 15 cm and height 13 cm. Is she correct or not?

10 How many of the small boxes below can you fit into the large box?

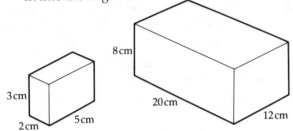

11 A packing case is 50 cm long, 40 cm wide and 30 cm high. How many boxes 5 cm long, 4 cm wide and 3 cm high can you fit into it?

12 Washing powder packets measure 20 cm by 8 cm by 30 cm. The wholesaler receives them in a large box 1 metre by 60 cm by 40 cm. How many packets come in each box?

13 (a) Explain how would you fit the washing powder packets in question **12** into the box.

(b) Is it possible to pack this box in more than one way?

(c) Explain your answer to (b) with a sketch.

The volume of a more complicated shape can often be found by **dividing** the shape up into **separate cuboids**.

14 Find the volume of each of these shapes.

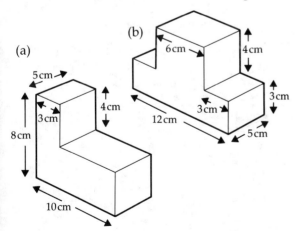

15 Find the volume of plastic in the window frame moulding shown.

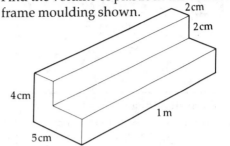

16 A duct is made by drilling a square hole through the centre of a length of wood. Find the volume of wood in the duct.

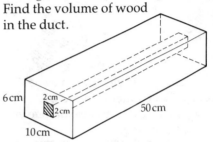

17 Wooden picture-frame moulding can be obtained in different cross-sections. Find the volume of a 3 metre length of each moulding.

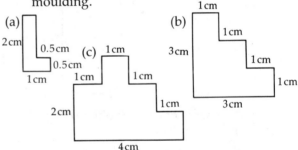

18 A square hole of edge 2 cm is drilled right through each face of a cube of edge 8 cm. Find the volume of wood remaining.

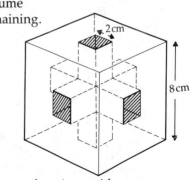

19 Find the volume of a prism, with cross-sectional area 4.8 cm², and length 32 cm.

3.12 *Volume: cylinders and prisms*

The **volume of a prism** is found by multiplying the area of its cross-section by its length.

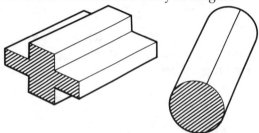

The cross-section of a cylinder is a circle.
The **volume of a cylinder** is found by multiplying the area of the circle on one end by the length of the cylinder.

1 (a) Find the area of the end of each cylinder.
       (Use $\pi = 3.14$)

   (i)
   (ii)

   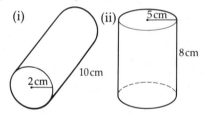

   (b) Now write down the volume of each cylinder in (a).

2 Find the area of the end of each cylinder and then the volume. (Use $\pi = 3.14$)

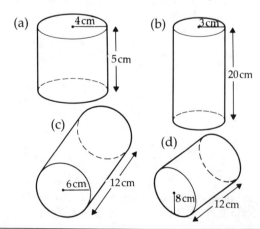

3 Write down, in your own words, how you would find the volume of a cylinder, when you are given its radius and its length.

4 Find the volume of each cylinder.
   (a) radius   2 cm, length 10 cm
   (b) radius   3 cm, length   8 cm
   (c) radius   5 cm, length 12 cm
   (d) radius   7 cm, length 5.6 cm
   (e) radius 1.2 cm, length 8.5 cm

5 Which cylinder has the larger volume?
   (a) radius 10 cm, length   5 cm
   (b) radius 5 cm, length 10 cm

6 Mr Heinz is trying to choose the best tin can for his baked beans.
   For each of the cans shown below which:
   (a) holds the most beans
   (b) uses the least tin?

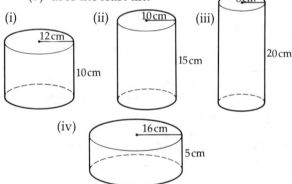

7 Find the volume of each of the shapes below.

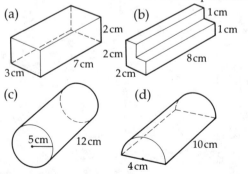

8 Each of the diagrams below represents the cross-section of a prism.
Find the volume of the prism if its length is 25 cm.

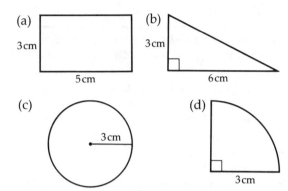

(a)
3 cm
5 cm

(b)
3 cm
6 cm

(c)
3 cm

(d)
3 cm

9 A concrete drainpipe has an outer radius of 2 metres and an inner radius of 1.8 metres. Find the volume of concrete in a 10 metre length of pipe.

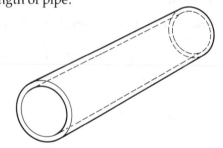

10 Each of the diagrams below represents the end of a railway tunnel.
Find the total volume of each tunnel if the length is 150 metres.

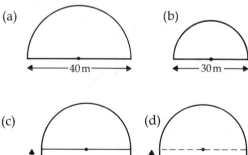

(a)
←—40 m—→

(b)
←—30 m—→

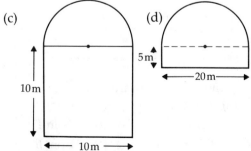

(c)
10 m
10 m

(d)
5 m
←—20 m—→

11 A manufacturer is making a batch of 1000 plastic knitting needles. Each one is 6 mm in diameter and 32 cm long.
What volume of plastic will be needed altogether?

12 The local supermarket wishes to display its new brand of Caff coffee as shown below. Each tin has diameter 20 cm and height 15 cm.
What is the total volume of coffee in the display?
Ignore any air space in the tins.

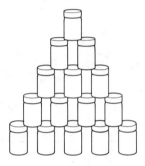

13 Say, in your own words, how you would find:
(a) the length of a cylinder, if you knew its volume and the diameter of its circular end
(b) the radius of the circular end of a cylinder, if you knew its volume and its length.

14 The volume of a cylinder is 502.4 cm³.
Using $\pi = 3.14$ find:
(a) the length of the cylinder, if its radius is 10 cm
(b) the radius of the cylinder, if its length is 10 cm.

15 A cylindrical can holds 5 litres of paint. Its height and radius are the same.
Find the radius of the can.

16 (a) Make your own collection of as many different cylindrical containers as you can find.
(b) Find the volume of each container as accurately as you can.
(c) Arrange your containers in order of size.

3.13 *Bearings*

Bearings or **directions** are often given using points of the compass.

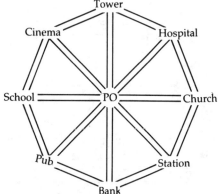

N: north   NE: north east
E: east   NW: north west
S: south   SE: south east
W: west   SW: south west

1   Look at the map below.

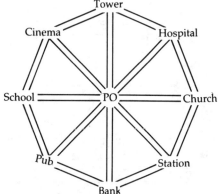

Adam is standing at the Post Office. Write down the bearing of:
(a) the church   (b) the school
(c) the station   (d) the pub.

2   Look again at the map in question **1**. What can Adam see on each bearing?
(a) N   (b) SE   (c) NW   (d) S

3   Find out which direction is due north when you are at home.
Imagine you are standing outside your home.
Draw a sketch map to show what you can see on each of the bearings:
    N, E, S, W, NE, SE, SW and NW.

4   If you are standing outside your home, in what compass direction does the road go:
(a) to the left   (b) to the right?

5   Write down the bearings which are in the opposite directions to:
(a) W   (b) S   (c) NE   (d) SE

Bearings can also be given using an angle measured in a **clockwise direction** from due north.
Bearings are given using **three** numbers.

    A from O is 090°
    B from O is 135°
    C from O is 270°
    D from O is 000°

6   Look at the diagram above. Write down the bearing of:
(a) P from O   (b) Q from O
(c) R from O   (d) S from O

7   Look again at the diagram above. Which point has a bearing from O of:
(a) 180°   (b) 045°   (c) 225°   (d) 315°?

8   For each diagram write down:
(a) the bearing of Q from P
(b) the bearing of P from Q.

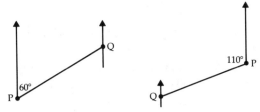

9   Look at the diagram below.

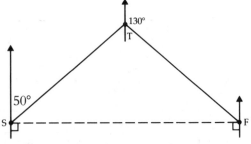

Write down the bearing of:
(a) T from S   (b) F from S
(c) S from F   (d) S from T
(e) F from T   (f) T from F.

10 If the bearing of a water tower from a church is 050°, what is the bearing of the church from the water tower?

11 A ship sails 6 km due north, then changes direction and sails 3 km due east, and then changes direction a second time and sails 3 km due south.
(a) Make a scale drawing of the ship's course.
(b) Find the bearing of the ship now, from its starting point.

The diagram below shows the course of a plane immediately after leaving Heathrow Airport.

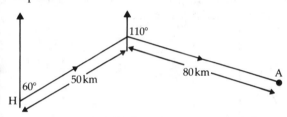

The plane flies on a bearing of 060° for 50 km, and then on a bearing of 110° for 80 km.

---

12 Make an accurate drawing for the above diagram.
Find how far the plane is now:
(a) to the east of its starting point
(b) to the south of its starting point.

13 Using your diagram from question **12**, find the bearing and distance of the plane from its starting point.

14 Tim likes to take long walks across the moor. On Sundays he usually walks 5 miles on a bearing of 030°, followed by 3 miles on a bearing of 140°, and finally straight back to his starting point.
(a) Make a scale drawing of his route.
(b) How far has he to walk on the final leg of his route?
(c) What is the bearing of this final leg?

15 What difference would it make to your answers to question **14** if Tim started by walking 3 miles on a bearing of 140° and then followed this with 5 miles on a bearing of 030° before heading for home?

16 Two ships set sail from port together. One sails 12 km on a bearing of 045°, the other 12 km on a bearing of 135°.
(a) Make a scale drawing of the paths of the two ships.
(b) Find the distance between the ships.
(c) Find the bearing of the second ship from the first ship.

17 Anne and Joe agree to meet at the Smuggler's Cafe.
Anne leaves her house and walks on a bearing of 315°. Joe's house is 4 miles due west of Anne's house and he sets off on a bearing of 050°.
(a) Make a scale drawing to show where the Smuggler's Cafe is located.
(b) How far is Anne's house from the cafe?
(c) How much further than Anne did Joe have to walk?

18 The diagram below shows the layout for the race course for the local windsurfing championships.

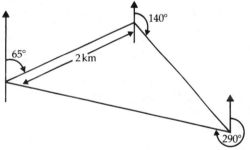

(a) Using the bearings shown, make a scale drawing of this race course.
(b) Find the lengths of the second and third legs of the course.
(c) If Chris averages 10 km per hour, find how long he takes to complete the course.

19 The points A, B and C have coordinates (0, 0), (3, 3) and (2, 5) respectively.
(a) Show these points on a grid.
(b) Find the bearing of B from A.
(c) Find the bearing of C from A.
(d) Find the bearing of C from B.
(e) Find the distance between B and C.

3.14 *Pythagoras*

In the diagram below squares are drawn on each side of a right-angled triangle.

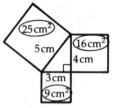 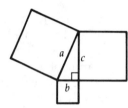

The area of the square on the longest side is equal to the sum of the areas of the squares on the two shorter sides.

$$5^2 = 3^2 + 4^2 \qquad a^2 = b^2 + c^2$$

1   The three sides of a right-angled triangle are 5 cm, 12 cm and 13 cm.
   (a) What is the area of a square drawn on the longest side?
   (b) What are the areas of the squares drawn on the two shorter sides?
   (c) Do you agree that: $13^2 = 5^2 + 12^2$?

2   A triangle has sides a cm, b cm and c cm. For which triangle is $a^2 = b^2 + c^2$?
   (a) $a = 10$, $b = 6$ and $c = 8$
   (b) $a = 12$, $b = 8$ and $c = 10$
   (c) $a = 25$, $b = 7$ and $c = 24$

3   Which of these is a right-angled triangle?

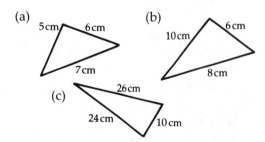

4   A semicircle is drawn on each side of a 3 cm, 4 cm, 5 cm right-angled triangle.
   (a) Find the area of each semicircle.
   (b) What can you say about these areas?

You can find the third side of a right-angled triangle as follows.

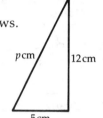

Now $p^2 = 5^2 + 12^2$
so $p^2 = 25 + 144$
         $= 169$
so $p = \sqrt{169} = 13$

5   Use the method above to find the longest side of each right-angled triangle.

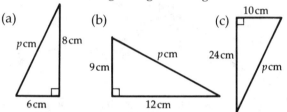

6   Find the longest side of a right-angled triangle if the two shorter sides are 12 cm and 16 cm.

7   A ladder is leaning up against a wall. The foot of the ladder is 1 metre from the wall. The top of the ladder is 3 metres above the ground. How long is the ladder?

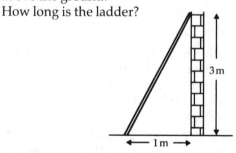

8   Michael is watching a yacht from the top of a vertical cliff 50 metres above the level of the sea. The yacht is 120 metres from the foot of the cliff. How far is Michael from the yacht?

9   Ian leaves his house and cycles 4.5 miles due north and then 6 miles due west. How far is Ian from his house now?

You can also find the other sides of a right-angled triangle using Pythagoras' theorem.

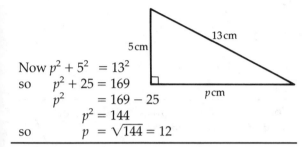

Now $p^2 + 5^2 = 13^2$

so $\quad p^2 + 25 = 169$

$\quad\quad p^2 \quad\quad = 169 - 25$

$\quad\quad p^2 = 144$

so $\quad\quad p = \sqrt{144} = 12$

10 Use the above method to find the shortest side of each right-angled triangle.

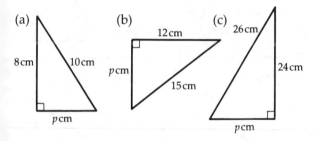

11 Find the shortest side of a right-angled triangle if the other two sides are 20 cm and 16 cm.

12 A ladder is leaning up against the top of a tree. The foot of the ladder is 1 metre from the tree. The length of the ladder is 3 metres. How far is the top of the tree above the ground?

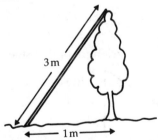

13 Suzy is watching a swimmer from the top of a vertical cliff. The swimmer is 200 metres from the foot of the cliff. The distance between Suzy and the swimmer is 250 metres.
How far is the top of the cliff above the level of the sea?

14 Find the length of the diagonal of each rectangle.

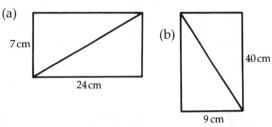

15 Find the length of the sides of each rhombus.

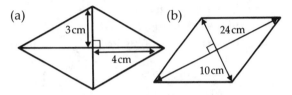

16 Find the length of the sides of each square.

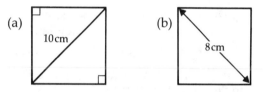

17 (a) Find the height of the triangle ABC.
(b) Find the length of the side BC.
(c) Is ABC a right-angled triangle?

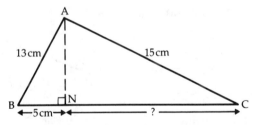

18 In the diagram below, P is a telegraph pole which serves two houses A and B.

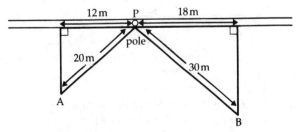

Find how far each house is from the road.

3.15 *Trigonometry: tangent*

In the right-angled triangle ABC:
AC is the longest side.
It is called the **hypotenuse**.
BC is **opposite** the angle BAC.
AB is **adjacent** to the angle BAC.

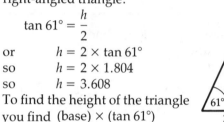

$$\tan A = \frac{BC}{AB} = \frac{\text{opposite}}{\text{adjacent}}$$

You can use tangents to find the sides of a right-angled triangle.

$$\tan 61° = \frac{h}{2}$$

or $\quad h = 2 \times \tan 61°$

so $\quad h = 2 \times 1.804$

so $\quad h = 3.608$

To find the height of the triangle you find (base) × (tan 61°)

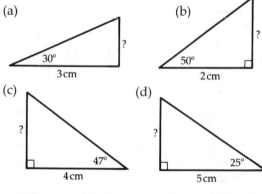

1   Use tables, or your calculator, to find:
(a) tan 20°   (b) tan 35°   (c) tan 78°
(d) tan 41.5°   (e) tan 67.2°   (f) tan 71.1°

2   Use tables, or your calculator, to find the angle whose tangent is:
(a) 0.700   (b) 0.900   (c) 0.249   (d) 1.000
(e) 1.600   (f) 0.637   (g) 0.759   (h) 1.834

3   (a) For each triangle write down tan *A*.
(b) Now find angle *A*.

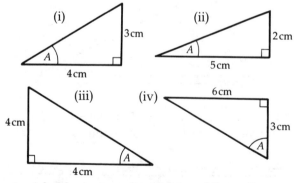

4   A ladder just reaches the top of the wall which is 4 metres high. The foot of the ladder is 1 metre from the wall.
Draw a sketch for this.
Write down the tangent of the angle the ladder makes with the ground.
Now find this angle.

5   Find the height of each triangle.

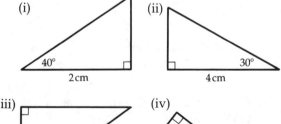

6   (a) Copy each triangle and show which side is opposite the given angle.
(b) Now find the length of this side.

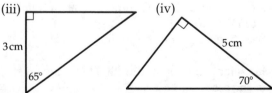

There are two possible ways of finding the base of a right-angled triangle.

Method 1 $\tan 58° = \dfrac{16}{b}$

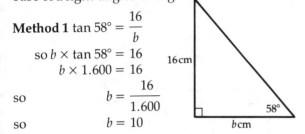

so $b \times \tan 58° = 16$

$b \times 1.600 = 16$

so $\qquad b = \dfrac{16}{1.600}$

so $\qquad b = 10$

Method 2 The third angle of the triangle is

$90° - 58° = 32°$

so $\quad b = 16 \times \tan 32°$

$= 16 \times 0.625$

$= 10$

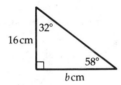

---

7 Use the method you prefer to find the base of each triangle.

(a)   (b)

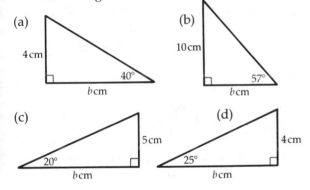

(c)   (d)

8 Find the length of the unknown side.

(a)   (b)

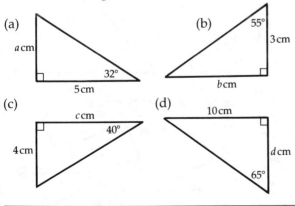

(c)   (d)

9 The string attached to a kite makes an angle of 67° with the ground. The shadow of the string on the ground is 20 metres long. How high is the kite in the sky?

10 A rectangle has sides of 4 cm and 5 cm. Draw the rectangle and mark in one of the diagonals. Find the angle the diagonal makes with the longer side.

11 A flagpole is supported by two wires as shown below.

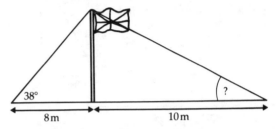

(a) Find the height of the flagpole.

(b) Find the angle that the second wire makes with the ground.

12 In New York a skyscraper can be seen from each of two roads. One of the roads goes east, the other goes south as shown below.

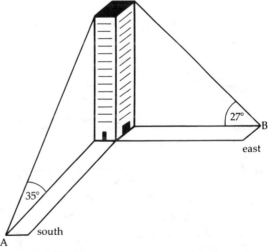

From the point A, the angle of sight to the top of the skyscraper is 35°. From B the angle to the top is 27°. If the height of the skyscraper is 120 metres, find:

(a) the distance from A to the building

(b) the distance from B to the building

(c) the bearing of B from A.

3.16 *Trigonometry: sine and cosine*

In the right-angled triangle ABC:
AC is the **hypotenuse**.
BC is **opposite** the angle BAC.
AB is **adjacent** to the angle BAC.

$$\sin A = \frac{BC}{AC} = \frac{\text{opposite}}{\text{hypotenuse}}$$

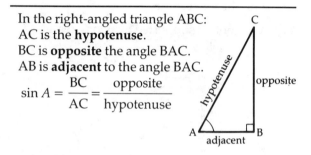

1 Use tables, or your calculator, to find:
  (a) sin 20°   (b) sin 35°   (c) sin 78°
  (d) sin 41.5°   (e) sin 67.2°   (f) sin 71.1°

2 Use tables, or your calculator, to find the angle whose sine is:
  (a) 0.500 (b) 0.906 (c) 0.866 (d) 0.035
  (e) 0.600 (f) 0.637 (g) 0.759 (h) 0.434

3 (a) For each triangle write down sin A.
  (b) Now find angle A.

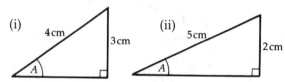

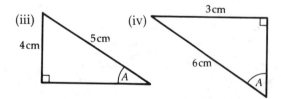

4 A ladder, which is 4 metres long, just reaches the top of a wall. The top of the wall is 3 metres above the ground.
Draw a sketch for this.
Write down the sine of the angle the ladder makes with the ground.
Now find this angle.

You can use sines to find the sides of a right-angled triangle.

$$\sin 61° = \frac{h}{2}$$

or $h = 2 \times \sin 61°$
so $h = 2 \times 0.875$
so $h = 1.75$

To find the height of the triangle you find
  (hypotenuse) × (sin 61°)

5 Find the height of each triangle.

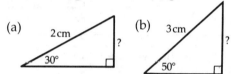

6 (a) Copy each triangle and show which side is opposite the given angle.
  (b) Now find the length of this side.

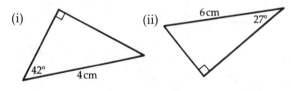

To find the hypotenuse of triangle ABC:

$$\sin A = \frac{BC}{AC}$$

$$AC \times \sin A = BC$$

$$AC = \frac{BC}{\sin A}$$

7 Find the hypotenuse of each triangle.

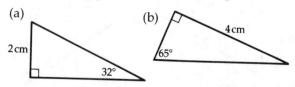

8 Find the length of the unknown side.

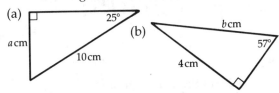

(a)

25°

a cm

10 cm

(b)

b cm

57°

4 cm

In the right-angled triangle ABC:

$$\cos A = \frac{AB}{AC} = \frac{\text{adjacent}}{\text{hypotenuse}}$$

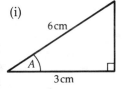

C

hyp   opp

A   adj   B

---

9 Use tables, or your calculator, to find:
(a) $\cos 20°$ (b) $\cos 35°$ (c) $\cos 78°$
(d) $\cos 41.5°$ (e) $\cos 67.2°$ (f) $\cos 71.1°$

10 Use tables, or your calculator, to find the angle whose cosine is:
(a) 0.500   (b) 0.906
(c) 0.866   (d) 0.035
(e) 0.600   (f) 0.637
(g) 0.759   (h) 0.434

11 (a) For each triangle write down $\cos A$.
(b) Now find angle A.

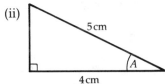

(i)

6 cm

A

3 cm

(ii)

5 cm

4 cm

A

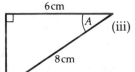

6 cm

A

8 cm

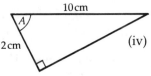

(iii)

A

2 cm

10 cm

(iv)

12 A ladder, which is 4 metres long, just reaches the top of a wall. The foot of the ladder is 1 metre from the wall.
Draw a sketch for this.
Write down the cosine of the angle the ladder makes with the ground.
Now find this angle.

13 A kite string is 20 metres long. The shadow of the string on the ground is 15 metres long. Find the angle between the string and the ground.

You can use cosines to find the sides of a right-angled triangle.

$$\cos 61° = \frac{b}{2}$$

or $b = 2 \times \cos 61°$
so $b = 2 \times 0.485$
so $b = 0.97$

2 cm

61°

b cm

To find the base of the triangle you find
(hypotenuse) $\times$ ($\cos 61°$)

---

14 Find the base of each triangle.

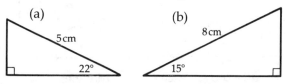

(a)

5 cm

22°

(b)

8 cm

15°

15 (a) Copy each triangle and show which side is adjacent to the given angle.
(b) Now find the length of this side.

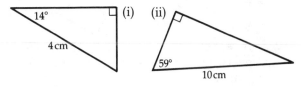

14°

4 cm

(i)   (ii)

59°

10 cm

You can use cosines to find the hypotenuse of triangle ABC.

$$\cos A = \frac{AB}{AC}$$

$$AC \times \cos A = AB$$

$$AC = \frac{AB}{\cos A}$$

C

hyp   opp

A   adj   B

---

16 Find the hypotenuse of each triangle.

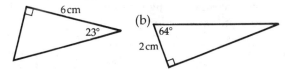

6 cm

23°

(b)

64°

2 cm

17 Find the length of the unknown side.

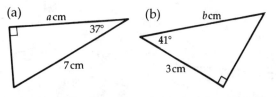

(a)

a cm

37°

7 cm

(b)

b cm

41°

3 cm

3.17 *Trigonometry: applications*

In the right-angled triangle ABC:

$$\sin A = \frac{\text{opp}}{\text{hyp}} = \frac{BC}{AC}$$

$$\cos A = \frac{\text{adj}}{\text{hyp}} = \frac{AB}{AC}$$

$$\tan A = \frac{\text{opp}}{\text{adj}} = \frac{BC}{AB}$$

1   On a copy of each of these triangles mark the hypotenuse and the sides opposite and adjacent to the angle.

(a) 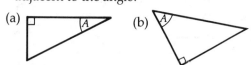 (b)

2   For each of the triangles, first write down a sine, cosine or tangent statement, then find the angle.

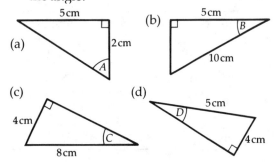

(a)   (b)   (c)   (d)

3   For each of the triangles, first write down a sine, cosine or tangent statement, then find the unknown side.

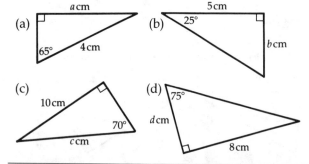

(a)   (b)   (c)   (d)

4   PQN and PRN are right-angled triangles.

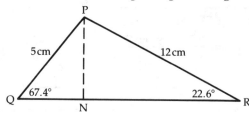

(a)   Find the length of PN.
(b)   Find the length of QN.
(c)   Find the length of RN.
(d)   What is the length of QR?
(e)   Find QR².
(f)   Now find PQ² + PR².
(g)   Check your answers for (e) and (f) by showing that   PQ² + PR² = QR².
(h)   Explain why angle QPR is 90°.

5   A ship sails 50 km on a bearing of 060°, followed by 40 km on a bearing of 070°, as shown below.

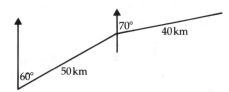

(a)   How far has the ship travelled north?
(b)   How far has the ship travelled east?

6   Sam drives 4 miles due north and then 3 miles due east.
Find the bearing of his current position from his starting point.

7   Jake starts at the same time as Sam, drives 5 miles due east and then 8 miles due north.
Find Jake's bearing from his starting point.

8   (a)   Make a drawing of Sam's and Jake's routes in questions **6** and **7**.

(b)   Use your drawing in (a) to find the bearing of Jake's position from Sam's position at the end.

9 A grandfather clock has a pendulum 1 metre long. At the end of its swing it makes an angle of 20° with the vertical. Find, for this point:

(a) how far it has moved to one side

(b) how far it has risen above the lowest point on its swing.

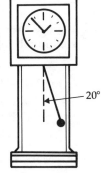

10 Carl tries to row straight across a river which is 60 metres wide. He actually finishes up on the other bank 30 metres downstream. Find the angle which his path makes with the bank.

11 The diagram below shows the big wheel at a fairground. The radius of the wheel is 25 metres.

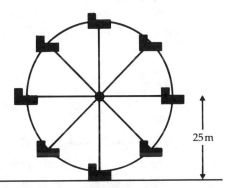

Fatima and Becky start at the bottom. Find how high they are, above the ground, when the wheel has turned through:
(a) 30° (b) 60° (c) 120° (d) 240°

12 The height of the mast of a yacht, above the deck, is 35 metres. Four wires are attached to the top of the mast and fixed to the deck. Two of these wires are 40 metres long, one is 70 metres long, and the fourth is 50 metres long.
Find the angle each wire makes with the deck.

13 A cross-section of a road is shown below. The road slopes up at 25°.

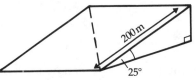

(a) Find the height of the top of the road.

(b) Would it be easier to cycle straight up the hill or along the dotted line? Why?

14 The diagram shows a stepladder. Each side is 2 metres long and makes an angle of 65° with the ground. A safety rope is used halfway up.

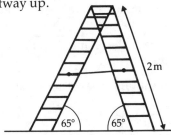

(a) Find the height of the top of the steps above the ground.

(b) Find the length of the safety rope.

15 The diagram below shows the cuboid ABCDPQRS. The diagonals AC, AQ and AR have been marked with dotted lines.

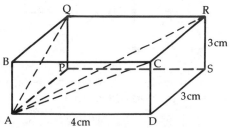

(a) Draw triangle ADC and find the angle CAD.

(b) Draw triangle APQ and find the angle QAP.

(c) Draw triangle ASR and find the angle RAS.

16 (a) Use your answers for question **15** to help you to find the lengths of the diagonals AC, AQ and AR.

(b) Use Pythagoras' theorem to check your answers to (a).

3.18 *Problems and investigations*

1 (a) Cut out a square and then divide it into two identical triangles as shown.

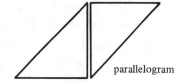

(b) Make as many different shapes as you can with the two triangles.

(c) Draw each shape and then name it, as above.

2 Repeat question 1 for two identical triangles which are:

(a) right-angled but not isosceles

(b) isosceles but not right-angled

(c) scalene but not right-angled.

3 Look at the diagram below. Two triangles have been drawn on a grid with nine dots.

(a) Copy the grid and make as many more triangles as you can which are different from the two shown above.

(b) Write down all you can about the sides and angles of each of your triangles.

(c) If your triangles are drawn on a grid of 1 cm squares, find the area of each.

4 (a) Make as many different quadrilaterals as you can on a grid with nine dots, like the one in question 3.

(b) Write down all that you can about the sides and angles of each of your quadrilaterals.

(c) If your quadrilaterals are drawn on a grid of 1 cm squares, find the area of each.

5 Look at the diagram below. The pattern has rotational symmetry of order four.

(a) Make some more patterns with rotational symmetry on the nine dot grid.

(b) Make some patterns on a nine dot grid which have one or more lines of symmetry.

(c) Record your results in (a) and (b) showing any lines of symmetry and the order of rotational symmetry.

6 Look at the diagram below. Two identical right-angled triangles have been drawn on a grid with nine dots. If you traced over one of the triangles, you could rotate the tracing paper about the centre dot until the first triangle fitted exactly onto the other one.

(a) How many more identical triangles can you draw in different positions on the grid?

(b) Investigate how you could move the top triangle onto each of the other possible positions.

(c) Draw a diagram for each pair of triangles and say how you would do the move.

7 (a) What other shapes can you make on a grid with nine dots?

(b) Which of these shapes has most sides?

(c) Investigate shapes which you can make on a grid with: (i) 16 dots   (ii) 25 dots.

8 Each of the shapes below has been drawn on a grid of 1 cm squares.

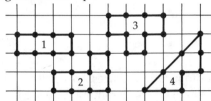

(a) Find the area of each shape.

(b) Count the dots on the edge of each shape.

(c) For each shape complete the table:

	Dots on edge	Dots ÷ 2	Area
Shape 1	8	4	3
Shape 2			

(d) Write down a rule connecting the area of each shape and the number of dots on the edge of the shape.

(e) Make up some other shapes like the ones above. Does your rule still work?

(f) What happens when there are dots inside the shape, as well as on the edge? (*Hint*: Add another column to your table in (c) for **Dots inside**.)

9 (a) Make as many different shapes as you can with:

    (i)   three squares   (triominoes)

    (ii) four squares   (quadrominoes)

    (iii) five squares   (pentominoes)

    (iv) six squares   (hexominoes).

(b) Describe the symmetry of each shape.

(c) Which of the pentominoes can be folded up to make a box without a lid?

(d) Which of the hexominoes can be folded up to make a cube?

(e) You should have found twelve pentominoes. Cut them out and use them to make a rectangle.

10 (a) Repeat question **9** (a) and (b) for shapes made with equilateral triangles.

(b) Which of your shapes, using four triangles, can be folded up to make a tetrahedron (pyramid)?

11 Any polygon can be divided into triangles.

(a) Copy and complete this table for regular polygons.

	Number of edges	Number of triangles	Sum of angles	Size of each angle
Triangle	3	1	180°	60°
Square	4	2	360°	90°
Pentagon	5			
Hexagon	6			
Octagon	8			
Decagon	10			
Duodecagon	12			

(b) Write down a rule connecting the sum of the angles of a regular polygon with its number of edges.

(c) How would you find the size of each angle of a regular polygon if you know the number of edges?

(d) Can you find a rule to give the number of diagonals of any polygon?

12 You can make a tessellation with squares and octagons as shown below.

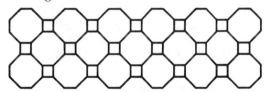

(a) Use you results from question **11** to check that the sum of the angles at each point in the tessellation is 360°.

(b) Investigate other combinations of regular polygons which will fit together at a point without leaving any gaps. (*Hint*: One possibility is two squares and three triangles since $(2 \times 90°) + (3 \times 60°) = 360°$)

(c) For each of your combinations in (b) make a tessellation pattern. Does this always work?

3.19 *Oral test*

1   How many degrees are there in a complete turn?

2   What is the sum of all of the angles in a triangle?

3   Two angles of a triangle are 30° and 50°. What is the third angle?

4   One angle of a right-angled triangle is 35°. What is the third angle?

5   How many lines of symmetry has:
(a) a kite
(b) a parallelogram?

6   What order of rotational symmetry has:
(a) a rhombus
(b) a parallelogram?

7   A square has a side of 7 cm. What is the perimeter of the square?

8   A rectangle has sides of 5 cm and 8 cm. What is the perimeter of the rectangle?

9   A rectangle has sides of 7 cm and 12 cm. What is the area of the rectangle?

10 A cuboid has edges of 3 cm, 4 cm and 5 cm. What is the volume of the cuboid?

11 Five centimetre squares are arranged in the shape of a cross.
(a) What is the area of the shape?
(b) What is the perimeter of this shape?

12 Imagine a box of matches.
(a) How many faces does it have?
(b) How many edges does it have?
(c) How many corners does it have?

13 Imagine a model of one of the Egyptian pyramids.
(a) How many faces will it have?
(b) How many vertices will it have?
(c) How many edges will it have?

14 A photograph measures 7 cm by 9 cm.
(a) What length of framing is needed?
(b) What area of glass is needed?

15 A photographic print measures 5 cm by 7 cm, and includes a 1 cm border.
What is the area of:
(a) the actual photograph
(b) the border?

16 A circular boating pond at the park has a diameter of eight metres.
The pond has a concrete edging strip.
(a) What is the approximate length of this strip?
(b) What is the approximate surface area of the water?

17 The minute hand of Big Ben is two metres in length.
Approximately how far does the end move in one hour?

18 I walk three miles due north, then four miles due east and finally three miles due south. Where do I finish my walk?

19 Jem cycles three miles due south and then four miles due west.
How far, as the crow flies, is he from his starting point?

20 What is the longest straight line that I can draw on a piece of paper which measures 5 cm by 12 cm?

21 Imagine a right-angled triangle with sides of 3 cm, 4 cm and 5 cm.
(a) Which of these sides is opposite the smallest angle?
(b) What is the sine of the smallest angle?

22 A square has a side of 4 cm.
(a) What is the length of its diagonal?
(b) What are the angles in each triangle?
(c) What is the tangent of 45°?

3.20 *Fact sheet III*

Area of triangle

$\frac{1}{2} \times$ (base) $\times$ (height)

$A = \frac{1}{2}bh$

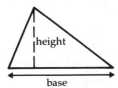

Area of parallelogram

(base) $\times$ (height)

$A = b \times h$

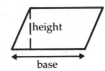

Area of trapezium

$\frac{1}{2} \times (a + b) \times h$

$A = \frac{1}{2}(a + b)h$

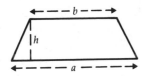

Circumference of circle

$\pi \times$ (diameter)
or $2 \times \pi \times$ (radius)

$C = \pi D$ or $2\pi r$

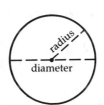

Area of circle

$\pi \times$ (radius)^2
where $\pi = 3.14$ (2 dec. pl.)

$A = \pi r^2$

Volume of cuboid

(length) $\times$ (width) $\times$ (height)

$V = lwh$

Volume of prism

(area of cross-section) $\times$ (length)

$V = Al$

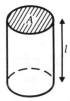

Volume of cylinder

$\pi \times$ (radius)$^2 \times$ (height)

$V = \pi r^2 h$

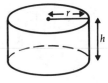

Pythagoras' theorem

Area on hypotenuse
= sum of areas on other two sides

$h^2 = a^2 + b^2$

$a^2 = h^2 - b^2$

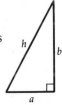

Trigonometry

$\tan A = \dfrac{\text{opp}}{\text{adj}}$ opp = adj $\times \tan A$

$\sin A = \dfrac{\text{opp}}{\text{hyp}}$ opp = hyp $\times \sin A$

$\cos A = \dfrac{\text{adj}}{\text{hyp}}$ adj = hyp $\times \cos A$

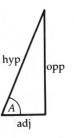

4.1 *Coordinates and graphs*

Look at the graph below.
The point P has **coordinates** (6, 3).
The point Q has **coordinates** (−4, −2).

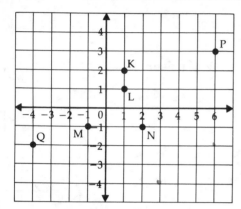

Note: Any point on the *x*-axis has a **second** coordinate of 0.
Any point on the *y*-axis has a **first** coordinate of 0.

4   Write down the coordinates of the points P, Q, R and S shown below.

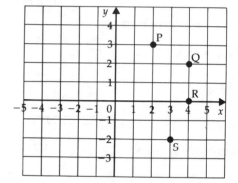

1   Look at the graph above.
  (a) Write down the coordinates of the points K and L.
     What do you notice about the first coordinate in each?
  (b) Write down the coordinates of the points M and N.
     What do you notice about the second coordinate in each?
  (c) Join KL and join MN.

2   (a) Copy the graph above.
  (b) On your graph mark the points S (−4, 3) and T (6, −2).
  (c) Join PS, SQ, QT and TP.
     What shape have you made?

3   Use the same graph as in questions **1** and **2**.
  (a) Mark the points (3, 2), (5, 2), (5, 0) and (3, 0).
     Now join these points in order.
     What shape have you made?
  (b) Mark the points (−3, 2), (−1, 2), (−1, 0) and (−3, 0)
     Now join these points in order.
     What shape have you made?

5   Look at the diagram in question **4**.
  The *y*-axis is a mirror line.
  (a) Do you agree that the reflection of P in this line is (−2, 3)?
  (b) Write down the coordinates of the reflections of the three points Q, R and S.
  (c) What do you notice about the coordinates of the reflected points?
  (d) What shape can you make by joining the points P, Q, R and S and their images?

6   Look again at the diagram in question **4**. The points P, Q, R and S are to be reflected in the *x*-axis.
  (a) Do you agree that the image of P is (2, − 3)?
  (b) Write down the coordinates of the images of Q and S.
  (c) What do you notice about the coordinates of the reflected points?
  (d) What is special about the image of the point R?

7 Look at the graph below.

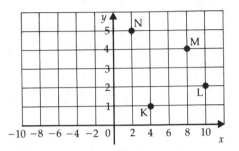

(a) What does each interval on the *x*-axis represent?

(b) What does each interval on the *y*-axis represent?

(c) Write down the coordinates of K, L, M and N.

(d) Write down the coordinates of the images of each of these points when they are reflected in the *y*-axis.

8 Use a grid like the one in question **7**. Mark your axes in the middle of the grid. On both axes take each interval to represent 2 units.

(a) Mark the points (8, 4), (4, 2), (6, 3), (2, 1), (−8, −4), (−6, −3), (−4, −2) and (−2, − 1).

(b) What can you say about each pair of coordinates for these points?

(c) Join your points with a straight line.

(d) These points also lie on your line. Complete their coordinates:

  (i)   (5,)
  (ii) (3,)
  (iii) (, 0.5)

(e) Check that your answers to (d) and (b) agree.

9 For each set of points write down a statement showing the connection between each pair of coordinates.

(a) (0, 1), (1, 2), (2, 3), (3, 4), (4, 5)

(b) (7, 5), (6, 4), (5, 3), (4, 2), (3, 1)

(c) (1, 2), (2, 4), (3, 6), (4, 8), (5, 10)

(d) (1, 3), (2, 6), (3, 9), (4, 12), (5, 15)

(e) (0, 3), (1, 5), (2, 7), (3, 9), (4, 11)

(f) (1, 1), (2, 4), (3, 9), (4, 16), (5, 25)

10 Look at the graph below.
The point P has coordinates (2.4, 1.2).
The point Q has coordinates (3.6, 1.8).

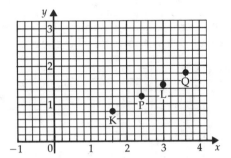

(a) Write down the coordinates of points K and L.

(b) What can you say about each pair of coordinates for the four points P, Q, K and L?

(c) Do you agree that the four points lie on a straight line?

(d) Write down the coordinates of two other points which lie on this line.

(e) Is the second coordinate equal to half the first coordinate for each of your points in (d)?

11 Use a grid like the one in question **10**.

(a) Mark the points (1.5, 0.5), (2.4, 0.8), (3, 1), (3.6, 1.2), (4.2, 1.4) and (6, 2).

(b) What do you notice about the positions of these six points?

(c) What can you say about each pair of coordinates for these points?

(d) Write down the coordinates of two more points which lie on this line.

12 Use a grid like the one in question **7**.
Draw your axes in the middle of the grid.
Use a scale of 1 cm to represent 1 unit on the *x*-axis and a scale of 1 cm to represent 5 units on the *y*-axis.

(a) Show each of the sets of points in question **9** on your grid.

(b) Which of these sets of points lie on a straight line?

(c) Can you draw a smooth curve through the set of points in **9** (f)?

4.2 *Reading and drawing graphs*

Mrs Stuart sets out on a journey at 9.00 a.m.
The graph below shows how far she travelled.

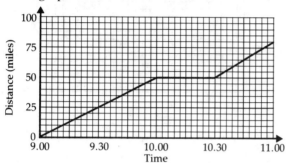

1   Look at the graph above.
  (a) How far had Mrs Stuart travelled by
       10.00 a.m?
  (b) What was the length of her journey?
  (c) By what time had she travelled:
       (i) 25 miles   (ii) 70 miles?
  (d) How far did she travel in the last half hour?
  (e) What do you think she was doing
       between 10.00 and 10.30 a.m.?

2   Jackie and Sally recorded the temperature
    during part of a summer day.
    The graph below shows their findings.

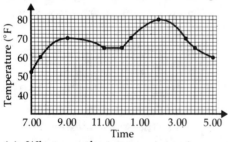

  (a) What was the temperature at:
       (i) 9.00 a.m. (ii) 11.00 a.m. (iii) 3.30 p.m?
  (b) At what time, or times, was the
       temperature:
       (i) 50°F (ii) 60°F (iii) 70°F (iv) 65°F?
  (c) What was the highest temperature
       during the day?
  (d) At what time was the highest
       temperature recorded?

3   The graph below shows how much petrol
    Mrs Stuart had in her car during her journey.

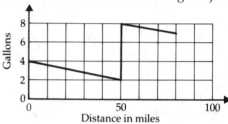

  (a) How much petrol did Mrs Stuart have in
       her car:
       (i)   at the start of her journey
       (ii)   at the end of her journey
       (iii) when she had done 25 miles
       (iv)   when she had just bought petrol?
  (b) How many miles had she travelled when
       the petrol gauge showed:
       (i) 3 gallons (ii) 5 gallons?

4   Mr Patel wants to know how much his
    electricity bill is going to be.
    The graph below shows the cost of using a
    given number of units of electricity.

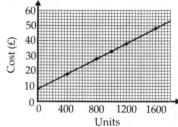

  (a) How much does he have to pay for
       using:
       (i)   400 units   (ii) 1000 units
       (iii) 500 units   (iv) 1350 units?
  (b) How many units has he used if the bill is:
       (i) £28 (ii) £33 (iii) £15.50 (iv) £26.75?
  (c) Can you explain why the graph starts at
       £8 on the cost axis?
  (d) If you exclude the fixed charge of £8,
       what is the cost of 100 units?
  (e) What is the cost of the electricity per
       unit?

5 The graph started below shows the distance travelled by a car when it is doing 80 m.p.h. on a motorway in Italy.

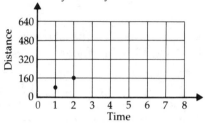

(a) What does each interval represent:

    (i) on the time axis

    (ii) on the distance axis?

(b) Copy and complete the table to show the distance travelled.

Time (hours)	1	2	3	4	5	6	7	8
Distance (miles)	80	160						

(c) Copy the graph above and use your results from (b) to complete it.

(d) Use your graph to find:

    (i) how far the car has travelled in 3.5 hours

    (ii) how long it takes to travel 420 miles on this motorway.

6 The table shows the temperature of a cake when it is being cooked in an oven.

Time (minutes)	10	20	30	40	50	60
Temperature (°C)	15	30	50	75	105	140

(a) Draw a graph to show this information. You will need to show 0–60 on the time axis and 0–150 on the temperature axis.

(b) Join the points on your graph with a smooth curve.

(c) Use your graph to estimate:

    (i) what the temperature of the cake will be after 45 minutes

    (ii) how long the cake will need to cook for the temperature to rise to 100°C.

(d) How much does the temperature rise during each ten minute period between 30 minutes and one hour?

(e) After how long will the temperature be 150°C?

7 The table below shows the size of each angle of a regular polygon.

Number of sides	3	4	5	6	7	8
Size of angle	60°	90°	108°	120°		135°

(a) Draw a graph to show this information.

(b) Join your points with a smooth curve.

(c) Which points on your curve represent sensible information?

(d) Why is there no part of the graph to the left of the point (3, 60)?

(e) Use your graph to estimate the size of the angle in a regular 7-sided polygon.

(f) What happens to the angles as the number of sides increases?

8 The table below shows the areas of circles with given radii.

Radius (cm)	1	2	3	4	5
Area (cm²)	3.14	12.56	28.26	50.24	78.50

(a) Draw a graph to show this information.

(b) Join your points with a smooth curve.

(c) Do all of the points on your curve represent sensible information?

(d) Use your graph to estimate:

    (i) the area of a circle of radius 2.5 cm

    (ii) the radius of a circle of area 40 cm².

9 The arm of Jill's metronome swings from side to side. The table below shows the angles it makes with the vertical during the first second.

Time (seconds)	0	0.25	0.5	0.75	1.0
Angle	0°	15°	20°	15°	0°

(a) Draw a graph to show this information.

(b) Join your points with a smooth curve.

(c) Use your graph to estimate:

    (i) the angle after 0.6 seconds

    (ii) the times when the angle is 10°.

(d) What would your graph look like for the next second? And the third second?

4.3 *Graphs and formulae*

1 Barry delivers 50 newspapers every day.
   (a) Copy and complete the table below to show how many papers he will deliver in a given number of days.

Number of days	1	2	3	4	5	6	7
Number of papers	50						

   (b) Complete the statement:
   number of papers is . . . number of days

   (c) Draw a graph to show this information.
   (You will need one axis for the number of days, showing 0 – 7, and the other axis for the number of papers showing 0 – 400.)

   (d) Barry gets paid 1p for each paper. How much will he earn in four weeks?

2 Anne and Mary decide to cycle 60 miles whilst on holiday.
   (a) How can they work out how long it will take them at a particular cycling speed?

   (b) Copy and complete the table below.

Speed (m.p.h.)	2	4	6	8	10	12
Time (hours)					6	

   (c) Copy and complete the graph below to show this information.

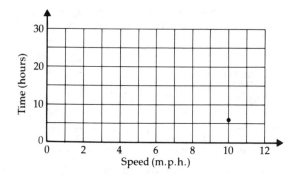

   (d) Do you agree that the formula:
   $$\text{time} = \frac{60}{\text{speed}} \quad \left(\text{or } t = \frac{60}{s}\right)$$
   describes this graph?

3 (a) Copy and complete the table which shows the area of a square for given sides.

Side (cm)	1	2	3	4	5	6	7	8
Area (cm²)	1	4						

   (b) Copy and complete the statement:
   the area of a square is . . .

   (c) Draw a graph to show this information.
   (You will need to be able to show areas from 0 to 80 cm².)

   (d) Use your graph to find:
   (i) the area of a square of side 4.5 cm
   (ii) the side of a square of area 40 cm².

4 The graph below shows the cost of hiring a television per week.

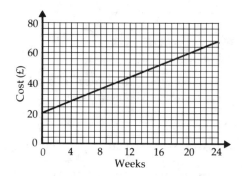

   (a) Use the graph to copy and complete this table.

Number of weeks	4	8	12	16	20	24
Cost of hiring (£)	28	36				

   (b) Why does the graph start at the point (0, 20)?

   (c) How much will it cost to hire the TV for:
   (i) 6 weeks (ii) 15 weeks
   (iii) 52 weeks?

   (d) Say how you can work out the cost of hiring the TV for any given number of weeks.

5 Jim owns a Porsche which can do 120 m.p.h.
   (a) How can you find the distance he can travel in a given number of hours?
   (b) Make up a table to show how far Jim can travel in a given number of hours.
   (c) Draw a graph to show this information.

6 (a) How do you find the volume of a cube when you know the length of its edge?
   (b) Copy and complete the table below.

Edge (cm)	1	2	3	4	5	6	7	8	9	10
Volume (cm³)										

   (c) Draw a graph to show this information. (It will show volumes from 0 to 1000 cm³.)
   (d) Use your graph to estimate:
      (i) the volume of a cube of edge 4.5 cm
      (ii) the edge of a cube of volume 200 cm³.

7 The graph below shows the cost of hiring a car for one day. The cost depends on the number of miles covered.

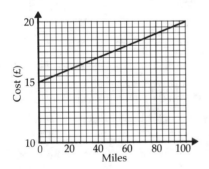

   (a) Use the graph to copy and complete this table.

Number of miles	20	40	60	80	100
Cost of hiring (£)	16	17			

   (b) How much will it cost to use the car for:
      (i) 50 miles   (ii) 70 miles   (iii) 85 miles?
   (c) What is the fixed charge irrespective of mileage?
   (d) What is the extra cost for each mile?
   (e) How can you work out the cost of hiring the car for any given number of miles?

8 Taxi fares are based on a fixed charge of 50p plus 15p for each whole mile, or part of a mile, covered. Draw a graph to show the cost of using a taxi for any mileage up to 50 miles.

9 The graph shows the cost of posting a second class letter. The cost depends on the weight.

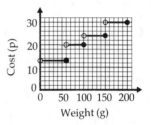

   (a) Use the graph to copy and complete this table.

Weight of letter (g)	50	75	100	125	175
Cost of letter (pence)					

   (b) How much will it cost to post a second class letter weighing:
      (i) 37 g   (ii) 89 g   (iii) 151 g?
   (c) What are the possible weights of letters which will require a 20p stamp?
   (d) Would it be cheaper to post two separate 55 g letters to the same address, or to put them in the same envelope?

10 The cost of posting a first class letter is given by the table below.

Letter weighing up to	60 g	100 g	150 g	200 g
Cost		18p	26p	32p	40p

   (a) How much will it cost to post a first class letter weighing:
      (i) 37 g   (ii) 89 g   (iii) 151 g?
   (b) What are the possible weights of letters which will require a 32p stamp?
   (c) Draw a graph to show this information.

11 (a) Find the cost of making telephone calls:
      (i) at various times of the day
      (ii) for various distances.
   (b) Draw two separate graphs to show the information in (a) for a 3-minute call.

4.4 *x-y graphs*

Draw the graph of $y = x + 3$.
(i)   First complete a table of values:

x	0	1	2	3	4	5	6	7	8
$x + 3$	3	4	5					10	11

(ii)   Now draw your axes:

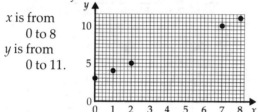

x is from
0 to 8
y is from
0 to 11.

(iii) Now plot the points using your table of
values.

1 (a) Copy and complete the table of values in
(i) above.

(b) Copy and complete the graph in (iii)
above.

(c) Do you agree that the points on the
graph of $y = x + 3$ lie on a straight
line?

(d) Where does this line cross the *y*-axis?

2 (a) Copy and complete the table for the
given values of *x*.

x	0	1	2	3	4	5	6	7	8
$x + 1$	1			4			7		

(b) Make a copy of the graph above.

(c) Now use your table of values to draw the
graph of $y = x + 1$.

(d) How does this graph compare with the
one you drew in question 1?

(e) Where does this graph cross the *y*-axis?

3 Repeat question **2** for $y = x + 2$.

4 Look again at the graphs you drew for:
   $y = x + 1$, $y = x + 2$ and $y = x + 3$.
Can you say what the graph of $y = x + 4$
will look like?

5 (a) Copy and complete the table for the
given values of *x*.

x	0	1	2	3	4	5	6	7	8
$4x$	0			12					32

(b) Now draw your axes:
   x is from 0 to 8   *y* is from 0 to 32.

(c) Use your table of values to draw the
graph of $y = 4x$.

6 Repeat question **5** for:
(a) $y = 2x$ (b) $y = 3x$ (c) $y = 0.5x$.

7 Look again at the graphs you drew of:
   $y = 4x$, $y = 3x$, $y = 2x$ and $y = 0.5x$.
Can you say what the graph of $y = 5x$ will
look like?

8 (a) Copy and complete the table for the
given values of *x*.

x	0	1	2	3	4	5	6	7	8
$2x + 3$	3				11				19

(b) Now draw your axes:
   x is from 0 to 8   *y* is from 0 to 19.

(c) Use your table of values to draw the
graph of $y = 2x + 3$.

(d) Where does this line cross the *y*-axis?

9 Repeat question **8** for:
(a) $y = 2x + 5$ (b) $y = 3x + 2$ (c) $y = 4x - 1$.

10 Look at the graphs you drew in questions **8**
and **9**.
(a) Do you agree that:
   (i)   the number multiplying the *x* term
tells you how steep the graph is
   (ii) the number that is added on tells you
where the graph crosses the *y*-axis?

(b) Can you say what the graph of $y = 3x + 5$
will look like?

(c) What can you say about the graphs of
   $y = 3x + 2$, $y = 3x + 4$ and $y = 3x - 2$?

Draw the graph of $y = 3x^2$.
(i) First complete a table of values.

x	0	1	2	3	4	5	6	7	8
x^2	0	1	4					49	64
$3x^2$	0	3	12					147	192

(ii) Now draw your axes:
  x is from 0 to 8
  y is from 0 to 200.

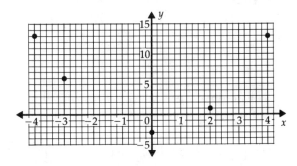

(iii) Now plot the points using your table of values.

11 (a) Copy and complete the table above.
  (b) Copy and complete the graph above.
  (c) Do you agree that the points on the graph of $y = 3x^2$ lie on a smooth curve?
  (d) Use your graph to write down:
    (i) the value of y when $x = 4.5$
    (ii) the value of x when $y = 100$.

12 (a) Copy and complete the table for the given values of x.

x	0	1	2	3	4	5	6	7	8
x^2	0			9					64
$2x^2$	0			18					128

  (b) Make a copy of the graph above.
  (c) Now use your table of values to draw the graph of $y = 2x^2$.
  (d) How does this graph compare with the one you drew in question **11**?
  (e) Use your graph to write down:
    (i) the value of y when $x = 4.5$
    (ii) the value of x when $y = 100$.

13 (a) Repeat question **12** for:
    (i) $y = 2x^2 + 1$ (ii) $y = 3x^2 + 1$
  (b) What effect does the number multiplying the x^2 have on the shape of the graph?
  (c) What effect does adding the 1 have?

14 Part of the graph of $y = x^2 - 3$ is drawn below.

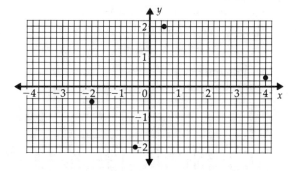

  (a) Copy and complete the table of values for $y = x^2 - 3$.

x	−4	−3	−2	−1	0	1	2	3	4
x^2	16	9			0		4		16
$x^2 - 3$	13	6			−3		1		13

  (b) Now use your table of values to copy and complete the graph.
  (c) Join the points on your graph with a smooth curve.
  (d) What is the value of y when:
    (i) $x = 2.5$ (ii) $x = -3.5$?
  (e) What are the values of x when $y = 0$?

15 Part of the graph of $y = \dfrac{1}{x}$ is drawn below.

  (a) Copy and complete the table of values.

x	−4	−2	−1	−0.5	0.5	1	2	4
$\dfrac{1}{x}$		−0.5		−2	2			0.25

  (b) Copy and complete the graph.

4.5 *Gradients*

The graph of $y = 3x + 2$ is drawn below.
It crosses the y-axis where $y = 2$.

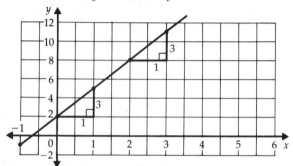

Look at the two triangles.
As the value of x increases by 1, the value of y increases by 3.
This tells us how **steep** the graph is.
The line $y = 3x + 2$ has a **gradient** of 3.

1 (a) On a grid like the one above draw the graph of $y = 2x + 3$.
   (b) What are the values of y when:
      (i) $x = 1$ and $x = 2$ (ii) $x = 3$ and $x = 4$?
   (c) Is the difference between the y values equal to 2 in each part of (b)?
   (d) Do you agree that the gradient of $y = 2x + 3$ is 2?

2 The graphs of $y = 4x + 1$ and $y = 5x + 1$ are shown below.

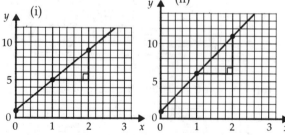

   (a) Use the triangles drawn to help you to write down the gradient of each line.
   (b) Write, in your own words, how you would find the gradient of a straight line.

The **gradient** of a straight line is given by finding the increase in the value of y for an increase of 1 in the value of x.
In the graph of $y = mx + c$ the gradient is m.

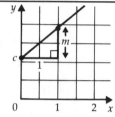

Look at the line joining the points $(1, 2)$ and $(4, 8)$. The **gradient** of this line is given by finding:
$$\frac{8 - 2}{4 - 1} = \frac{6}{3} \quad \text{i.e. } 2$$

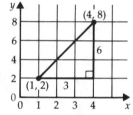

3 (a) Find the gradient of these lines:

(i) 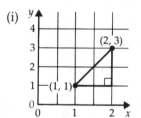   (ii)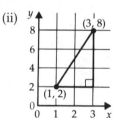

   (b) Find the gradient of the line joining:
      (i) $(0, 0)$ and $(1, 5)$ (ii) $(1, 1)$ and $(2, 5)$
      (iii) $(3, 2)$ and $(5, 8)$ (iv) $(1, 4)$ and $(4, 7)$.

Look at the line joining the points $(1, 9)$ and $(3, 5)$. The gradient of this line is given by finding:
$$\frac{5 - 9}{3 - 1} = -\frac{4}{2} \quad \text{i.e. } -2$$

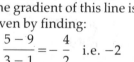
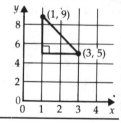

4 (a) Draw a diagram, as above, for:
      (i) $(1, 5)$ and $(2, 2)$ (ii) $(2, 5)$ and $(3, 1)$
      (iii) $(3, 8)$ and $(5, 2)$ (iv) $(1, 7)$ and $(4, 4)$.
   (b) Find the gradient of the line joining each pair of points.

Note: When the gradient of a line is **positive** the line is sloping up to the right.

When the gradient of a line is **negative** the line is sloping down to the right.

5 A party of Venture Scouts are walking in the Black Mountains. The graph below shows one of the mountains.

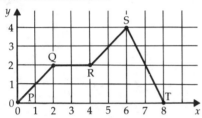

(a) Which of the lines PQ, QR, RS and ST have a gradient which is:

(i) positive   (ii) negative?

(b) What is special about the gradient of the line QR?

(c) Find the gradient of the lines:

(i) PQ   (ii) RS   (iii) ST.

6 Look at the graph below. Sandra left home at 9.00 a.m. on her bicycle. She went to visit her friend Tani for an hour and then cycled home.

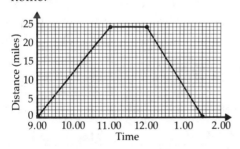

(a) How far away did Tani live?

(b) How long did it take Sandra to cycle there?

(c) What was Sandra's speed?

(d) How long was Sandra's return journey?

(e) What was her speed on her return journey?

7 Look again at the graph in question **6**.
(a) Find the gradient of each of the three parts of this graph.

(b) How do your answers for (a) compare with the speeds you found in question **6**?

(c) What do you think the negative gradient tells you about the speed?

$$speed = \frac{distance\ travelled}{time\ taken}$$

So you can find the **speed** from the **gradient** of a distance-time graph.

8 The travel graph below shows Carl's evening run.

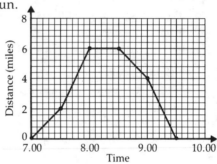

(a) Find the gradient of each part of the graph.

(b) Use your answers to find Carl's speed on each part of his run.

(c) When was Carl running fastest?

(d) When was Carl running slowest?

(e) When was Carl resting?

(f) Describe Carl's run in your own words.

9 The graph below shows how a sports car moves between two sets of traffic lights.

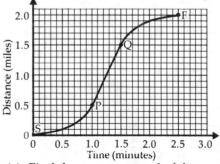

(a) Find the average speed of the car for each of the three sections of the graph.

(b) Describe the journey from S to F.

4.6 *Number patterns*

1 (a) Complete the table below using the rule:
double the number and add three

Number	1	2	3	4	5	6	7	8
Result	5	7						

(b) Describe any patterns you can see in your results for (a).

(c) What result would you get using the number:

(i) 10   (ii) 20   (iii) 100   (iv) n?

2 (a) Complete the table below using the rule:
multiply by five and subtract three

Number	1	2	3	4	5	6	7	8
Result	2	7						

(b) Describe any patterns you can see in your results for (a).

(c) What result would you get using the number:

(i) 10   (ii) 20   (iii) 100   (iv) n?

3 (a) Complete the table below using the rule:
add four and then multiply by three

Number	1	2	3	4	5	6	7	8
Result	15	18						

(b) Describe any patterns you can see in your results for (a).

(c) What result would you get using the number:

(i) 10   (ii) 20   (iii) 100   (iv) n?

4 Do you agree that the results when using the above rules with the number n are:

(i) $2n + 3$ (ii) $5n - 3$ (iii) $(n + 4) \times 3$?

5 The result of using the number n with a rule is $4n + 7$.
Say, in your own words, what the rule is.

When the result of using a rule is $2n + 5$
the rule is: *double and add 5*
When the result of using a rule is $2(n + 5)$
the rule is: *add 5 and then double*

6 Look at the dotty shape patterns below.

(i)

(ii)

(iii)

(iv)

For each of the above carry out these steps.
(a) Draw the next dotty shape.
(b) Say how many dots there will be in the:

(i)   next but one shape   (ii)   10th shape

(iii) 100th shape   (iv) nth shape.

(c) Say, in your own words, how you can find the number of dots in the nth shape.

7 Copy and complete the table:

Number n	1	2	3	4	5	6	7
Result $3n - 2$	1			10			

8 Copy and complete the table:

Number n	1	2	3	4	5	6	7
Result $5(n+1)$	10			25			

9 Can you find another way of describing the rule in question 8?

10 For each of the number sequences find:
(a) the next number (b) the 10th number
(c) the 20th number (d) the *n*th number.

    (i)   2, 4, 6, 8, 10, 12, 14, 16, . . .

    (ii) 3, 5, 7, 9, 11, 13, 15, 17, . . .

    (iii) 3, 6, 9, 12, 15, 18, 21, 24, . . .

    (iv) 4, 7, 10, 13, 16, 19, 22, 25, . . .

    (v) 1, 4, 9, 16, 25, 36, 49, 64, . . .

    (vi) 2, 5, 10, 17, 26, 37, 50, 65, . . .

11 Look at this pattern of shapes.

(a) How many dots will there be in:
    (i)   the next shape
    (ii) the next but one shape
    (iii) the 10th shape?

(b) How you can find the number of dots in:
    (i) the 20th shape (ii) the *n*th shape?

(c) Write, using *n*'s, the number of dots in the *n*th shape.

12 Look at this pattern of rectangles.

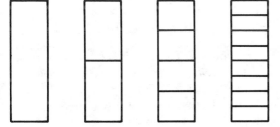

(a) How many rectangles will there be in:
    (i)   the next shape
    (ii) the next but one shape
    (iii) the 10th shape?

(b) Say, in your own words, how you can find the number of rectangles in:
    (i) the 20th shape (ii) the *n*th shape.

13 A chess board has 64 squares on it. One grain of rice is put on the first square, twice as many on the second square and so on. How many grains of rice will there be on:
(a) the 10th square (b) the 64th square?

14 The pattern below contains black and white triangles.

(a) Find the number of triangles in each row.

(b) How many triangles will there be in:
    (i) the 10th row (ii) the *n*th row?

(c) How many triangles are there altogether in the first:
    (i) 2 rows (ii) 3 rows (iii) *n* rows?

15 Repeat question **14** for:
(a) the black triangles

(b) the white triangles.

16 You will need a strip of paper.
(a) Fold it in half and then unfold it.
How many small rectangles can you see?
How many fold lines can you see?

(b) Fold it in half again and then unfold it as shown below.

How many small rectangles can you see?
How many fold lines can you see?

(c) Continue folding the paper in half and record your results in a table like this.

Number of folds	Number of small rectangles	Number of fold lines
1	2	1
2	.	.
3	.	.

(d) Try to predict the number of rectangles and fold lines for 20 folds.

4.7 *Use of letters and formulae*

In **algebra** we often use letters to stand for numbers.

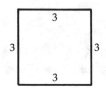 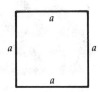

$$3 + 3 + 3 + 3 = 4 \times 3 \qquad a + a + a + a = 4a$$

The **perimeter** of each **square** is:
four times the length of one side.

1 Write down the perimeter of each shape and then simplify your answer.

(a)   (b)

2 Simplify:
(a) $a + a + a$ (b) $p + p + p + p + p$
(c) $3a + 5a$ (d) $7p + 4p$
(e) $a + 2a + 3a$ (f) $7p + 5p + 3p$

Often more than one letter is used.

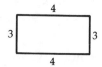 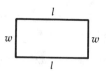

$$4 + 3 + 4 + 3 = 8 + 6 \qquad l + w + l + w = 2l + 2w$$

The **perimeter** of a **rectangle** is:
twice the length plus twice the width.

3 Write down the perimeter of each shape and then simplify your answer.

(a) 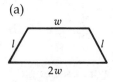 (b)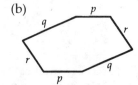

4 Simplify:
(a) $a + a + b + b$ (b) $p + p + q + q + q$
(c) $a + b + a + b + a$ (d) $7p + 3p + 4q$
(e) $a + 2b + 3a$ (f) $7p + 5q + 3p + 2q$

5 Simplify:
(a) $p + p + p - p - p$ (b) $7a - 4a$
(c) $2p + 6p - 3p$ (d) $a + a - b - b - b$
(e) $4p + 5q - 2q$ (f) $8a - 5a + 7b - 4b$

The **area** of a rectangle is found by multiplying the length by the width.

area is $4 \times 3 \text{ cm}^2$ area is $l \times w \text{ cm}^2$

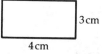 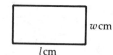

This result can be written as the formula:
area = length × width   or $A = l \times w$

6 The area of a square can be found by
   multiplying the length of one side by itself.
   Write this result as a formula:
   (a) in words: area is . . .
   (b) using letters: $A = $ . . .

7 The volume of a cube can be found by
   multiplying the length of one side by itself and by itself again.
   Write this result as a formula:
   (a) in words: volume is . . .
   (b) using letters: $V = $ . . .

8 The volume of a cuboid can be found by
   multiplying the length by the width and by the height.
   Write this result as a formula:
   (a) in words: volume is . . .
   (b) using letters: $V = $ . . .

9 The area of a triangle can be found from
   half the base multiplied by the height.
   Write this result as a formula:
   (a) in words: area is . . .
   (b) using letters: $A = $ . . .

You can find the value of an unknown letter in a formula by **substituting** the values of the other letters.

Example 1   Find p in the formula $p = 2a + 2b$ when $a = 5$ and $b = 6$.

$$p = 2 \times 5 + 2 \times 6$$
$$p = 10 + 12$$
$$\text{so } p = 22$$

Example 2   Find b in the formula $p = 2a + 2b$ when $p = 28$ and $a = 5$.

$$28 = 2 \times 5 + 2b$$
$$28 = 10 + 2b$$
$$18 = 2b$$
$$\text{so } b = 9$$

10   (a) Use the formula $A = l \times l$ to find A when:

(i) $l = 7$ (ii) $l = 12$ (iii) $l = 2.5$ (iv) $l = 3.1$.

(b) Say how you would find l if you know A.

(c) Find l when $A =$:

(i) 36   (ii) 169   (iii) 441   (iv) 1.44.

11   (a) Use the formula $A = l \times w$ to find A when:

(i) $l = 7, w = 5$ (ii) $l = 12, w = 8$

(iii) $l = 4, w = 3.5$ (iv) $l = 16, w = 2.5$.

(b) How can you find w if you know A and l?

(c) Find w when:

(i) $A = 32, l = 4$ (ii) $A = 30, l = 2.5$.

12   (a) If $V = l \times l \times l$, find V when $l =$:
(i) 7   (ii) 12   (iii) 2.5   (iv) 3.1.

(b) Find l when:

(i) $V = 8$ (ii) $V = 27$.

13   (a) If $V = l \times w \times h$, find V when:
(i) $l = 7, w = 5, h = 2$

(ii) $l = 12, w = 8, h = 5$

(iii) $l = 4, w = 3, h = 2.5$

(iv) $l = 16, w = 7, h = 1.4$.

(b) Find h when:

(i) $V = 60, l = 5, w = 6$

(ii) $V = 48, l = 8, w = 12$.

14   (a) If $A = \frac{1}{2} \times b \times h$, find A when:

(i) $b = 8, h = 5$ (ii) $b = 12, h = 8$

(iii) $b = 4, h = 3.5$ (iv) $b = 16, h = 2.5$.

(b) Find b when:

(i) $A = 8, h = 2$ (ii) $A = 54, h = 9$.

15   (a) Find the average speed of a car which travels:

(i)   120 miles in two hours

(ii)   357 miles in seven hours

(iii) 40 miles in half an hour.

(b) Write, in your own words, how you found the speeds in (a).

(c) Write this result as a formula:

(i) in words: average speed is . . .

(ii) using letters: $s = \ldots$

16   (a) Write a formula for a car doing s m.p.h. to give the distance travelled in t hours.

(b) Use your formula to find the distance when:

(i) $s = 60$ m.p.h. and $t = 4$ hours

(ii) $s = 44$ m.p.h. and $t = 3.5$ hours

(iii) $s = 36$ m.p.h. and $t = 20$ minutes.

17   The formula for calculating the cost of using electricity is given as:

15 pence for each unit used plus a fixed charge of £8.

(a) Find the cost of using:

(i)   100 units

(ii)   240 units

(iii) 428 units.

(b) Write a formula for this using letters. (Use £T for the total cost and n for the number of units used.)

(c) If the total cost comes to £38 how many units have been used?

18   Saleem is trying to find how much income tax he has to pay. The formula for calculating the amount of tax for a single person is:

subtract the single person's allowance from the gross salary and then find 27% of the result.

(a) If the single person's allowance is £2425 find the tax payable on a gross salary of:

(i) £4245   (ii) £5925   (iii) £8500.

(b) Write a formula for this using letters. (Use £T for the tax payable and £G for the gross salary.)

(c) If Saleem paid £270 in tax what was his income?

4.8 *Formulae from tables*

1 The coordinates of a set of points are given in the table below.

Value of x	0	1	2	3	4	5	6	7	8
Value of y	0	2	4		8	10		14	16

   (a) Fill in the missing values in the table.
   (b) What is the value of y when $x = 10$?
   (c) What is the value of y when $x = 100$?
   (d) Complete the formula: $y = \ldots x$
   (e) Check that your formula works for $x = 7$.

2 The coordinates of a set of points are given in the table below.

Value of x	0	1	2	3	4	5	6	7	8
Value of y	3	4		6	7	8		10	11

   (a) Fill in the missing values in the table.
   (b) What is the value of y when $x = 10$?
   (c) What is the value of y when $x = 100$?
   (d) Complete the formula: $y = x \ldots$
   (e) Check that your formula works for $x = 7$.

3 The coordinates of a set of points are given in the table below.

Value of x	0	1	2	3	4	5	6	7	8
Value of y	3		7	9	11		15	17	

   (a) Fill in the missing values in the table.
   (b) What is the value of y when $x = 10$?
   (c) What is the value of y when $x = 100$?
   (d) Complete the formula: $y = \ldots x \ldots$
   (e) Check that your formula works for $x = 7$.

4 The coordinates of a set of points are given in the table below.

Value of x	0	1	2	3	4	5	6	7	8
Value of y	1	6	11		21	26	31		41

   (a) Fill in the missing values in the table.
   (b) What is the value of y when $x = 10$?
   (c) What is the value of y when $x = 100$?
   (d) Complete the formula: $y = \ldots x \ldots$
   (e) Check that your formula works for $x = 7$.

5 The table below shows the cost of buying chocolate bars.

Number of bars	1	2	3	4	5	6	7
Cost in pence	12	24		48		72	84

   (a) Fill in the missing values in the table.
   (b) Write down how you can find the cost when you are told the number of bars.
   (c) What would be the cost of buying ten chocolate bars?
   (d) Using C for the cost in pence, and N for the number of bars required, copy and complete the formula:
   $C = \ldots N$

6 Look again at the table in question **5**.
   (a) How many bars can you buy for 156p?
   (b) Write down how you can find the number of bars if you are told the total cost.
   (c) Now copy and complete the formula:
   $N = \ldots C$

7 When driving at a constant speed on the motorway the distance travelled after a given time is given in the table below.

Time in hours	1	2	3	4	5	6	7
Distance in miles	55	110		220		330	385

   (a) Fill in the missing values in the table.
   (b) Write down how you can find the distance travelled after any given time.
   (c) How far will the car travel in ten hours?

(d) Using D for the distance in miles, and T for the time in hours, copy and complete the formula:
$D = \ldots T$

8 Look again at the table in question **7**.

(a) How long will it take to travel 660 miles?

(b) Write down how you can find the time taken for any given distance.

(c) Now copy and complete the formula:
$T = \ldots D$
for the time taken for any given distance.

9 Look at the shapes below.

Number of shape	1	2	3	4	5	6
Perimeter of shape	4					

(a) Copy and complete the table to show the perimeter of each shape.

(b) Write down how you can find the perimeter of any given shape in this set.

(c) What is the perimeter of the tenth shape?

(d) Using P for the perimeter, and N for the number of the particular shape, copy and complete the formula:
$P = \ldots + \ldots$

10 Repeat question **9** for the set of shapes below.

Number of shape	1	2	3	4	5	6
Perimeter of shape	6					

11 The cost of developing and printing films is shown in the table below.

Number of prints	1	2	12	20	24	36
Cost in pence	85	95	195	275	315	435

(a) What would be the cost for 10 prints?

(b) What would be the cost for 30 prints?

(c) If N is the number of prints and C pence is their cost, copy and complete the formula:
$C = \ldots N + \ldots$

(d) What would be the cost for 50 prints?

12 The cost of photographic enlargements is quoted as 50p each, plus a fixed charge of £1. Write this information as a formula and show as a table the cost in pence of up to 10 enlargements.

13 Look at the patterns of dots below.

Number of pattern	1	2	3	4	5	6
Number of dots	2					37

(a) Copy and complete the table to show the number of dots in each pattern.

(b) Write down how you can find the number of dots in any given pattern.

(c) How many dots are in the tenth pattern?

(d) Using N for the number of dots and P for the particular pattern, copy and complete the formula:
$N = \ldots + \ldots$

14 The coordinates of a set of points are given in the table below.

Value of x	0	1	2	3	4	5	6	7	8
Value of y	5		11	14		20	23		29

(a) Fill in the missing values in the table.

(b) What is the value of y when $x = 10$?

(c) What is the value of y when $x = 20$?

(d) Write down how you can find each value of y when you know the value of x.

(e) Complete the formula: $y = \ldots + \ldots$

(f) Check that your formula works for $x = 7$.

15 The cost of using any given number of units of electricity is quoted as 5p per unit, plus a fixed charge of £3.50.

(a) Write this information as a formula to show the cost in pence, and then complete the table below to show the cost of using up to 600 units.

Number of units	100	200	300	400	500	600
Cost in £						

(b) Find the cost of using 750 units.

4.9 *Linear equations*

1 When four is added to a number the result is 12.
What is the number?
How did you work this out?

2 When seven is subtracted from a number the result is 8.
What is the number?
How did you work this out?

3 When a number is multiplied by three the result is 24.
What is the number?
How did you work this out?

4 When a number is divided by four the result is 9.
What is the number?
How did you work this out?

5 Find the missing number:
(a) $\square + 5 = 7$ (b) $3 + \square = 8$
(c) $\square + 19 = 53$ (d) $7 - \square = 3$
(e) $\square - 9 = 5$ (f) $\square - 19 = 53$

6 Find the missing number:
(a) $\square \times 5 = 30$ (b) $8 \times \square = 72$
(c) $\square \times 12 = 84$ (d) $8 \div \square = 2$
(e) $\square \div 4 = 30$ (f) $\square \div 12 = 30$

7 Find the value of the letter:
(a) $x + 3 = 9$ (b) $8 + y = 23$
(c) $a + 15 = 31$ (d) $x - 5 = 2$
(e) $8 - b = 3$ (f) $p - 21 = 47$

8 Find the value of the letter:
(a) $2x = 8$ (b) $7y = 35$
(c) $12p = 72$ (d) $q \div 5 = 6$
(e) $s \div 4 = 13$ (f) $42 \div t = 7$

9 When a number is doubled and 5 is added the result is 25.
What is the number?
How did you work this out?

10 Find the missing number in:
$2 \times \square + 3 = 25$.

Two methods of solving the equation
$2x + 7 = 19$ are shown below.

(i) $2x + 7$ means *double a number and add seven*

so $x \rightarrow \boxed{\times 2} \rightarrow \boxed{+ 7} \rightarrow 19$
$6 \leftarrow \boxed{\div 2} \leftarrow \boxed{- 7} \leftarrow 19$

so working backwards $x = 6$

(ii) $2x + 7 = 19$

subtract 7 from each side $2x + 7 - 7 = 19 - 7$
so $2x = 12$

divide each side by 2 $\dfrac{2x}{2} = \dfrac{12}{2}$
so $x = 6$

11 Which of the two methods above did you use to answer questions **9** and **10**?

12 Solve the equations:
(a) $2x + 3 = 23$ (b) $2x + 7 = 13$
(c) $2x - 3 = 11$ (d) $3x + 5 = 26$
(e) $3x - 2 = 19$ (f) $4x + 7 = 23$

13 When a number is doubled and nine is added, the result is 17.
(a) Write this information as an equation.
(b) Solve your equation to find the number.

14 When a number is multiplied by three and five is added, the result is 17.
(a) Write this information as an equation.
(b) Solve your equation to find the number.

15 When a number is multiplied by four and five is subtracted, the result is 19.
(a) Write this information as an equation.
(b) Solve your equation to find the number.

16 Find the value of the letter:
(a) $5x + 3 = 48$ (b) $7p + 5 = 54$
(c) $6a - 7 = 41$ (d) $4y - 5 = 23$
(e) $2b + 3 = 12$ (f) $4q - 9 = 21$

17 The angles on a straight line add up to 180°.
  (a) Write an equation for each diagram.
  (b) Solve your equation to find the value of the letter.

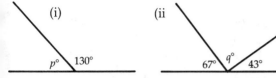

18 The sum of the angles in a triangle is 180°.
  (a) Write an equation for each diagram.
  (b) Solve your equation to find the value of the letter.

(i)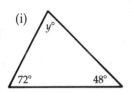

(ii)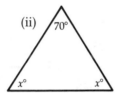

19 The perimeter of each rectangle below is 28 cm.
  (a) Write an equation for each diagram.
  (b) Solve your equation to find the value of the letter.

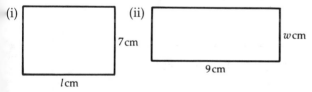

20 The perimeter of each triangle below is 36 cm.
  (a) Write an equation for each diagram.
  (b) Solve your equation to find the value of the letter.

(i)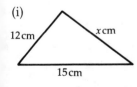

(ii)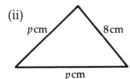

21 John is five years older than Tim. Their combined age is 31 years.
  (a) If Tim's age is a years, how old is John?
  (b) Write an equation for this information.
  (c) Solve your equation to find their ages.

22 Tina is three years younger than Tim. The sum of their ages is 25 years.
  (a) If Tim's age is a years, how old is Tina?
  (b) Write an equation for this information.
  (c) Solve your equation to find their ages.

23 The perimeter of a rectangle is 42 cm. The length of the rectangle is 3 cm longer than the width.
  (a) If the width is w cm, what is the length?
  (b) Write an equation for this information.
  (c) Solve your equation to find the length and the width.

24 Jane and her three children want to go to the cinema. The cost of an adult's ticket is twice the cost of a child's ticket.
  (a) If £c is the cost of a child's ticket, what is the total cost of taking the family to the cinema?
  (b) Jane has to pay £3.50. Write an equation to show this information.
  (c) Solve your equation to find the cost of each ticket.

25 The sum of two consecutive numbers is 93.
  (a) Write this information as an equation.
  (b) Solve your equation to find the two numbers.

26 The sum of three consecutive numbers is 93.
  (a) Write this information as an equation.
  (b) Solve your equation to find the two numbers.

27 In a competition Ann scored three more than Jane and five more than David. If their combined marks came to 64 find their individual marks.

28 The angles of a triangle are $x°$, $2x°$ and $3x°$.
  (a) Find the size of each angle of the triangle.
  (b) What kind of triangle is this?

29 The angles of a quadrilateral are $x°$, $2x°$, $3x°$ and $4x°$.
  (a) Find the size of each angle of the triangle.
  (b) Now sketch this shape.

4.10 Linear inequalities

Sets of numbers can be shown on a number line.

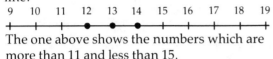

The one above shows the numbers which are more than 11 and less than 15.

1 What number is four more than 7?

2 What number is three less than 9?

3 What numbers are less than nine but more than six?

4 What numbers are more than four but less than seven?

5 What numbers are more than three but less than or equal to nine?

6 Show on number lines, like the one below, the numbers which are:

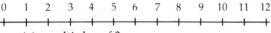

   (a) multiples of 3
   (b) more than 9
   (c) less than 5
   (d) more than 6 but less than 9
   (e) less than 8 but more than 3

7 What numbers from 1 to 10 fit?
   (a) $\square + 5 < 7$ (b) $3 + \square < 8$
   (c) $\square + 1 > 8$ (d) $7 - \square > 5$
   (e) $8 - \square > 3$ (f) $\square - 4 > 2$

8 What numbers from 10 to 20 fit?
   (a) $\square + 5 < 17$ (b) $3 + \square < 18$
   (c) $\square + 11 > 28$ (d) $17 - \square > 5$
   (e) $18 - \square > 3$ (f) $\square - 14 > 2$

9 What numbers from 1 to 10 fit?
   (a) $\square \times 5 < 10$ (b) $3 \times \square > 24$
   (c) $\square \times 2 > 8$ (d) $8 \times \square > 64$
   (e) $\square \times 4 < 20$ (f) $\square \times 12 > 84$

10 What numbers from 1 to 10 could the letter represent?
   (a) $x + 3 > 9$ (b) $8 + y < 13$
   (c) $a + 15 > 22$ (d) $x - 5 > 2$
   (e) $8 - b > 3$ (f) $p - 2 > 7$

11 What numbers from 1 to 10 could the letter represent?
   (a) $2x < 8$ (b) $3y > 21$
   (c) $5p > 30$ (d) $7q < 35$
   (e) $12s > 72$ (f) $4t < 36$

Inequalities can be solved in a similar way to equations, so long as great care is taken when multiplying or dividing by negative numbers.

$$2x + 7 < 19$$

subtract 7 from each side $2x + 7 - 7 < 19 - 7$

so $2x < 12$

divide each side by 2 $\dfrac{2x}{2} < \dfrac{12}{2}$

so $x < 6$

---

12 Solve the inequality:
   (a) $x + 5 < 9$ (b) $x + 7 > 15$
   (c) $3x < 24$

13 Solve the inequality:
   (a) $x - 5 > 9$ (b) $x - 7 < 15$
   (c) $4x > 24$

14 Solve the inequality:
   (a) $2x + 3 < 23$ (b) $2x + 7 > 13$
   (c) $2x - 3 < 11$ (d) $3x + 5 > 26$
   (e) $3x - 2 < 19$ (f) $4x + 7 > 23$

15 When a number is doubled and nine is added, the result is less than 17.
   (a) Write this information as an inequality.
   (b) Solve your inequality to find the possible values of the number.

16 When a number is multiplied by three and five is added, the result is more than 17.
  (a) Write this information as an inequality.
  (b) Solve your inequality to find the possible values of the number.

17 When a number is multiplied by four and five is subtracted, the result is less than 19.
  (a) Write this information as an inequality.
  (b) Solve your inequality to find the possible values of the number.

18 Find the possible values of the letter:
  (a) $5x + 3 < 48$ (b) $7p + 5 > 54$
  (c) $6a - 7 < 41$ (d) $4y - 5 > 23$
  (e) $2b + 3 < 12$ (f) $4q - 9 > 21$

19 What number is three less than 0?

20 What number is four less than 1?

21 What number is four less than -1?

22 What number is four more than -4?

23 What number is four more than -1?

24 What number is three more than -5?

25 What numbers are less than 3 but more than -2?

26 What numbers are more than -5 but less than -2?

27 Show on number lines, like the one below, the numbers which are:

$-7 \ -6 \ -5 \ -4 \ -3 \ -2 \ -1 \ \ 0 \ \ 1 \ \ 2 \ \ 3 \ \ 4 \ \ 5 \ \ 6 \ \ 7 \ \ 8$

  (a) less than 0
  (b) less than -2
  (c) more than -5
  (d) less than 3 but more than -4
  (e) more than -6 but less than -2
  (f) less than -3 but more than -5

28 What numbers from -5 to 5 fit?
  (a) $\square < 2$ (b) $\square < -2$
  (c) $\square > -3$ (d) $\square > -5$
  (e) $\square \leqslant 3$ (f) $\square \geqslant -2$

29 What numbers from -5 to 5 fit?
  (a) $\square + 2 < 4$ (b) $\square + 5 > 2$

  (c) $\square - 2 < 0$ (d) $\square \times 3 < 6$
  (e) $\square \times 4 < 8$ (f) $\square \times 2 > -6$

30 Solve the inequality:
  (a) $x + 3 < 1$ (b) $x + 5 > 2$
  (c) $3x < -6$

31 Solve the inequality:
  (a) $x - 5 > -2$ (b) $x - 2 < -5$
  (c) $4x > -12$

32 Solve the inequality:
  (a) $2x + 8 < 2$ (b) $2x + 7 > -3$
  (c) $2x + 3 < -1$ (d) $3x + 5 > -4$
  (e) $3x - 2 < -8$ (f) $4x + 7 > -1$

33 When a number is doubled and seven is added, the result is less than three.
  (a) Write this information as an inequality.
  (b) Solve your inequality to show the possible values of the number.

34 When a number is multiplied by three and four is added, the result is more than -2.
  (a) Write this information as an inequality.
  (b) Solve your inequality to show the possible values of the number.

35 What positive whole numbers have squares which are less than twenty?
  Show this information as an inequality.

36 What numbers have squares which lie between forty and eighty?
  Show this information as an inequality.

37 A bride's parents are prepared to spend £500 on the wedding reception.
  If the hotel charges £7 per person for a buffet meal how many guests could be invited?
  Show this information as an inequality.

38 There are 640 pupils at St Joseph's College. No class must have more than 30 pupils.
  How many classrooms could be used?
  Show this information as an inequality.

39 A taxi can hold four people and their luggage.
  How many taxis could be used to take 25 people to the airport?
  Show this information as an inequality.

4.11 *Brackets and factors*

1 Ian wants to save time so he buys two identical shirts and ties. Each shirt costs £4.99 and each tie £1.99.
(a) Find the total cost.
(a) How did you work this out?
(c) Can you find the total cost in more than one way?

2 (a) Find the cost, in question **1**, of:
(i) three shirts   (ii) three ties.
(b) Add your answers for (a).
(c) Find the cost of one shirt and one tie.
(d) Find three times your answer for (c).

Your answers for **2**(b) and **2**(d) above should be the same.
$3 \times £4.99 + 3 \times £1.99 = 3 \times (£4.99 + £1.99)$

---

3 Find:
(a) $3 \times 5 + 3 \times 6$ and $3 \times (5 + 6)$. Are they the same?
(b) $2 \times 4 + 2 \times 5 + 2 \times 6$ and $2 \times (4 + 5 + 6)$.
(c) Are your answers to (b) the same? Which was quicker to work out?

4 A shirt costs £6.99, a tie £1.50 and a pair of trousers £8.99.
Find the cost of:
(a) two shirts, two ties and two pairs of trousers
(b) three shirts, three ties, and three pairs of trousers.

The **perimeter** of a rectangle can be found in two different ways.

Double the length, double the width, add these results.		Add the length to the width and then double this result.
$2l + 2w$	$\leftarrow$ **or** $\rightarrow$	$2(l + w)$

---

5 Write without brackets:
(a) $2(p + q)$ (b) $3(x + y)$ (c) $4(r + s)$
(d) $2(x - y)$ (e) $5(a - b)$ (f) $7(p - q)$

6 Write without brackets:
(a) $2(p + q + r)$ (b) $3(x + y + z)$
(c) $4(r + s + t)$ (d) $2(x + y - z)$
(e) $5(a - b + c)$ (f) $7(p - q - r)$

7 Do you agree that:
(a) $7 \times 16 + 7 \times 24 = 7 \times (16 + 24)$
(b) $3 \times 78 + 3 \times 22 = 3 \times (78 + 22)$?

8 Look again at question **7**.
Use this idea to write down the answers for:
(a) $7 \times 16 + 7 \times 24$ (b) $3 \times 78 + 3 \times 22$
(c) $4 \times 37 + 4 \times 63$ (d) $2 \times 3.1 + 2 \times 6.9$

9 Find the answer by first re-writing the numbers using brackets:
(a) $6 \times 47 + 6 \times 13$
(b) $3 \times 12 + 3 \times 13 + 3 \times 15$
(c) $7 \times 2.4 + 7 \times 1.6$
(d) $4 \times 53 + 4 \times 21 + 4 \times 26$

When each term has a number or letter in common it is called a **common factor**.
So $3a + 3b + 3c$ can be written as $3(a + b + c)$
$pa + pb + pc$ can be written as $p(a + b + c)$

---

10 Write, using brackets:
(a) $3p + 3q$ (b) $5a + 5b$ (c) $7x + 7y + 7z$
(d) $3a - 3b$ (e) $6x - 6y$ (f) $4p + 4q - 4r$

11 Write, using brackets:
(a) $px + py$ (b) $ax + ay$ (c) $sx + sy + sz$
(d) $xa - xb$ (e) $px - py$ (f) $kp + kq - kr$

12 Write, without brackets:
(a) $2(4x + y)$ (b) $3(p + 7q)$ (c) $4(2a + 3b + 5c)$
(d) $3(5a - b)$ (e) $7(x - 4y)$ (f) $5(2p + 7q - 8r)$

Each term below has $2p$ as a **common factor** so we can re-write
$6px + 8py + 2pz$ as $2p(3x + 4y + z)$

---

13 Write, using brackets:
(a) $2px + py$ (b) $3ax + 3ay$ (c) $4sx + 2sy$
(d) $xa - 2xb$ (e) $4px - 4py$ (f) $3kp + 6kq$

14 Write, using brackets:
(a) $2px + py + 3pz$ (b) $3ax + 3ay + 3az$
(c) $4sx + 2sy - sz$ (d) $9kx + 6ky + 3kz$
(e) $4xa - 2xb + 8xc$ (f) $4px - 4py + 6pz$

15 (a) Write down the area of a square of side:
     (i) $7\,cm$ (ii) $p\,cm$ (iii) $2x\,cm$ (iv) $3y\,cm$
(b) Write down the volume of a cube of
     edge:
     (i) $7\,cm$ (ii) $p\,cm$ (iii) $2x\,cm$ (iv) $3y\,cm$

Indices can be used to shorten your answers to
question **15**.

$7 \times 7 = 7^2$ $p \times p = p^2$ $(3y) \times (3y) = 9y^2$

$7 \times 7 \times 7 = 7^3$ $p \times p \times p = p^3$

$(3y) \times (3y) \times (3y) = 27y^3$

---

16 Write, using indices:
(a) 5×5 (b) $6 \times 6 \times 6$
(c) $3 \times 3 \times 3 \times 3$ (d) $7 \times 7 \times 7 \times 7 \times 7$
(e) $n \times n$ (f) $w \times w \times w$
(g) $y \times y \times y \times y$ (h) $z \times z \times z \times z \times z$

17 Simplify:
(a) 5^2 (b) 4^3 (c) 3^4 (d) 2^5 (e) $(3 \times 4)^2$

18 Simplify:
(a) $p^2 \times p^3$ (b) $y^4 \times y^5$ (c) $(3p)^2$ (d) $(2y)^3$

19 Simplify:
(a) $(2p) \times (5p)$ (b) $(3y) \times (4y)$
(c) $(2n) \times (3n) \times (4n)$

20 Write without brackets and then simplify:
(a) $2x(y + 3z)$ (b) $p^2(2p + q)$
(c) $3y^2(2y^2 + 6z)$

Each term below has p^2 as a **common factor** so we
can re-write
$2p^2b^2 + 3p^2ab + p^3$ as $p^2(2b^2 + 3ab + p)$

---

21 Factorise:
(a) $p^2b + p^2c$ (b) $2p^2b + 3p^2c$ (c) $p^3 + p^2y$

22 Factorise:
(a) $2p^2x + 3p^2y + 4p^2z$ (b) $2p^2b + 3p^2c + p^3$

23 Factorise:
(a) $p^2b^2 + p^3c^2$ (b) $2p^3b + 3p^2b^2 + pb^3$

24 Look at the large rectangles below.
Write down the area of each of the small
rectangles.

(a) 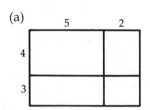   (b)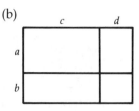

The areas of the large rectangles above can be
written in two different ways.

$$(4 + 3) \times (5 + 2)$$
$$\text{or } 4 \times 5 + 4 \times 2 + 3 \times 5 + 3 \times 2$$

$$(a + b) \times (c + d)$$
$$\text{or } a \times c + a \times d + b \times c + b \times d$$

---

25 Look at the large rectangle below.
Write down the area of each of the small
rectangles.

(a) 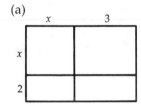   (b)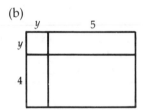

26 (a) Do you agree that the area of the large
rectangles in question **25** can also be
written as:
$(x + 2) \times (x + 3)$ and
$(y + 4) \times (y + 5)$?
(b) Use your answers for question **25** to
write each of the expressions in (a)
without brackets.

27 Use the given rectangles and the above
ideas, to write the expressions without
brackets:
(a) $(2x + 3) \times (x + 6)$ (b) $(3y + 5) \times (2y + 9)$

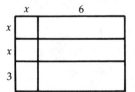   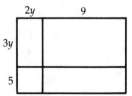

4.12 *Transformation of formulae*

1   Mrs Jones earns £10 a week more than her husband.
    (a) If Mrs Jones earns £95 a week, how much does Mr Jones earn?
    (b) If Mr Jones earns £124 a week, how much does Mrs Jones earn?

2   Mr Green earns £25 a week more than his wife.
    (a) If Mr Green earns £h a week, how much does Mrs Green earn?
    (b) If Mrs Green earns £w a week, how much does Mr Green earn?

The results in question **2** could be shown as a **formula.**
If £h is used for the husband's weekly wage, and £w is used for the wife's weekly wage:
in **2** (a) $w = h - 25$ or in **2** (b) $h = w + 25$
If you were told the value of h you would use the first formula but if you knew w you would use the second.

---

3   Do you agree that:
    (a) if you add 25 to each side of the formula
        $w = h - 25$ it becomes
        $h = w + 25$?
    (b) if you subtract 10 from each side of the formula
        $p = q + 10$ it becomes
        $q = p - 10$?

4   Do you agree that:
    (a) if you divide each side of the formula
        $w = 5h$ by 5 it becomes $h = \dfrac{w}{5}$?
    (b) if you multiply each side of the formula
        $p = \dfrac{q}{8}$ by 8 it becomes $q = 8p$?

5   Re-write each formula as $h = \ldots$
    (a) $w = h + 4$ (b) $w = h - 7$ (c) $p = h + q$
    (d) $w = 3h$ (e) $w = \dfrac{h}{6}$ (f) $A = bh$

When you re-write a formula as you did in question **5**, it is called **changing the subject** of the formula. In each part you made h the subject.

---

6   The formula for finding the area of a rectangle is $A = l \times w$.
    (a) How would you find the width, w, if you know the area, A, and the length, l?
    (b) Re-write the formula as $w = \ldots$
    (c) Use this formula to find the width of a rectangle of area 56 cm² and length 7 cm.

7   The formula for finding the average speed of a car which does d miles in t hours
    is $s = \dfrac{d}{t}$.
    (a) How would you find the distance, d, if you know the speed, s, and the time, t?
    (b) Rewrite the formula as $d = \ldots$
    (c) Use this formula to find the distance travelled by a car in 3 hours, when it is doing 52 m.p.h.

8   The formula for finding the volume of a cuboid is $V = l \times w \times h$.
    (a) How would you find the height, h, if you know the volume, V, the length, l, and the width, w?
    (b) Re-write the formula as $h = \ldots$
    (c) Use this formula to find the height of a cuboid of volume 72 cm³, length 6 cm and width 4 cm.

Changing the subject of a formula is rather like solving an equation.

If $h = 3p + 2$ then $h - 2 = 3p$
and $p = \dfrac{(h - 2)}{3}$

If $y = 5(x - 3)$ then $\dfrac{y}{5} = x - 3$

and so $x = \dfrac{y}{5} + 3$

9 Re-write each equation as $p = \ldots$
(a) $h = 5p + 2$ (b) $h = 4p + 3$ (c) $w = 5p + b$
(d) $h = 5p - 3$ (e) $w = 7p - 4$ (f) $k = 2p - a$

10 Re-write each equation as $x = \ldots$
(a) $y = 3x + 2$ (b) $y = 4x - 3$
(c) $y = 5(x + 2)$ (d) $w = 5x + u$
(e) $t = 7x - s$ (f) $k = 2(x - 3)$

11 The cost of printing a reel of film is given as eight pence for each good picture plus 75 pence for the developing.
(a) Find the cost of printing:
   (i) 15 pictures   (ii) 24 pictures.
(b) Do you agree that the formula to give C the cost, in pence, of printing n pictures is $C = 8n + 75$?
(c) Re-write this formula as $n = \ldots$
(d) How many pictures can be done for:
   (i) £2.75   (ii) £1.55   (iii) £2.35?

12 The cost of hiring a chain saw is given as £5 plus £2 for each day.
(a) Find the cost of hiring the saw for:
   (i) 3 days   (ii) 5 days   (iii) 12 days.
(b) Write a formula to give C, the cost of hiring the saw for n days.
(c) How many days can you hire the saw for if you are willing to spend:
   (i) £9   (ii) £13   (iii) £33?
(d) Re-write the formula in (b) as $n = \ldots$
(e) Use this to check your answers for (c).

13 In a restaurant Mario works out the bill as follows. He adds a £1 cover charge to the cost of the meal and then multiplies by the number of people at the table.
(a) Find the total bill for four people if the price of the set menu is:
   (i) £4.50   (ii) £6.95   (iii) £p.
(b) Write a formula to give £B, the total bill for a set menu, priced at £p, for a party of four people.
(c) Find the price of the menu if the bill for four people is:
   (i) £24   (ii) £18   (iii) £35.80.
(d) Rewrite the formula as $p = \ldots$.

14 (a) What is the total bill in question **13**, for a party of n people, when the price of the set menu is:
   (i) £4   (ii) £6.50   (iii) £p?
(b) Write your answer to (iii) as a formula.
(c) Re-write your formula in (b) as:
   (i) $n = \ldots$ (ii) $p = \ldots$

15 The formula for converting temperatures from Fahrenheit to Centigrade is
   $C = (F - 32) \times \frac{5}{9}$
(a) Use this formula to find the equivalent temperature in Centigrade to:
   (i) 50°F   (ii) 68°F   (iii) 86°F   (iv) 32°F.
(b) Say, in your own words, how you would try to find the Fahrenheit temperature for a given Centigrade temperature.
(c) What is the Fahrenheit equivalent of 5°C?

The formula in question **16** can be thought of as
$$F \to \boxed{-32} \to \boxed{\times 5} \to \boxed{\div 9} \to C$$
so working backwards:
$$F \leftarrow \boxed{+32} \leftarrow \boxed{\div 5} \leftarrow \boxed{\times 9} \leftarrow C$$
so $F = C \times 9 \div 5 + 32$ or $F = \frac{9}{5}C + 32$.

16 Use the formula $F = \frac{9}{5}C + 32$ to find the equivalent temperature in Fahrenheit to:
(a) 5°C   (b) 10°C   (c) 25°C
(d) 30°C   (e) 0°C.

17 (a) Copy and complete the flowchart for the formula $y = mx + c$.
   $$x \to \boxed{\times} \to \boxed{+} \to y$$
(b) Find the value of y when $x = 5$, $m = 7$ and $c = 8$.
(c) Use the flowchart backwards to re-write the formula as $x = \ldots$

18 The formula $v = u + at$ gives the value of the final velocity v, at time t, when the initial velocity u, and the acceleration a, are known.
(a) Find the value of v when:
   (i) $u = 5$, $a = 7$, $t = 2$
   (ii) $u = 3$, $a = 5$, $t = 8$.
(b) Re-write the formula with u as its subject.
(c) Re-write the formula with a as its subject.

4.13 *Flow charts*

1 (a) Use the flow chart below.
What do you notice about your answer?

(b) Now try using a different number.
What happens this time?

(c) Do you think this always happens?

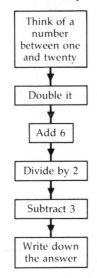

Think of a
number
between one
and twenty

↓

Double it

↓

Add 6

↓

Divide by 2

↓

Subtract 3

↓

Write down
the answer

2

Think of a number between 1 and 10	Think of a number between 1 and 10
↓	↓
Multiply by 5	Double it
↓	↓
Add 2	Add 1
↓	↓
Double it	Multiply by 5
↓	↓
Subtract 3	Subtract 4
↓	↓
Write down the answer	Write down the answer

(a) Use each of the flow charts above with
the same starting number.
What do you notice about your results?

(b) Repeat (a) using a different starting
number. What happens this time?

(c) Now try this again with another starting
number.

(d) Can you describe in a simpler way what
happens to your starting number?

3 (a) Use the flow chart below. What was the last
number you wrote down?

(b) What can you say about the set of ten
numbers you have written down?

(c) How would you change this flow chart to
give you the even numbers?

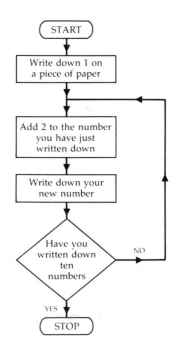

START

↓

Write down 1 on
a piece of paper

↓

Add 2 to the number
you have just
written down

↓

Write down your
new number

↓

Have you
written down
ten
numbers   NO →

YES ↓

STOP

4 (a) Use the flow chart which follows.
How many numbers did you write
down?

(b) What can you say about the numbers
you have written down?

(c) How would you change this flow chart to
give you the powers of 3?

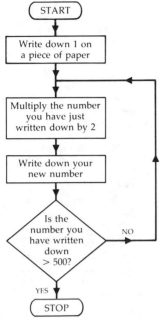

(a) Complete the last box in the flow chart.

(b) Use the four triangles with the above flow chart.
Do you agree with their descriptions?

6

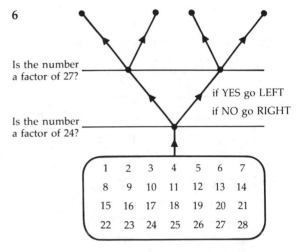

Is the number a factor of 27?

if YES go LEFT

if NO go RIGHT

Is the number a factor of 24?

1	2	3	4	5	6	7
8	9	10	11	12	13	14
15	16	17	18	19	20	21
22	23	24	25	26	27	28

(a) Use the sorting tree to find which of the numbers finish at the end of each branch.

(b) Which of these numbers are factors of both 24 and 27?

(c) Which of these numbers are neither factors of 24 nor 27?

5

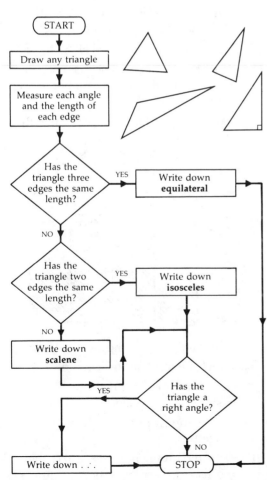

7

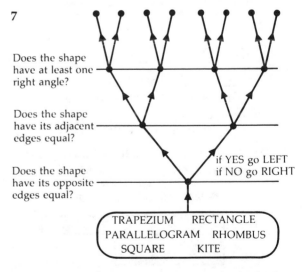

Does the shape have at least one right angle?

Does the shape have its adjacent edges equal?

if YES go LEFT

if NO go RIGHT

Does the shape have its opposite edges equal?

TRAPEZIUM   RECTANGLE
PARALLELOGRAM   RHOMBUS
SQUARE   KITE

(a) Use the sorting tree above to find which shapes finish at the end of each branch.

(b) You should have found that there were no shapes on two of the branches of the tree. Try to draw a shape to fill each of these two gaps.

4.14 *Bar charts and block graphs*

1 (a) Describe the information shown in the pictogram below.

January sales figures	
Ford	🚗 🚗 🚗 🚗
Vauxhall	🚗 🚗 🚗
Austin Rover	🚗 🚗 🚗
Peugeot Talbot	🚗

(b) If Peugeot Talbot sold 2000 cars, how many cars did each of the others sell?

(c) If Ford sold 20 000 cars how many cars did each of the others sell?

(d) How many more cars did Vauxhall sell than Peugeot Talbot?

2

	Council expenditure in £1 millions
Schools	24
Police	6
Fire service	3
Other services	12

Draw a pictogram to represent the above information.
(Use a £ sign for each £1 000 000 spent.)

3 (a) Describe the information shown in the bar chart below.

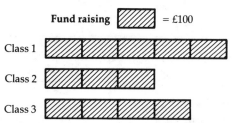

(b) How much did each class raise?

(c) How much was raised altogether?

(d) Show the amount each class raised as a fraction of the total amount raised.

4 (a) Describe the information shown in the block graph below.

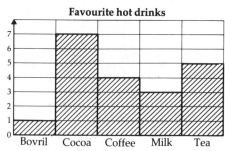

(b) Which was the most popular hot drink? How many chose this one?

(c) Which was the least popular drink? How many chose this one?

(d) How many people voted altogether?

(e) Write the number who chose each drink as a fraction of the total number of voters.

5 (a) Describe the information shown in the block graph below.

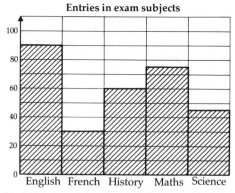

(b) Which subject had the most entries?

(c) Which subject had the least entries?

(d) How many more entries were there in History than in Science?

(e) How many entries were there altogether?

(f) Write the number of entries in each subject as a fraction of the total number.

6 A traffic survey was carried out between 9.00 a.m. and 10.00 a.m. The results are shown below in the form of a tally chart.

Vehicle	Tally	Frequency
Cars	⃫⃫⃫ ⃫⃫⃫ ⃫⃫⃫ ||	17
Lorries	⃫⃫⃫ |||	
Buses	|||	
Bicycles	⃫⃫⃫ ⃫⃫⃫ ⃫⃫⃫ ⃫⃫⃫		

(a) Copy and complete the frequency column.

(b) How many vehicles were counted?

(c) Draw a block graph to represent the information above.

7 (a) Draw a block graph to show the information in each frequency table.

	Number of hours of sunshine	Number of new births
Monday	8	12
Tuesday	9	14
Wednesday	11	17
Thursday	12	18
Friday	7	10
Saturday	4	6
Sunday	3	5

(b) How do your two graphs compare?

(c) Do you think there is a connection between the number of new births and the amount of sunshine on any particular day?

(d) Can you think of an explanation for the low weekend figures?

8 (a) What is the longest word you can find in this question?

(b) What is the shortest word you can find in this question?

(c) Make a tally chart for the number of letters in each word in this question.

(d) Copy and complete the table below.

Length of word	1	2	3	4	5	6	7	8	9
Frequency of word	2								

(e) Draw a block graph for your table.

9 The information below indicates the number of matches found in a sample of 120 matchboxes.

Number of matches	45	46	47	48	49	50
Number of boxes	12	18	35	28	16	11

(a) Draw a block graph for this information.

(b) How many boxes contained more than 47 matches?

(c) What fraction of the 120 boxes contained less than 47 matches?

10 (a) Describe the information shown in the block graph below.

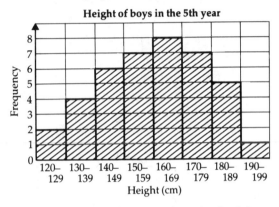

Height of boys in the 5th year

(b) How many boys are there in the 5th year?

(c) In which range of heights are there most boys?

(d) What can you say about the height of the tallest boy?

(e) How many boys are shorter than 150 cm?

11 (a) Draw a block graph for this information.

Girls' heights (cm)	120 –129	130 –139	140 –149	150 –159	160 –169	170 –179	180 –189
Frequency	3	4	6	9	5	2	1

(b) What can you say about the height of the tallest girl?

(c) How many girls are shorter than 150 cm?

(d) How does this graph compare with the graph for the boys in question **10**?

(e) In which ranges of heights are there more girls than boys?

4.15 *Pie charts*

1 (a) Describe the information shown in the pie chart below.

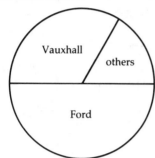

   (b) What fraction of the cars sold in February were made by:

      (i) Ford   (ii) Vauxhall   (iii) others?

   (c) If 3600 cars were sold altogether in February, how many were made by:

      (i) Ford   (ii) Vauxhall   (iii) others?

   (c) What angles in the pie chart are used to represent this information?

In the pie chart above 360° represents the 3600 cars sold, so **each degree** represents 10 cars. Ford sales are represented by half of the pie chart, so their 180° represents 1800 cars sold.

---

2 In March, another 3600 cars were sold. 1200 were sold by Ford, 900 by Vauxhall, and 600 by Austin Rover.

   (a) What fraction of the cars were sold by:

      (i) Ford   (ii) Vauxhall

      (iii) Austin Rover?

   (b) What angles in a pie chart would you use to represent this information?

   (c) Draw a pie chart for this information.

3 In April, in the Liverpool area, 720 cars were sold.

   (a) How many cars in the pie chart would one degree represent this time?

   (b) Draw a pie chart to show these sales:
      Ford: 360, Vauxhall: 240,
      Austin Rover: 90.

4 (a) Describe the information shown in the pie chart below.

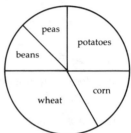

   (b) Measure the number of degrees in each section of the pie chart.

   (c) What fraction of the pie chart is represented by each section?

   (d) How many acres of each type of crop are there, if the total number of acres represented is:

      (i) 360   (ii) 720   (iii) 240   (iv) 900?

5 (a) Describe the information shown in the pie chart below.

   (b) Measure the number of degrees in each section of the pie chart.

   (c) What fraction of the pie chart is represented by each section?

   (d) How much is spent on each item, if the total expenditure represented is:

      (i) £1800   (ii) £1200   (iii) £4800?

6 (a) If the amount spent on rent in question **5** is £2400, how much is spent on each of the other items?

   (b) If the amount spent on food is £1200, how much is spent on each of the other items?

The pie chart started below shows how Mick spends his day.

The whole day:	24 hours is 360°
so sleeping:	12 hours is 180°
school:	6 hours is 90°
eating:	3 hours is 45°
and others:	3 hours is 45°

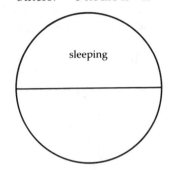

7 (a) Copy and complete the pie chart started above.

(b) How many degrees in the pie chart represent one hour?

(c) How many degrees would represent an activity lasting:

(i) 4 hours   (ii) 5 hours   (iii) 7 hours?

8 Mick's sister spends her day as shown in the table below:

sleeping	8 hours
eating	2 hours
working	9 hours
watching TV	2 hours

(a) Copy the table and add another column to show how many degrees are needed for each activity.

(b) Draw a pie chart to represent this information.

9 The seventy-two members of the local youth club choose to spend club night as follows.
table tennis:30, chess: 12, darts: 18, Trivial Pursuit: 8, Scrabble: 4

(a) Draw a pie chart to represent this information.

(b) Re-draw your pie chart to show this information if there had been 144 members.

10 In the pie chart below 120 students are taking Maths in the exams.

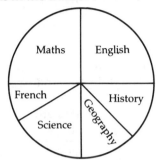

(a) Find the number of degrees in each section of the pie chart.

(b) How many exam entries are shown in all?

(c) Find the number of entries for each of the other subjects.

(d) Check your answer to (c): show that the total of the individual subject entries is the same as your answer for (b).

11 A poll was taken to find the most popular country for holidays. Draw a pie chart to show this information.

Country	Austria	France	Greece	Italy	Spain
Numbers	100	300	450	150	500

12 Another poll was taken to find out what type of holiday 600 families choose. Draw a pie chart to show this information.

Holiday	Boat	Caravan	Hotel	Tent	Others
Numbers	60	150	90	240	60

13 Draw a pie chart for each set of sales figures.

(a)

Days	Mon	Tues	Wed	Thurs	Fri
Sales	15	25	20	30	30

(b)

Days	Mon	Tues	Wed	Thurs	Fri
Sales	12	24	18	36	30

14 (a) Collect information about the number of sisters each of 12 of your friends has.

(b) Draw a pie chart to show your findings.

4.16 *Information from tables*

1 The tables below show the cost of different sized tyres and different exhausts.

Tyre size	Price	Make of car	Exhaust Price
145 × 10	£17.50	Mini	£14.50
135 × 12	£19.30	Fiesta	£16.50
135 × 13	£19.50	Allegro	£19.50
145 × 13	£20.50	Escort	£28.50
155 × 13	£21.50	Cavalier	£28.50

Find the cost of:
(a) four 135 × 13 tyres.
(b) a Fiesta exhaust and two 145 × 13 tyres.

2 The American elections were shown live on television all over the world. When the polls closed it was 11.00 p.m. in London.
New York time is 5 hours behind London time.
Moscow is 3 hours ahead of London time. This is shown in the first entry in the table below.

New York time	London time	Moscow time
6.00 p.m.	11.00 p.m.	2.00 a.m.
7.00 p.m.	. . .	. . .
. . .	7.00 p.m.	. . .
. . .	. . .	7.00 a.m.

(a) Fill in the missing values in the table.
(b) A clock in New York shows 11.00 p.m. What time is showing on a clock in Moscow?
(c) A clock in Moscow shows 11.00 p.m. What time is showing on a clock in New York?
(d) It takes 4 hours to fly from Moscow to London when leaving Moscow at 5.00 a.m.
What time will a clock in London show when the plane arrives?

3 Part of a train timetable is shown below.

London	1030	1100	1200	1300	1330	1400
Edinburgh	1453	1542	1626	1728	1825	1838
Glasgow	1618	1648	1748	1848		1948
Dundee	1610	1738	1817	1904		2035
Aberdeen	1724		1940	2028		
Stirling	1604	1704	1716	1837	1937	2004
Perth	1648		1755	1927	2023	2129
Inverness	1910		2005			

(a) How long does the 1453 from Edinburgh take to get to Aberdeen?
(b) What time does the train which reaches Dundee at 1904 leave Edinburgh?
(c) What is the latest time you can leave London to get to Inverness?
(d) Which is the quickest train from London to Edinburgh?

4 The table below shows the cost, in £, of buying different types of motor fuel.

Litres	Diesel	Unleaded	2–star	4–star
1	33.4p	34.6p	35.8p	36.6p
5	1.67	1.73	1.79	1.83
10	3.34	3.46	3.58	3.66
15	5.01	5.19	5.37	5.49
20	6.68	6.92	7.16	7.32
25	8.35	8.65	8.95	9.15
30	10.02	10.38	10.74	10.98

(a) A taxi runs on diesel and uses 10 litres per 100 miles.
(i) What is the cost of 30 litres?
(ii) How much more would it have cost to use 2-star petrol?
(iii) What is the cost for 250 miles?
(b) The new Fiesta can run on unleaded fuel. Overall it uses 10 litres per 80 miles.
(i) What is the cost of 25 litres?
(ii) What will a 400 mile journey cost?
(iii) How much is saved by not using 4-star petrol?

5 The table below shows the prices, in £, for various types of skiing holiday in Austria.

	Berg (Apts) (SC)		Pension Dom (BB)		Hotel Post (HB)	
Date	7	14	7	14	7	14
Dec 17	152	302	188	379	240	485
Dec 24	226	311	262	398	314	504
Dec 31	257	252	293	329	345	435
Jan 7	172	231	208	308	260	414
Jan 14	152	231	188	308	240	414

(a) What do you think SC, BB and HB stand for?

(b) When is the most expensive week in the Berg apartments?

(c) How much would two weeks in the Pension Dom leaving on December 24th cost?

(d) What would be the minimum cost for a family of four, if they wanted to spend a week at the Hotel Post?
When could they travel?

6 It is possible to fly from four different airports for the holidays in question **5**. This information is shown below.

Airport	Take off	Home landing	Supplement
Birmingham	1500	1945	£11
Glasgow	1445	1345	£29
Leeds	1530	2015	£15
Newcastle	0800	1315	£19

(a) Which place is the cheapest to fly from?

(b) Which place has the earliest departure?

(c) Which place has the latest return home?

(d) How much will the family of four save by flying from Leeds rather than Glasgow?

(e) If the flight time from Glasgow is two and a half hours, and the time from the airport to the resort by coach is another one hour, what time should you arrive at the resort?

7 The prices of five different models of a car and the available options are shown below.

	1.3 (3-dr)	1.3 (5-dr)	1.4 (auto)	1.6 (5-dr)	1.6 (est.)
Basic cost	7252	7519	7694	7962	8472
Auto	345	345	n/a	345	345
5-speed	n/a	195	n/a	195	195
Met. paint	140	140	140	140	160
Extra cover	110	110	150	130	140

£350 is added to put the car on the road.

What is the cost on the road of:
(a) a basic 1.3 (3-door) car
(b) a 1.4 (auto) car with metallic paint
(c) a 1.6 (est.) car with a 5-speed gearbox and extra cover?

8 The table below shows the weekly repayments for borrowing different sums of money over 3, 5, 8 or 10 years.

LOANS

Borrow	Weekly repayments over			
	3 years	5 years	8 years	10 years
£ 500	4.28	3.05	2.47	2.22
£1000	8.55	6.09	4.94	4.43
£2000	17.11	12.18	9.88	8.86
£3000	25.66	18.27	14.82	13.29
£4000	34.21	24.35	19.76	17.72
£5000	42.76	30.44	24.70	22.15

(a) Use the above table to calculate the weekly repayments on a loan of:
(i) £1000 over 5 years
(ii) £4000 over 8 years
(iii) £1500 over 3 years
(iv) £7000 over 10 years.

(b) How much can I borrow if I am willing to pay back:
(i) £12.18 weekly over 5 years
(ii) £24.70 weekly over 8 years
(iii) £ 9.14 weekly over 5 years
(iv) £39.87 weekly over 10 years?

4.17 *Using statistics*

1 Look at the block graph below for Fixit. It shows a comparison of their profits in 1986 and 1987.

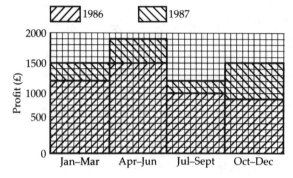

(a) Describe the pattern of profits in 1987.

(b) Compare the profits in 1987 with 1986.

(c) Which was the best quarter in 1986?

(d) Which was the worst quarter in 1987?

(e) Find the total profit in 1987.

(f) By how much did the profits in 1987 exceed those in 1986?

2 Look at the pie chart below for TV viewing. It compares the popularity of certain TV programmes among a sample of 5000 viewers.

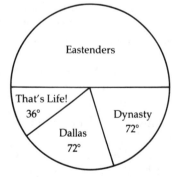

(a) Which is the most popular programme?

(b) Which is the least popular programme?

(c) What can you say about the popularity of these two programmes?

(d) How many people voted for each programme?

3 The line graph below shows the increase in attendance at the All Stars football club.

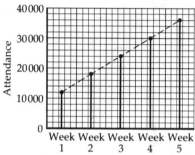

(a) How many saw the game in the first week?

(b) How many saw the game in the fifth week?

(c) In which week was the attendance 30 000?

(d) By how much had the attendance increased over these five weeks?

(e) If the attendance kept on increasing at the same rate, how many would you expect to watch the game in the sixth week?

4 The graph below shows the journeys of two cyclists between Coventry and Birmingham.

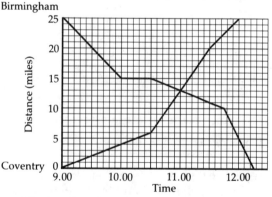

(a) At what time did:
  (i) the second cyclist reach Coventry
  (ii) the two cyclists pass each other?

(b) How far from Birmingham was each cyclist when they passed each other?

(c) When was each cyclist travelling fastest?

5 The weights, in kg, of twenty-one newborn babies are shown below:

2.5 2.7 2.7 2.8 2.9 3.0 3.1
3.1 3.1 3.3 3.5 3.6 3.7 3.8
4.0 4.2 4.2 4.2 4.3 4.5 4.7

(a) Which weight occurred most often?

(b) What is the average (mean) weight?

(c) Draw a block graph to show this information.

6 John, David, Andy and Mike are comparing their wages. They receive per week £93, £108, £87 and £104 respectively.

(a) Find their average (mean) wage.

(b) Bob joins the group.
His weekly wage is £103.
What is the new average (mean) wage?

(c) Their employer now wants to take on a sixth lad.
He can afford an average (mean) wage of £100.
What can he afford to pay this person?

7 Look at the pie-chart below. It shows the popularity of four car manufacturers amongst a sample of 100 drivers.

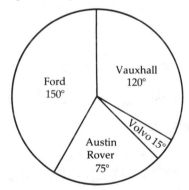

(a) How many of the above sample chose:
(i) a Ford car (ii) a Vauxhall car?

(b) If you asked 1000 drivers, how many would you expect to choose:
(i) a Ford car (ii) a Vauxhall car?

(c) Look in a car park and make a note of the makes of 120 cars.
Draw a pie chart to show this information.
How does it compare with the one above?

8 The block graph below shows the frequency of success in a test when hatching batches of five eggs.

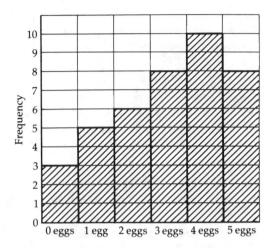

(a) How many times were no eggs hatched?

(b) Which number of eggs hatched occurred most often?

(c) What was the total number of successful hatchings?

(d) How many eggs were used altogether in this test?

(e) What fraction of the eggs were successfully hatched?

(f) What was the average (mean) number of eggs hatched in each batch of five?

9 Use your results for question **8** to predict:
(a) the number of times you would expect to hatch all five eggs in 1000 batches of five.

(b) the number of successful hatchings for a total of 2000 eggs.

10 A die was rolled a large number of times to see if it was biased. The results are shown below:

1 — 300   2 — 250   3 — 200   4 — 100
5 — 100   6 — 50

(a) Draw a block graph to show this information.

(b) Write down each result as a fraction of the total number of rolls.

(c) Do you think this die was biased or not? Explain your answer.

4.18 Mean, mode and median

1   Nine pupils in Matt's class decided to compare the number of rooms in each of their houses.
    The results were as follows:
      6, 8, 5, 11, 5, 3, 7, 4, 5
    (a) Re-write these numbers in order of size.
    (b) What was the smallest number of rooms? What type of house do you think this was?
    (c) What was the largest number of rooms? What type of house do you think this was?
    (d) What was the most common number of rooms? What type of houses do you think these were?

2   If you had to describe the houses in Matt's class to a stranger, what would you say was the average number of rooms?

Three types of average are often used to describe statistics:
    the **mode**, the **median** and the **mean**.
The item which occurs **most frequently** is called the **mode**.

---

3   Which item is the mode for the number of rooms in question **1**?

If the items are put in order of size, the **middle** item is called the **median**.

---

4   Which item is the median for the number of rooms in question **1**?

The **mean** of a set of items is the total of the items divided by the number of items.
For example the mean of 3, 4, 5 and 8 is 5.

---

5   What is the mean for the number of rooms in question **1**?

6   If, in question **1**, a tenth house with 16 rooms had been included, what would have been the mean number of rooms?

7   For each set of data write down:
    (a) the mode   (b) the median
    (c) the mean.
    (i)   2, 2, 5
    (ii) 2, 2, 4, 5, 7
    (iii) 2, 5, 7, 8, 9, 9, 9
    (iv) 4, 3, 4, 5, 4, 5, 3, 4

8   A group of children had their heights measured. These are shown below.
      136 cm, 142 cm, 133 cm, 153 cm, 127 cm, 137 cm, 128 cm, 144 cm, 133 cm.
    (a) Re-write these heights in order of size.
    (b) What was the height of:
        (i)   the shortest child
        (ii) the tallest child
        (iii) the middle child?

9   Use the information given in question **8**.
    (a) Write down which height is the mode.
    (b) Write down which height is the median.
    (c) Find the mean of these heights.

10 The table below shows the size of family in a sample of 235 households.

Size of household	1	2	3	4	5	6
Frequency	10	40	65	70	40	10

    (a) Draw a block graph to show this data.
    (a) Which size of household is the mode?
    (c) Which size of household is the median?

11 Look again at the information in question **10**. If you were a local council providing housing would you need more three-bedroomed houses or more one-bedroomed flats? Explain your answer.

The item which occurs most frequently is shown by the tallest block.
In question **10** the mode is 4.
In this example the median is also 4.

---

12 The table below shows the number of bookings made at the local Youth Hostel during a two month period.

Number of bookings	25	26	27	28	29	30
Frequency	5	11	15	23	4	3

(a) Draw a graph to show this information.

(b) Which number of bookings is the mode?

(c) Which number of bookings is the median?

13 Look again at the information in question **12**. If you were the Youth Hostel warden, how many meals would you aim to have available on any given night? Explain your answer.

14 Mrs Sandell runs a shoe shop. She collects information from a sample of her male customers as shown below.

Shoe size	6	7	8	9	10	11	12
Frequency	3	9	15	12	14	7	2

(a) Draw a graph for this information.

(b) Which shoe size is the mode?

(c) Which shoe size is the median?

(d) Which size shoe would you expect Mrs Sandell to stock most of?

15 Mrs Chick is an egg producer.

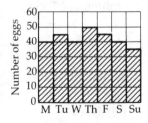 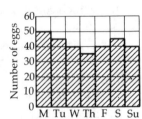

For each graph:
(a) Describe the information shown.

(b) Write down which number of eggs is:
(i) the mode   (ii) the median.

(c) Find the total number of eggs.

(d) If eggs sell at 4p each how much will Mrs Chick receive?

16 Mrs Sandell sold 20 pairs of shoes today. She made £3 profit on each of 12 pairs, £2 profit on each of 6 pairs and £1 profit on the others she sold.
(a) What was her total profit?

(b) Find the mean profit she made on each pair of shoes today.

17 The block graph below shows how often a given number of goals are scored in a football match.

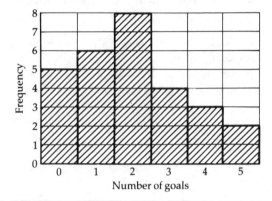

(a) Which number of goals is scored most frequently?

(b) Which number of goals is scored least frequently?

(c) If you were to visit your local football club, how many goals would you expect to see scored?

18 Use the information given in question **17** above.
(a) Find the median number of goals scored.

(b) Find the mean number of goals scored.

19 In a small firm each of the nine employees earns £100 per week. Their boss pays himself £600 per week.
(a) What is the modal wage for this firm?

(b) What is the mean wage for this firm?

(c) Which of these two averages gives the fairer indication of the total salary bill for the firm?

20 Find the mode, median and mean for the numbers of letters in:
(a) the days of the week

(b) the months of the year.

4.19 *Probability: equally likely events*

1   Say whether you think the event is certain to
happen, or impossible.
(a) Ice will melt if it is put in the sun.

(b) You can make gold out of sand.

(c) If I drop a stone it will fall to the ground.

(d) Water will flow uphill.

2   The scale below is used to show whether an
event is likely to happen or not.
Show, on a copy of the scale, whether you
think the event is likely to happen or not.

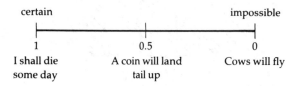

certain   impossible

1   0.5   0

I shall die   A coin will land   Cows will fly
some day   tail up

(a) It will rain some time next week.

(b) School will be cancelled tomorrow.

(c) A coin will land head up.

(d) If a die is rolled it will show a six.

(e) You will not get married next year.

Probability tells us how **likely** something is to
happen.

It is measured on a scale from 0 to 1.
A probability of 1 tells us an event is **certain** to
happen.

A probability of 0 tells us an event is
impossible.

---

3   (a) Write down the probability of each event
in question **1** happening.

(b) Estimate the probability of each event in
question **2** happening.

The probability that a coin will land head up is
one in two or 0.5.

Probability does not tell us what is *actually*
going to happen, only how *likely* or not it is.

4   Jill has a bag
containing three
white beans and
seven red beans.
John picks one bean
out without looking.

(a) Is John more likely to pick a white or a
red bean?

(b) Do you agree that the probability of John
picking a red bean is seven out of ten, i.e.
0.7?

(c) What is the probability that John picks a
white bean?

5   (a) How many of the beans in question **4** are
not white?

(b) What is the probability that John picks a
bean out that is not white?

6   Jill adds six black
beans to the bag in
question **4**.

(a) Which colour bean is John most likely to
pick?

(b) Which colour bean is John least likely to
pick?

(c) What is the probability that John picks a
bean out that is:

(i) red   (ii) white   (iii) black?

7   (a) How many of the beans in question **6** are
not white?

(b) What is the probability that John picks
out a bean that is not white?

(c) How many of the beans in question **6** are
not red?

(d) What is the probability that John picks
out a bean that is not red?

When John picks a bean in question **6**, a red bean is more likely to be chosen since there are more of them.

The possible outcomes are not equally likely.

When a coin is tossed there are only two possible outcomes.

A head or a tail is equally likely.

---

8 The chairperson of the school council is to be chosen at random from a committee of five girls and three boys.
   (a) Is it more likely that the chairperson will be a boy or a girl?
   (b) What is the probability that a boy will be chosen?

9 The captain of the club tennis team is to be chosen from the four men and four women who make up the team.
   (a) If one of the team members is selected at random is it more likely to be a man or a woman?
   (b) What is the probability that a woman will be the captain?

10 A die is rolled.
   (a) Are you more likely to score a number less than four or a number greater than four?
   (b) What is the probability of getting:
     (i) a number greater than four
     (ii) a number less than four?

11 A die is rolled.
   (a) Are you more likely to get a six or a five?
   (b) What is the probability of getting:
     (i) a six
     (ii) a five?

12 A die is rolled.
   (a) Are you more likely to get a result which is an even number, or a multiple of three?
   (b) What is the probability of getting:
     (i) an even number
     (ii) an odd number
     (iii) a multiple of three?

When a die is rolled there are three possible even numbers:
   2, 4 and 6.

There are only two possible multiples of three: 3 and 6.

The probability of getting an even number is 3 out of 6
   i.e. $\frac{3}{6} = \frac{1}{2}$

The probability of getting a multiple of three is 2 out of 6
   i.e. $\frac{2}{6} = \frac{1}{3}$

---

13 The die in a particular game is marked with three 1's, two 2's and one 3.
   (a) Which number are you most likely to get when you roll this die?
   (b) What is the probability of getting:
     (i) a 1 (ii) a 2
     (iii) a 3?

14 (a) Are you more likely to get an even number, or an odd number, when using the die in question **13**?
   (b) What is the probability of getting:
     (i) an even number
     (ii) an odd number?

15 Look at the spinner shown below.

What is the probability of getting:
(a) a six
(b) a number less than four
(c) a number greater than four
(d) a multiple of three
(e) a multiple of four?

16 Another spinner like the one in question **15** has three 1's, two 2's, two 3's and one 6. What is the probability of getting:
(a) a six
(b) a number less than four
(c) a multiple of three?

4.20 *Probability: listing outcomes*

The probability of a particular event taking place is:

$$\frac{\text{the number of the particular outcome}}{\text{the number of all the possible outcomes}}$$

An ordinary pack of playing cards has 52 cards, excluding any jokers. If one card is selected at random the possible outcome could be any one of the 52 cards.
So the probability of picking a heart is:

$$\frac{\text{the number of hearts}}{\text{the number of cards}} = \frac{13}{52} = \frac{1}{4}$$

---

1 A card is chosen at random from a pack of 52 playing cards.
   (a) Write down the number of ways of picking:
      (i) a spade   (ii) a red card   (iii) an ace.
   (b) Write down the probability of picking:
      (i) a spade   (ii) a red card   (iii) an ace.

2 A card is selected at random from the diamonds in a pack of playing cards.
   (a) How many diamonds are there?
   (b) How many picture cards are there among the diamonds?
   (c) What is the probability of picking a picture card from the diamonds?

3 A card is selected at random from the red cards in a pack of playing cards.
   (a) How many red cards are there?
   (b) How many red picture cards are there?
   (c) What is the probability of picking a picture card from the red cards?

4 A card is chosen at random from a pack of 52 playing cards.
   (a) How many picture cards are there?
   (b) How many red aces are there?
   (c) What is the probability of picking:
      (i) a picture card   (ii) a red ace?

5 One day of the week has to be chosen.
   (a) List the possible outcomes.
   (b) How many of the outcomes begin with the letter T?
   (c) What is the probability that the chosen day begins with the letter:
      (i) T   (ii) S   (iii) M   (iv) P?
   (d) What is the probability that the chosen day does not begin with the letter:
      (i) T   (ii) S   (iii) M   (iv) P?

6 One month of the year has to be chosen at random.
   (a) How many of the outcomes begin with the letter J?
   (b) What is the probability that the chosen month begins with the letter:
      (i) J   (ii) M   (iii) A   (iv) P?
   (c) What is the probability that the chosen month does not begin with the letter:
      (i) J   (ii) M   (iii) A   (iv) P?

7 A bag contains a number of counters, as shown below. One counter is picked at random from the bag.

   (a) How many counters are there altogether?
   (b) How many of the counters are multiples of three?
   (c) What is the probability of picking a counter which is:
      (i)   a multiple of three
      (ii) an even number
      (iii) a multiple of four
      (iv) a square number?

8 A bag contains one £20 note, four £10 notes and twenty £5 notes. One note is chosen at random for testing.
   (a) How many notes are there altogether?
   (b) What is the probability that the note chosen is a:
       (i) £5 note   (ii) £10 note   (iii) £50 note?

9 A spinner is made from a regular pentagon. The five edges are coloured red, yellow, green, blue and black.

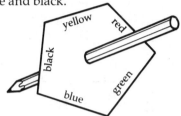

If you used the spinner 1000 times, on how many occasions would you expect the outcome to be:
   (a) red   (b) a colour beginning with 'b'?

10 A spinner similar to the one in question **9** has two of its edges coloured red, two coloured blue and one brown. In 500 spins how many would you expect to be:
   (a) red   (b) a colour beginning with 'b'?

11 Larry Wade makes light bulbs. After each batch he tests a sample of 100. In one batch, five are found to be faulty. What is the probability that one chosen at random is:
   (a) faulty   (b) not faulty?

12 One in twenty of Larry's bulbs are known to be faulty. What number are likely to be faulty in a batch of:
   (a) 100   (b) 500   (c) 2000   (d) 4800?

13 At a by-election, a sample of 1000 people were asked who they voted for.
The findings were as follows:

Conservatives	390
Labour	320
Alliance	280
Others	10

If one of this sample were picked out at random, what is the probability that he voted:
   (a) Conservative   (b) Labour   (c) Alliance?

14 Assuming that 10 000 people voted in the by-election in question **13**, approximately how many would you expect to have voted:
   (a) Conservative   (b) Labour   (c) Alliance
   (d) for none of these three parties?

15 The diagram shows four groups of children in Mrs Jones's infant class.

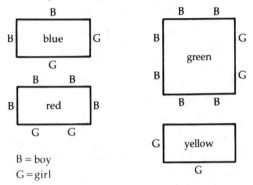

B = boy
G = girl

One child is selected at random each day to ring the break bell. What is the probability that the child chosen to ring the bell is:
   (a) a boy   (b) a girl
   (c) from the blue table
   (d) from the green table
   (e) from the red table
   (f) from the yellow table?

16 A survey of washing powders was carried out amongst a sample of users. The results are shown below in the graph.

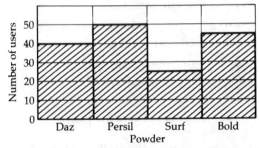

   (a) How large was the sample questioned?
   (b) If one of these was picked out at random, what is the probability that they used:
       (i) Daz   (ii) Persil   (iii) Surf   (iv) Bold?
   (c) If the Cleano supermarket used the results of this sample for their stock of 8000 packets, how many of each should they buy?

4.21 *Probability: combined events*

1 A 5p and a 10p coin are tossed together.
   (a) Write down the four possible outcomes.
        Are they all equally likely?
   (b) In how many of these outcomes is there a
        head on one coin and a tail on the other
        coin?
   (c) If two coins are tossed, what is the
        probability of getting:
        (i) a head and a tail   (ii) two heads?

2 A 2p, a 5p and a 10p coin are tossed together.
   The table started below shows the possible
   outcomes.

2p	h	h	h	h	t	t	t	t
5p	h	h						
10p	h	t						

   (a) Copy and complete the table to show the
        eight possible outcomes.
        Are they all equally likely?
   (b) In how many of these do you get:
        (i) just one head   (ii) two heads?
   (c) If three coins are tossed, what is the
        probability of getting:
        (i)   just one head   (ii) two heads
        (iii) just one tail   (iv) three tails?

3 A coin is tossed and a die is rolled at the
   same time. One of the possible outcomes is a
   head and a six.
   (a) Write down all the other possible
        outcomes.
        How many are there altogether?
        Are they all equally likely?
   (b) In how many of these outcomes does a
        six occur?
   (c) What is the probability of getting:
        (i)   a head and a six
        (ii) a tail and a five
        (iii) a head and an even number?

One way of showing the possible outcomes of
tossing a coin and rolling a die is shown
below.

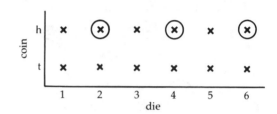

The circled x's represent a head and an even
number.

4 (a) Copy the diagram above and show
        which outcomes represent a tail with a
        multiple of three.
   (b) What is the probability of getting a tail
        and a multiple of three?

5 The diagram below represents all the
   possible outcomes when two dice are rolled
   together.

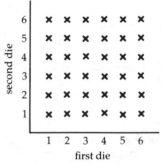

   (a) How many outcomes are possible when
        two dice are rolled together?
   (b) On a copy of the diagram above show
        the outcomes which represent:
        (i)   a total score of seven
        (ii) a total score of four
        (iii) doubles
        (iv) a total score of more than ten.

6 Using your diagrams for the two dice from question **5** on page 164, write down the probability of getting:

(a) a total score of seven

(b) a total score of four

(c) doubles

(d) a total score of more than ten.

7 On a copy of the diagram from question **5**, show the outcomes which represent:

(a) at least one six

(b) an even number on each die

(c) one six and an odd number

(d) no sixes.

8 Using your diagrams for the two dice from question **7**, write down the probability of getting:

(a) at least one six

(b) an even number on each die

(c) one six and an odd number

(d) no sixes.

An alternative way of finding the probabilities for combined events is to use a tree diagram.

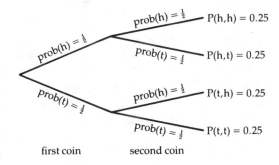

first coin   second coin

To find the combined probability you multiply the individual probabilities.

9 Using the tree diagram above, do you agree that the probability of getting:

(a) a head on both coins is 0.25

(b) a tail on both coins is 0.25?

10 (a) Do you agree that there are two ways of getting a head and a tail and that the probability of doing this is 0.25 + 0.25?

(b) Why is the sum of the four individual probabilities one?

The tree diagram started below shows the probabilities of getting sixes with two dice.

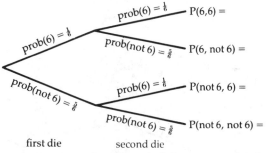

first die   second die

To find the combined probability you multiply the individual probabilities.

11 (a) Using the tree diagram above, find the probability of getting:

(i) a six on both dice

(ii) a six on only one of the dice

(iii) a six on neither die.

(b) Use the diagram from question **5** to check.

12 Check that the sum of the probabilities in question **11**(a) is one. Do these represent all possible outcomes when rolling two dice?

The tree diagram started below shows the probabilities when three coins are tossed.

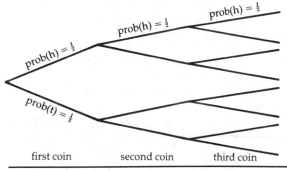

first coin   second coin   third coin

13 Copy and complete the tree diagram above, and find the probability of getting:

(a) a head on all three coins

(b) a tail on the first coin and a head on the other two coins

(c) a tail and two heads in any order.

14 Three dice are rolled.
Find the probability of getting:
(a) three sixes   (b) only one six.

4.22 *Problems and investigations*

1 Look at the cuboid below.
One route from A to G is shown.

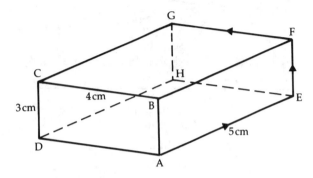

(a) What is the length of the given route
from A to G?

(b) How many other routes, of the same
length from A to G, can you find?

(c) Describe each of these routes.

(d) Investigate routes from A to G of other
lengths.

(e) What is the longest route you can find
from A to G, without going over any
edge a second time?

2 You can make two different patterns with
the colours red (r) and blue (b).
These are rb and br as shown below.

2 colours	3 colours	4 colours	5 colours
rb	rby	rbyg	rbgyp
br	ryb	rbgy	?
	?	?	

(a) Make as many different patterns as you
can by adding yellow (y) to your colours.
Two are shown above.

(b) Make as many different patterns as you
can by now adding the colour green (g)
as well. Two are shown above.

(c) Can you predict how many patterns you
can make using five colours?

3 (a) How many different 'words' can you
make with the letters:
(i) a,t (ii) a,e,t (iii) a,e,r,t?

(b) Can you predict how many 'words' you
can make with the letters: a,e,h,r and t?

4 Using + signs only, the four different ways
of making the number three are shown below.
3, 1+2, 2+1, 1+1+1

(a) Write down the two different ways of
making the number 2.

(b) Write down all the different ways of
making the number 4.

(c) Write down all the different ways of
making the number 5.

(d) Can you predict how many ways there
will be of making the number 6?

(e) Now copy and complete the table below.

Number	1	2	3	4	5	6
Number of ways		2	4			

(f) Check your prediction for the number of
ways of making the number 6. Were you
right?

5 Look again at your results for question **4**. Of
the four ways of making the number three:
1 way uses only one number: 3
2 ways use two numbers: 1+2   2+1
1 way uses three numbers: 1+1+1

(a) Now copy and complete this table.

		Number				
	1	2	3	4	5	6
Number of ways using 1	1	1	1			
2	–	1	2			
3	–	–	1			
4	–	–	–			
5	–	–	–	–		
6	–	–	–	–	–	

(b) Can you see any patterns in this table?

(c) Can you predict what will happen for the
different ways of making the number 7?

6 How many different 'words' can you make with
  (a) four r's and one u
  (b) three r's and two u's
  (c) other numbers of r's and u's?

7 Look at the grid below.
  One route from S to F is shown.
  We could call this route   rruru
  where r stands for go right and u for go up.

  (a) If you are only allowed to go right or up, how many other routes are there from S to F? Describe each route.
  (b) How many routes, of this sort, are there from S to G?
  (c) Find the number of routes from S to each of the other points on the grid.
  (d) On a copy of the grid mark the number of routes to each point.
  (e) What patterns can you find in these route numbers?

8 (a) Copy and complete the next few rows of Pascal's triangle:

1	. . . 1
1 1	. . . 2
1 2 1	. . . ?
1 3 3 1	. . . ?
? 4 ? ? ?	. . . ?

  (b) Find the total of the numbers in each row of the triangle. Is there a pattern here?
  (c) Investigate other patterns in your completed triangle.

9 Using your calculator, copy and complete the table on the right. What do you notice? Does the pattern continue?

$$11 = 11$$
$$(11)^2 =$$
$$(11)^3 =$$
$$(11)^4 =$$

10 (a) Toss two coins 100 times. Record the number of times you get:
    two heads
    a head and a tail
    two tails.
  (b) Draw a graph to show this information.
  (c) If you tossed the coins 4000 times, how many of each result would you expect?
  (d) If you toss two coins, what is the probability of getting a head and a tail?

11 (a) Toss three coins 160 times. Record the number of times you get:
    three heads
    two heads and one tail
    one head and two tails
    three tails.
  (b) Draw a graph to show this information.
  (c) If you tossed the coins 8000 times, how many of each result would you expect to get?
  (d) If you toss three coins, what is the probability of getting three heads?

12 (a) Roll two dice 360 times. Record the number of times you get:
    no sixes
    one six
    two sixes.
  (b) Draw a graph to show this information.
  (c) If you rolled two dice 3600 times, how many of each result would you expect to get?
  (d) If you roll two dice, what is the probability of getting two sixes?

13 A sequence of numbers is formed as follows:
    Choose any number.
    If it is even divide it by 2.
    If it is odd multiply it by 3 and add 1.
  Now repeat this to find the next number.
  For example:
  $$13 \rightarrow 40 \rightarrow 20 \rightarrow 10 \rightarrow 5 \rightarrow 16 \rightarrow 8 \rightarrow 4 \rightarrow 2 \rightarrow 1$$
  Do all these sequences finish with one?

14 Another sequence is formed as follows:
    Choose any number
    If it is a multiple of 3 divide it by 3.
    If it is not, multiply it by 2 and add 1.
  Now repeat this to find the next number.
  Find which starting numbers have sequences which finish with one.

4.23 *Oral test*

1 What is the next number in the sequence: 2, 4, 8, 16 . . .

2 What is the next number in the sequence: 2, 5, 8, 11, . . .

3 A number plus five is seventeen. What is the number?

4 Eight times a number is fifty-six. What is the number?

5 When a number is doubled and five is added the result is seventeen. What is the number?

6 When a number is halved and five is subtracted the result is seventeen. What is the number?

7 How do you find the area of a rectangle?

8 How do you find the volume of a cuboid?

9 How do you find the average speed of a car which covers d miles in t hours?

10 How do you find the distance travelled by a car in t hours, when it is doing s m.p.h.?

11 What does the gradient of a straight line measure?

12 My watch says 10.27. It is 15 minutes fast. What is the correct time?

13 My watch says 12.32. It is 15 minutes slow. What is the correct time?

14 Greek time is two hours ahead of UK time. What time will it be in Athens if the time in Manchester is 11 a.m.?

15 The time in New York is five hours behind the time in Scotland. What time will it be in Edinburgh if the time in New York is 11.30 p.m.?

16 What is the cost of three pairs of jeans which are £12 each and three pairs of trainers which are £8 each?

17 What is the cost of ten chocolate bars which are 22p each and ten tubes of mints which are 12p each?

18 Write with brackets: $3x + 3y$

19 Write with brackets: $8a - 8b$

20 Simplify: $5x + 3x + 2y$

21 Simplify: $3x + 7y - 3y$

22 For what value of x is $2x + 3 = 15$?

23 For what value of y is $3y - 5 = 10$?

24 How many whole numbers are less than 17 but more than 12?

25 Which integers are less than 3 but more than -4?

26 How do you find the *average* of three numbers? What is the average of 5, 10 and 15?

For GCSE only:

27 What is the *mode* of a set of numbers?

28 How do you find the *median* of a set of numbers?

29 What is the mean of five and nine?

30 What is the mean of 14, 20 and 26?

31 What is the probability of getting a multiple of three when you roll a die?

32 A bag contains four white beads, five red beads and one blue bead. What is the probability of picking a blue bead?

33 What does a probability of one tell us?

34 What does a probability of zero tell us?

35 If you roll two dice how many different possible outcomes are there?

36 If you roll two dice what is the probability of getting two sixes?

4.24 *Fact sheet IV*

Number sequences

Evens	2, 4, 6, 8, . . . , $2n$
Odds	1, 3, 5, 7, . . . , $2n - 1$
Multiples	4, 8, 12, 16, . . . , $4n$
Squares	1, 4, 9, 16, . . . , n^2
Cubes	1, 8, 27, 64, . . . , n^3

Expressions

$a + a + a = 3 \times a = 3a$ $a \times a \times a = a^3$
$5x + 4x + 3y = 9x + 3y$
$2a - 3b + 4a = 6a - 3b$
$5c + 2d - c = 4c + 2d$
$2x + 5$ means double x and then add 5
$2(x + 5)$ means add 5 and then double

Equations

$x + 3 = 5$	so $x = 5 - 3$	so $x = 2$
$4x = 12$	so $x = 12/4$	so $x = 3$
$2x + 5 = 11$	so $2x = 6$	so $x = 3$
$2(x + 5) = 16$	so $x + 5 = 8$	so $x = 3$

Inequalities

$x + 3 > 5$	so $x > 5 - 3$	so $x > 2$
$4x < 12$	so $x < 12/4$	so $x < 3$
$2x + 5 \leqslant 11$	so $2x \leqslant 6$	so $x \leqslant 3$
$2(x + 5) > 16$	so $x + 5 > 8$	so $x > 3$

Brackets, factors

$2(a + b) = 2a + 2b$
$5(x - 4) = 5x - 20$
$4(a + 3b) = 4a + 12b$
$6x - 3 = 3(2x - 1)$
$10x + 15y = 5(2x + 3y)$

Straight lines

$y = 3x + 2$ is the graph of a straight line.

x	0	1	2	3	4	5
y	2	5	8	11	14	17

The line $y = 3x + 2$ has a **gradient** of 3.
The line meets the y-axis at the point $(0, 2)$.

Formulae from tables

n	0	1	2	3	4	5	equal steps of size 1
c	3	5	7	9	11	13	equal steps of size 2

To get c, multiply n by 2 and add 3
So $c = 2n + 3$

FOR GCSE ONLY:

Statistics

Using the set $\{2, 2, 3, 3, 3, 4, 5, 5, 7, 8, 8\}$
Mode: the most common item is 3
Median: the middle item when in order is 4
Mean: $\dfrac{\text{total of items}}{\text{number of items}} = \dfrac{50}{11} = 4.55$

Probability

A die has 6 outcomes, each equally likely.
Probability of getting an odd number $= \frac{3}{6}$
A bag with two red, three white and five blue beads gives a probability of picking a red bead as $\frac{2}{10} = \frac{1}{5}$.

GCSE

PAPER 1 (level 1)

Time allowed 2 hours   Calculators may be used
Answer all questions

BILL

Potatoes
Carrots
Cauliflower ____

____

1   Copy and complete the
    following greengrocer's
    bill:
    8 lb potatoes at 9p per lb
    3 lb carrots at 15p per lb
    2 cauliflower at 35p each.

(4marks)

2   Alan, Brian, Carol and Deborah had a meal
    in a restaurant. Alan's meal cost £3.50,
    Brian's cost £4.25, Carol's cost £3.85 and
    Deborah's cost £3.20. Alan paid the bill for all
    the meals with a £20 note.
    (a) How much change did he get?
    (b) The four friends then decided to share
         the cost of the meals equally. How much
         did they each pay?

(5 marks)

3   The dials below represent the reading on an
    electricity meter on a given date when the
    meter is read. Three months previously the
    meter reading had been 2427 units.

    (a) State the reading shown by the dials
         above.
    (b) How many units have been used in the
         past 3 months?
    (c) The price per unit is 5.2p and there is a
         standing charge of £6.50 per quarter.
         What will be the total amount owing for
         the past 3 months?

(8 marks)

4   A field in the shape of a rectangle is 250 m
    long and 180 m wide.
    (a) Calculate its perimeter in metres.
    (b) Calculate its area in square metres.
    (c) Calculate its area in hectares, given that
         1 hectare = $10000 \, \text{m}^2$.

(4 marks)

5

	Distance from Coventry (miles)		
Coventry	0	dep. 09.40	dep. 19.30
Northampton	31	arr. 10.30	
		dep. 10.35	
Bedford	54	arr. 11.17	
		dep. 11.20	
Cambridge	85	arr. 12.15	

The timetable shows the times and distances
of a bus journey from Coventry to Cambridge
passing through Northampton and Bedford.
Use the table to find:
    (a) the distance from Northampton to
         Bedford
    (b) the time taken for the journey from
         Bedford to Cambridge
    (c) the time of arrival in Cambridge of the
         19.30 bus from Coventry, given that the
         journey times are the same
    (d) the price of a return ticket from Coventry
         to Cambridge if the cost is calculated at a
         rate of 5p per mile.

(7 marks)

6   The owner of a car drove 12000 miles in a
    year. The car averaged 30 miles to a gallon
    and throughout the year the cost of petrol
    was £1.80 per gallon. Other expenses
    involved in using the car were:
        tax £100, insurance £165, servicing £95.
    (a) Calculate the cost of petrol for the year.
    (b) Calculate the total cost of running the car
         for the year.
    (c) Calculate the average cost per mile of
         running the car for a year.

(7 marks)

7 The formula for calculating the area of a right-angled triangle is $A = \frac{1}{2}bh$.

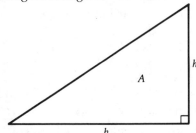

Calculate the area of each of the triangles below.

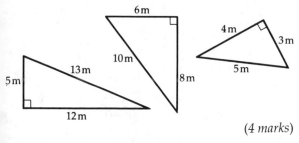

(4 marks)

8 A 6 lb turkey costs £4.68.
  (a) How much per lb is the turkey?
  (b) For how long should it be cooked in a microwave oven, if the instruction book tells you to allow 13 minutes per lb?
  (4 marks)

9 The diagram shows a brick in the shape of a cuboid with length 20 cm, breadth 9 cm and height 5 cm.

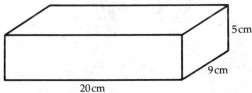

  (a) How many faces does the brick have?
  (b) How many edges does the brick have?
  (c) Find the volume of the brick.
  (4 marks)

10 At noon John sets out from a point P and walks in a straight line at constant speed of 6 km per hour. At 1.00 p.m. Mary sets out from P and walks in a straight line in the opposite direction at 4 km per hour. How far apart are they at 2.00 p.m?
  (4 marks)

11 The diagram is a scale drawing of a rectangular plot of ground. With the scale used 1 cm represents 4 m. By measuring the sides of the drawing, find the actual measurements of the plot in metres. What is the area of the plot?
  (5 marks)

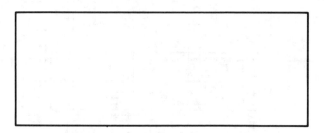

12 The pie chart represents the cost of a weekend in London. The hotel cost was £60.

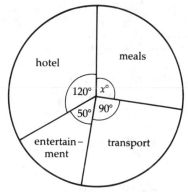

  (a) Calculate the angle marked $x°$.
  (b) Calculate the cost of entertainment.
  (c) Calculate the total cost of the weekend.
  (d) Express the hotel cost as a fraction of the total cost of the weekend.
  (7 marks)

13 A park has a circular flower-bed of radius 5 metres. Tulip bulbs are planted 20 cm apart all round the edge of the bed. Calculate the number of bulbs planted. (Take $\pi = 3.14$)
  (4 marks)

14 A van travels a distance of 174 kilometres along a motorway in 3 hours.
  (a) Calculate its average speed, in kilometres per hour.
  (b) How far does it travel in $2\frac{1}{2}$ hours if it maintains this speed?
  (4 marks)

15   The graph converts miles to kilometres.

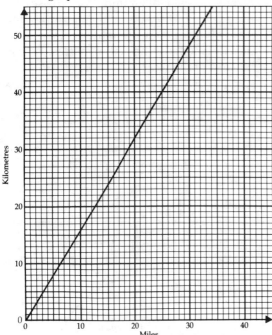

(a)   Use the graph to convert:
   (i)   25 miles to kilometres
   (ii)   28 kilometres to miles.

(b)   Find the number of miles which are equal to 120 kilometres.

(7 marks)

16   The drawing shows the pattern on the top half of a wall tile. The pattern on the tile is symmetrical about the line AB.

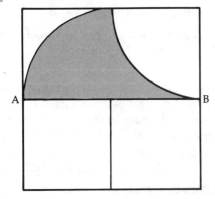

(a)   Draw a picture of the whole tile.

(b)   If the tile measures 10 cm by 10 cm, how many will be needed to cover a part of a wall measuring 2 m by 15 m?

(4 marks)

17   In an experiment on word size, the number of letters in each word of a piece of writing was counted. The numbers were:

  1 4 1 5 5 2 3 6 3 7
  6 6 5 1 4 4 3 5 2 6
  4 6 1 7 3 5 6 5 4 4
  2 1 3 6 5 5 4 3 1 2
  2 4 5 5 6 5 1 3 7 6

(a)   How many words were counted?

(b)   Copy and complete the frequency table.

Number of letters in a word	1	2	3	4	5	6	7
Frequency (number of words)							

(c)   On graph paper, draw a bar chart to illustrate the above information.

(d)   Which length of word was met most frequently?

(9 marks)

18   A television set at Ron's TV shop costs £350 including VAT. A similar set at Syd's shop costs £280 plus VAT. VAT is 15%.

(a)   Calculate the total cost price, including VAT, at Syd's.

(b)   Ron gives a discount of 5% off his total price, for cash. Calculate the price paid by a customer who pays cash at Ron's.

(c)   Syd offers his set on hire purchase. The terms are:
   deposit £60 plus 12 monthly payments of £24.
  Calculate the total hire purchase price of the set at Syd's.

(9 marks)

PAPER 2 (level 1 & 2)

Time allowed 2 hours   Calculators may be used
Answer all questions

1 Use your calculator to find the value of $2\frac{3}{4}$ metres of material at £1.89 per metre. Give your answer to the nearest 1p.

(4 marks)

2 A box holds 5.25 kg of oranges. Find how many oranges there are in the box if the average weight of an orange is 75 g.

(3 marks)

3 Washing powder is sold in three sizes. The 3 kg size costs £2.75, the 2 kg size costs £1.74 and the 800 g size costs 88p.
(a) What is the cost of 1 kg of powder if you buy the largest size?
Give your answer to the nearest 1p.
(b) What is the cost of 100 g of powder if you buy the smallest size?
(c) Which size is the best value for money? Show all your working.

(5 marks)

4 The shape ABDC is drawn on centimetre squared paper.

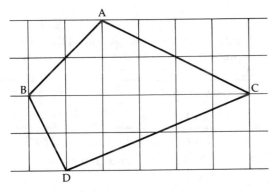

(a) Write down the length BC.
(b) Calculate the area of triangle ABC.
(c) Calculate the area of the quadrilateral ABDC.
(d) Use your ruler to measure the length AC.
(e) Use your protractor to measure the size of angle BDC.

(9 marks)

5 On a cold evening the temperature outside a house is −4°C. Inside the house the temperature is 12°C.
(a) How much higher is the temperature inside than outside?
(b) What will the outside temperature be if, during the night, it falls by 3°C?

(4 marks)

6 The ages of five children in a group are 13 years 1 month, 13 years 7 months, 12 years 6 months, 11 years 2 months and 13 years 5 months.
(a) Calculate the average (mean) age of the children.
(b) When a sixth child joins the group the average age becomes 12 years 10 months. What is the age of the sixth child?

(6 marks)

7 A bicycle wheel has diameter 70 cm.
(a) Calculate its circumference. (Take π = 3.14)
(b) Calculate how far the bicycle goes when the wheel turns through 20 complete revolutions.
(c) Through how many complete revolutions does the wheel turn when the bicycle travels 1 kilometre?

(7 marks)

8 John leaves home at the same time each day to walk to school. On most days, when he walks at his normal speed, the walk takes 20 minutes and he arrives at school 5 minutes early.
(a) How long does it take him on a day when he walks at twice his normal speed?
(b) How long does it take him on a day when he walks at only half his normal speed? By how many minutes is he late arriving at school?

(4 marks)

9 A supermarket employs extra staff at the weekend to help to stock the shelves. It employs 8 extra staff who each work from 9.30 a.m. until 1.30 p.m. without a break and are paid at the rate of £2.30 per hour. Calculate the total amount of money paid out by the supermarket to these extra staff.

(4 marks)

10   A ship sets out from a point A and sails in a straight line due north for 10 kilometres to a point B. It then turns and sails in a straight line in a direction north west for 5 kilometres to a point C. It then turns again to sail directly back to A. On your paper make a scale drawing of the ship's journey, using a scale of 1 cm to represent 1 km. Measure the length AC on your drawing. Use this to find the total distance actually travelled by the ship on its journey.

(7 marks)

11   A survey was taken of the number of people in a district who were watching television at a given time one evening. The results are shown in the bar chart.

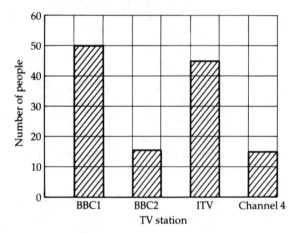

(a) How many people were watching television?

(b) How many more were watching ITV than Channel 4?

(c) What fraction of the total number of viewers in the survey was watching BBC 1? Express this as a decimal.

(5 marks)

12   A bag contains 10 white, 6 red and 4 blue beads. A bead is drawn at random. Find the probability that it is:
(a) white   (b) blue   (c) not red.
If the first two beads removed are both white and they are not replaced, find the probability that the third bead selected will be:
(d) white   (e) not white.

(7 marks)

13   A closed box is made of wood 3 cm thick. Its outside dimensions are:
    length 78 cm,   width 56 cm,
    height 56 cm.
(a) What are its inside dimensions?
(b) Calculate the total surface area of the inside of the box.

(5 marks)

14   The shaded shape shown below is the net of a solid. It is drawn on centimetre squared paper.
(a) What is the name of the solid whose net this is?

(b) On your centimetre squared paper, draw the same net as the one shown but with sides twice as long. Draw an axis of symmetry on your shape.

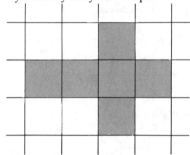

(c) If the net you have drawn was used to make a solid, what would be the volume of the solid?

(6 marks)

15   The children in a class measured their handspans. The results are shown in the table below.

Handspan (cm)	13	14	15	16	17	18
Number of children	2	3	9	7	6	2

(a) How many children were in the survey?

(b) Which length of handspan is the median?

(c) Which length is the mode?

(d) The children use their handspans to measure the width of a cupboard. A girl with a 15 cm handspan finds that the cupboard is 12 handspans wide. What would be the width of the cupboard in handspans if measured by the child with the largest handspan?

(8 marks)

16 The graph represents the journey of a bus. The bus travelled from Birmingham to Oxford via Warwick, a distance of 64 miles. Use the graph to find:

(a) the time at which the bus arrived at Warwick

(b) how long the bus stopped in Warwick

(c) how far the bus was from Oxford at 10.30 a.m.

At 10.00 a.m. a car left Oxford and travelled at a constant speed to Warwick, arriving in Warwick at 11.15 a.m.

(d) On your graph paper, draw a line to represent the journey of the car.

(e) Use your graph to find the time at which the car met the bus.

(8 marks)

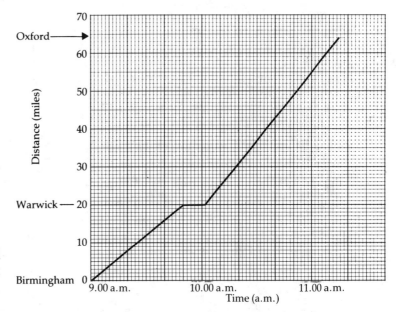

17 A man bought himself a second-hand car for £1600.

(a) For how much should he sell it if he wants to make a profit of 10% of his cost price?

(b) For how much did he sell it if he made a loss of 5% of his cost price?

(4 marks)

18 Before going to the USA on holiday, a man changed £120 into dollars. While on holiday he bought a book for $8.70. Given that the rate of exchange was $1.45 = £1, calculate:

(a) the number of dollars he received before going on holiday

(b) the cost of the book in pounds.

(4 marks)

PAPER 3 (level 2)

Time allowed 2 hours   Calculators may be used
Answer all questions

1 An engineer gives a measurement as 1.3×10^{-4}.
Write this as a decimal.

(2 marks)

2 The attendance at a football match is 46 570. Write this:

(a) to the nearest thousand

(b) in standard form.

(3 marks)

3 Jane went shopping. In the first shop she spent one third of her money. In the second shop she spent half of the money remaining. She then had 40p. How much did she have originally?

(3 marks)

4 On 1 January 1980 Susan invested £500 in a bank savings account. The interest paid by the bank is 8% on the sum of money in the account and this is added to the account on 31 December. On 1 January 1981 Susan invested another £500 in the same account. Calculate the amount in Susan's account on:

(a) 2 January 1981   (b) 2 January 1982.

(4 marks)

5 The formula to convert temperatures from degrees Celsius to degrees Fahrenheit is:
$$F = \tfrac{9}{5}C + 32$$
where F is the temperature in Fahrenheit and C is the temperature in Celsius.
(a) Find the temperature on the Fahrenheit scale which is the same as 30°C.
(b) Find the temperature on the Celsius scale which is the same as 50°F.
(c) Write an expression for C in terms of F.
(*5 marks*)

6 Three people share a sum of £4.20. How much do they each receive if:
(a) they share it equally
(b) they share it in the proportion 2:2:3?
(*4 marks*)

7 The equation of a line PQ is $y + 3x = 5$ and the equation of line RS is $5x - 3y = 8$.
Calculate:
(a) the gradient of PQ
(b) the gradient of RS
(c) the coordinates of the point where PQ cuts the y-axis
(d) the coordinates of the point where RS cuts the x-axis.
(*4 marks*)

8 A train leaves a station A at 11.36 a.m. and travels 112 miles to a station B in $1\tfrac{3}{4}$ hours.
Calculate:
(a) the arrival time at B
(b) the average speed of the train for this journey
(c) the price of a ticket for this journey, given that the price is calculated by charging 11.5p for each mile travelled.
(*5 marks*)

9 A box contains coloured rods of different lengths. The number of rods of each length is given in the following table:

Length of rod (cm)	1	2	3	4	5	6
Number of rods	2	3	11	1	5	9

(a) What is the median length of rod?
(b) Calculate the mean length of rod.
(*5 marks*)

10 (a) In the figure, ABC is a triangle in which BÂC = 90°, AB̂C = 25° and BC = 70 cm. Calculate the length of AB, correct to one decimal place.

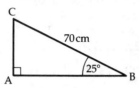

(b) A boat is 120 metres from the foot of a vertical cliff which is 45 metres high. Calculate the angle of elevation from the boat to the top of the cliff.

(c) In the figure, R is due north of Q, Q is due west of P, PR is 5.8 km and RP̂Q is 36°.

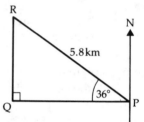

Calculate:
(i) the bearing of R from P
(ii) the distance QR, correct to two decimal places.
(*8 marks*)

11 The diagram shows the cross-section of a girder in which AB is parallel to DC, AC = DC, AB̂C = 105° and BĈA is 42°.

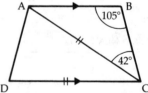

Calculate:
(a) the size of angle ACD
(b) the size of angle ADC.
(*4 marks*)

12 An engine part is made by drilling a cylindrical hole of diameter 3 cm through a solid cuboid of metal. The cuboid has length 8 cm, breadth 5 cm and height 4 cm and the axis of the cylindrical hole passes through the centres of the two largest faces. Calculate the volume of metal in the engine part.
(Take $\pi = 3.14$)
(*6 marks*)

13 Below is the pattern for a simple smock. Two of these shapes are needed to make the smock. A tailor has a piece of material which is in the shape of a rectangle 1 metre wide and 2 metres long and wants to use it to make two complete smocks (4 pieces).

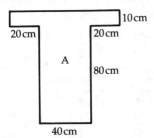

On centimetre squared paper, using a scale of 1 cm to represent 10 cm, draw a rectangle to represent the piece of material. On your rectangle, show how 4 pieces, each of shape A, could be cut from the material.

(a) What is the area of material unused?

(b) Express the unused area as a fraction of the total area of material.

(c) Express the unused area as a percentage of the total area of material.

(8 marks)

14 A lawn is in the shape of a rectangle 20 m long and 12 m wide. In the middle there is a circular rose-bed of diameter 5 m.

(a) Calculate the area of the rose-bed. (Take $\pi = 3.14$)

(b) Fertiliser is to be put on the lawn at the rate of 25 g per square metre. Calculate the number of 1 kg packets of fertiliser required.

(7 marks)

15 The probability that Rovers will beat United at football is $\frac{2}{3}$. The probability that United will win is $\frac{1}{4}$.

(a) What is the probability that there will be a draw?

(b) The two teams meet twice in a season. What is the probability that:
  (i) Rovers will win both games
  (ii) Rovers will win the first and United will win the second game
  (iii) Rovers will not win either game?

(8 marks)

16 AB is a diameter of a circle, centre O and radius 8 cm. The tangent to the circle at B passes through a point P. PB is 15 cm.

(a) Calculate the length OP.

(b) Calculate the length AP, giving your answer to 1 decimal place.

(4 marks)

17 The commission charged by an estate agent when selling a house is 3% of the first £20 000 of the selling price plus 2% of the remainder.

(a) Calculate the commission charged when a house is sold for £28 000.

(b) Find an expression in terms of x for the commission charged on a house sold for £x, where x is greater than 20 000.

(c) Calculate the selling price of a house on which the commission charged is £840.

(9 marks)

18 A manufacturer puts his product into packets whose base is a rectangle of area 36 cm². He decides to experiment by trying different rectangular shapes, each with an area of 36 cm². In the table below, x is the length of the rectangle in centimetres and y is the breadth in centimetres.

x	1	2	3	4	6	12	18	36
y	36							
Area	36	36	36	36	36	36	36	36

(a) Copy and complete the table.

(b) Explain why the breadth of the rectangle can be given by the formula $y = \dfrac{36}{x}$.

(c) On your graph paper, draw the graph of the equation $y = \dfrac{36}{x}$, using values of x from 1 to 36.

(d) Use your graph to find:
  (i) the breadth of the rectangle when the length is 5.5 cm
  (ii) the perimeter of the rectangle when the breadth is 3.5 cm
  (iii) the length and breadth of the rectangle with perimeter 26 cm.

(11 marks)

SCE Standard Grade

FOUNDATION LEVEL: PAPER 1

Assessing 'Knowledge and Understanding'
Time: 40 minutes
Calculators may be used

Aural Section

Questions 1–5 will be read by the teacher.

1 A school has one thousand four hundred and seventy pupils. Write this number in figures.
The number of girls is seven hundred and eight. Write this number in figures.

(2 marks)

2 Today is Tuesday April 30th. My birthday is on Wednesday next week.
What date is my birthday?

(2 marks)

3 How much change do you get from £10 if you buy two cassettes at £4.45 each?

(2 marks)

4 How many centimetres are there in $5\frac{1}{4}$ metres?

(1 mark)

5 My train is due to leave at 9.17 a.m. I arrive 20 minutes early.
When do I arrive?

(3 marks)

6 The school day ends at 15.45. Show this time on the clockface.

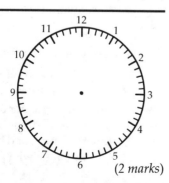

(2 marks)

7 On the grid, draw a triangle similar to this one, but with each side three times as long.

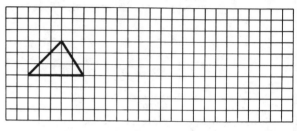

(3 marks)

8 A car's petrol tank holds 8 gallons of petrol. This is enough to drive 230 miles. How far can the car be driven on 1 gallon? (Answer to the nearest mile.)

(3 marks)

9 I buy 40 litres of petrol. How many gallons is this, to the nearest gallon?

CONVERSION CHART

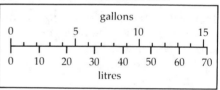

(2 marks)

10 The chart shows the distance in miles between various cities. John drives from Aberdeen to Perth and goes on from there to Edinburgh. How far does he drive altogether?

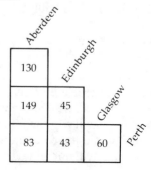

(3 marks)

11 Mr and Mrs Mackie drive to France with their two children, Peter aged 8 and Tracie aged 6.
How much does their return ferry crossing cost altogether?

CAR FERRY TARIFF	
Single journey prices	(£)
Adults	12.00
Senior Citizens	5.50
Children (under 14 yrs)	6.50
Cars	38.50

(3 marks)

12 A hi-fi system can be bought on special terms, as shown.
What would be the total cost on these terms?

SPECIAL OFFER

£80 deposit

plus £8.40 a week

for 52 weeks

(4 marks)

13 Calculate the size of the shaded angle.

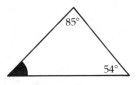

85°

54°

(3 marks)

14 A shop offers a discount of 15%. Complete this bill:

Records	8.46
Cassettes	5.63
Batteries	2.91
Less discount:	
Charge:	

(3 marks)

15 A floor is to be covered by carpet-tiles.
The floor measures 5 m by 4 m. Carpet tiles measure 50 cm by 50 cm.
How many tiles will be needed to cover the floor?

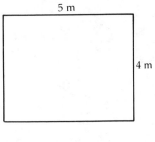

5 m

4 m

 50 cm

50 cm

(3 marks)

16 Patrick is paid £7.40 per hour. At weekends he is paid double time.
He works 40 hours during the week, plus 5 hours on Saturday and 1 hour on Sunday.
What is his gross pay?

(5 marks)

17 In sports trials, Kyle threw a javelin 6 times.
What was his average distance for these throws?

Throw	Distance (m)
First	30.8
Second	28.4
Third	30.5
Fourth	31.3
Fifth	29.8
Sixth	31.0

(4 marks)

18 In a smoking survey, 850 people were interviewed. 68% of them said they didn't smoke. How many didn't smoke?

(3 marks)

19

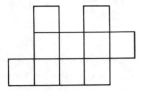

Which two of the small shapes fit together to make this large shape?

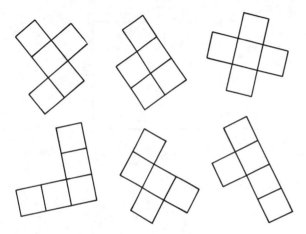

Show on this diagram how they fit together:

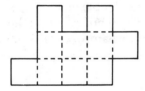

(*2 marks*)

20 This timetable shows the times of trains from Aberdeen to Inverness.

Aberdeen	0905	1117	1545	1755
Dyce	0914	1126	1554	1804
Elgin	1034	1244	1712	1927
Forres	1048	1258	1726	1941
Nairn	1100	1310	1738	1953
Inverness	1120	1330	1758	2013

(a) Jenny has to be in Inverness by 2.00 p.m. What times does the latest train she can catch leave Elgin?

(b) How long does this train take from Elgin to Inverness?

(*5 marks*)

21 A classroom measures 15 m by 8 m by 4 m. How many cubic metres of air will it contain?

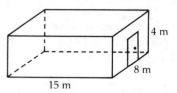

(*2 marks*)

22 The graph shows a journey made up of three parts:
(i) cycling from home to the shops
(ii) doing the shopping
(iii) cycling back home.

(a) How long did the return journey take?

(b) What was the total distance cycled?

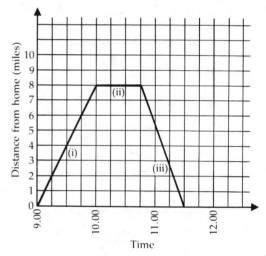

(*5 marks*)

FOUNDATION LEVEL: PAPER 2

Assessing 'Reasoning and Applications'
Time: 40 minutes
Calculators may be used

1 Glenda buys a kit to make a hanging mobile. The kit contains thread, rods and coloured diamonds. All the diamonds of the same colour are the same weight, but different colours are different weights. The mobile must balance:

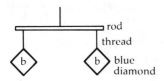

The thread is too light to make any difference to the balance. So are the rods.

Glenda knows that an apricot diamond weighs 10 grams. She wants to find out the other weights.

Colour	Weight (grams)
apricot (a)	10
blue (b)	
crimson (c)	
dun (d)	
emerald (e)	
purple (p)	

She finds that the following arrangements balance.

(a)

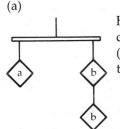

How much does a blue diamond weigh?
(Put your result in the table above.)

(b)

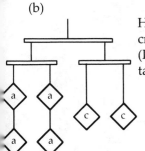

How much does a crimson diamond weigh?
(Put your result in the table above.)

(c)

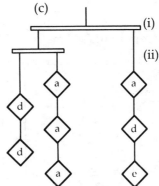

(i) How much does a dun diamond weigh?
(ii) How much does an emerald diamond weigh?
(Put your results in the table above.)

(d)

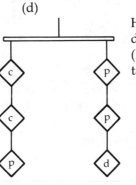

How much does a purple diamond weigh?
(Put your result in the table above.)

(12 marks)

2 Joanna is designing a T-shirt. She wants it to be symmetrical.
She has completed the left-hand side of the design.
Draw the right-hand side for her.

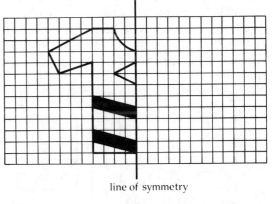

line of symmetry

(3 marks)

3   This is a map of a park.
The scale is:

1 centimetre represents 400 metres

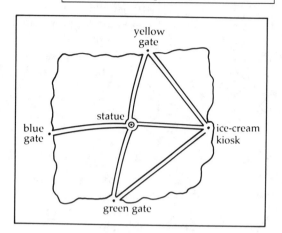

(a)   Use the map to find out how far it is from the ice-cream kiosk to the blue gate, via the statue.

(b)   Jan walks from the yellow to the green gate, buying an ice-cream on the way. How far does she walk?

(c)   Philip walks from the statue to the blue gate in half an hour.
How many metres would he walk in an hour, travelling at the same speed?

(*10 marks*)

4   Alan does a lot of driving.
The graph shows the number of gallons of petrol in his petrol tank during one of his journeys, from when he started to when he finished.
He had to fill up his tank twice on the journey.

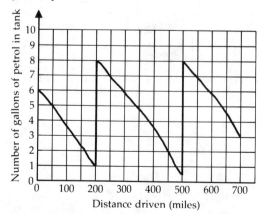

(a)   How much petrol was in the tank at the start of the journey?

(b)   How many gallons could the tank hold?

(c)   How much petrol did Alan buy on the second fill-up?

(d)   How much petrol did Alan use on the whole journey?

(*8 marks*)

5   Uncle William decided to give some money to each of his four young relatives, Ann, Billy, Carole and David.
He pays them a certain amount every year.

(a)   He gives Ann: £10 in the first year
£20 in the second year
£30 in the third year
£40 in the fourth year
and so on.
He gives Billy £50 each year.
Complete this table showing how much each receives in all:

	Total received (£)	
	Ann	Billy
after 1 year	10	50
after 2 years	30	100
after 3 years		
after 4 years		
after 5 years		

(b)   Work out how much Ann will have received after 10 years.

(c)   Work out how much Billy will have received after 10 years.

(d)   Uncle William gives Carole:
£5 in the first year
£10 in the second year
£20 in the third year
£40 in the fourth year
and so on.
He gives David:
£400 in the first year
£200 in the second year
£100 in the third year
£50 in the fourth year
and so on. (If there are pennies in the

sum due, he leaves them out.)
Complete this table showing how much each receives in all:

	Total received (£)	
	Carole	David
after 1 year	5	400
after 2 years	15	600
after 3 years		
after 4 years		
after 5 years		

(e) Work out how much Carole will have received after 10 years.

(f) Work out how much David will have received after 10 years.

(g) Which of the four children has the most generous gift?

(16 marks)

GENERAL LEVEL: PAPER 1

Assessing 'Knowledge and Understanding'
Time: 55 minutes
Calculators may be used

1 The chart shows rainfall in St Andrews for a 6-day period.
Calculate the average rainfall per day during this period.

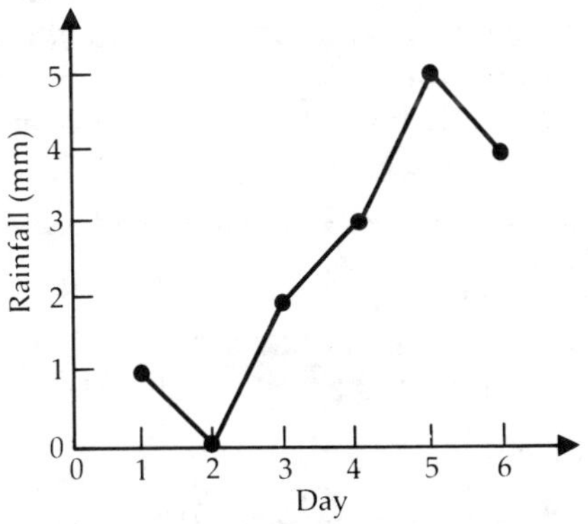

(3 marks)

2 150 people took part in a smoking survey. 112 didn't smoke. Of these, 61 were women. Of those who smoked, 25 were women.

(a) How many men smoked?

(b) What percentage of the women in the survey smoked?
(Answer to the nearest whole per cent.)

(4 marks)

3 The pie chart shows the places from which Britain imported butter in 1988.
Altogether 144 000 tonnes of butter were imported.
How much of this came from Australia?

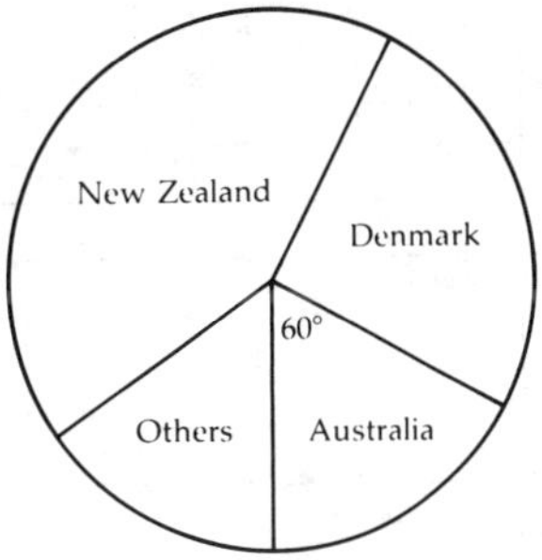

(2 marks)

4 Simplify $15x + 4(8 - 2x)$

(3 marks)

5 Mr and Mrs Galloway want to spend 1 week at Hotel Bergland. They can spend up to £520 on the hotel.
Between what dates could they begin their holiday?

HOTEL BERGLAND
Price per person in £

Departures on or between: No. of nights:	7	10	11	14
9 May–23 May	223	–	279	308
24 May–30 May	252	297	312	346
31 May–15 June	252	299	314	350
16 June–29 June	263	305	320	357
30 June–12 July	271	318	333	372
13 July–19 July	275	334	349	388
20 July–2 Aug	289	355	370	404
3 Aug–23 Aug	273	334	349	388
24 Aug–6 Sept	268	323	338	378
7 Sept–20 Sept	263	313	328	357
21 Sept–2 Oct	257	297	–	–

(3 marks)

6   Calculate the diameter, *D*, of this bolt:

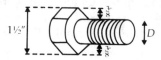

(*3 marks*)

7   Complete this diagram so that lines PQ and RS are both lines of symmetry.

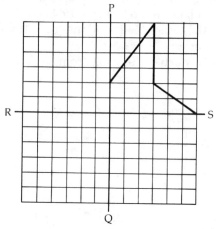

(*2 marks*)

8   This flowchart shows how to find the 2nd class postage for a letter weighing less than 150 grams.

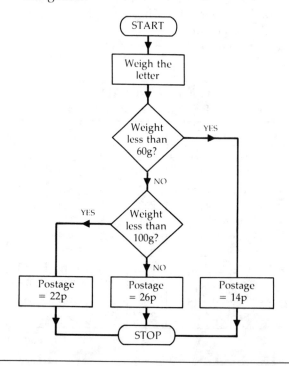

(a) How much would a letter weighing 75 g cost in postage?

(b) It is more expensive to post two letters weighing 55 grams each than one letter weighing 110 grams. How much more expensive?

(*3 marks*)

9   George Mitchell borrowed £15 000 from his bank. The bank charged $12\frac{1}{2}$ % interest each year.
    (a) How much did he owe at the end of 1 year?
    (b) He didn't pay anything back until the end of the second year.
          How much did he owe the bank by this time?

(*4 marks*)

10 A playground swing is supported on a metal frame like this.
    PQ and PR are the same length.
    Angle QPR = 40°.
    Calculate the size of angle SQR.

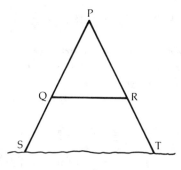

(*3 marks*)

11 A water tank is full of water as shown. Water is drained out of it until the depth of water remaining is 10 cm.
    (a) How much water, in cm³, has been drained out?
    (b) How many litres of water still remain in the tank?

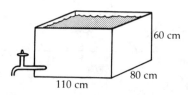

(*4 marks*)

12 Solve the inequality $3y - 8 > y - 4$

(*3 marks*)

13 In this magic square, all the rows, columns and diagonals have the same totals.
Fill in the missing numbers.

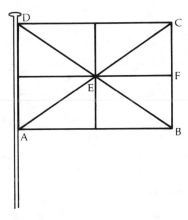

(*3 marks*)

14 A flag ABCD has a design as shown.
AB = 16 cm; BC = 10 cm.

(a) Calculate the area of triangle CEF.

(b) Calculate the size of angle ACB.

(c) A larger version of the flag is to be drawn, similar to this one. The height is to be 25 cm. What will be the new length?

(*8 marks*)

15 A stone is thrown up into the air.
After t seconds its height in metres is given by the formula
$h = ut - 5t^2$
where u is the speed of the throw, in metres per second.
If the speed of throw is 20 metres per second,

(a) what is the height of the stone after 3 seconds?

(b) what is the height of the stone after 4 seconds?
Explain the meaning of your second answer.

(*5 marks*)

GENERAL LEVEL: PAPER 2

Assessing 'Reasoning and Applications'
Time: 55 minutes
Calculators may be used

1 Cars may be hired from the Alpha or the Betta Car Hire Companies.

ALPHA CAR HIRE:

Hire of Metro for
a week: £100

PLUS

20p for each mile driven

BETTA CAR HIRE:

Hire of Escort for
a week: £60

No extra charge

for the first 100 miles driven.

40p for each mile thereafter.

(a) Jane expects to drive 250 miles in the week.
Calculate the total charge for each company.

(b) John can only afford to pay £140 altogether.
Calculate the distance he could drive a car hired from each company.

(c) Complete the following table:

Distance driven	Total charge (£)	
(miles)	Alpha	Betta
0		
50		
100		
150		
200		
250		
300		

(d) Use this grid to draw graphs showing the relationship between distance and charge for each company:

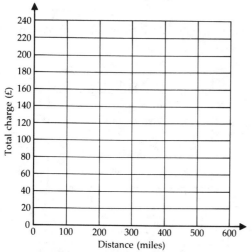

(e) For journeys more than a certain distance it is cheaper to use the Alpha Car Hire Company. What is this distance?

(12 marks)

2   A boat sails towards the foot of a cliff. The sailor looks up at a tree at the top of the cliff. The tree is 80 m above sea-level.

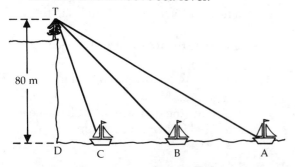

(a) When the boat is at A, angle DAT = 28°. How far is the boat from the foot of the cliff?

(b) From A to B is 60 m. Calculate angle DBT.

(c) Angle DCT = 58°. What is the distance from B to C?

(10 marks)

3   Ann and James took part in a general knowledge quiz. There were 3 rounds. Use the clues to complete the table.

	Ann	James
Points in Round 1	14	
Points in Round 2		
Points in Round 3		
TOTAL		30

Clues

(a) Ann got twice as many points in Round 1 as she got in Round 3.

(b) James got twice as many points in Round 3 as he got in Round 1.

(c) James beat Ann by 6 points in Round 2.

(d) Ann and James had the same final total.

(6 marks)

4   Gavin makes a design with a row of black and white tiles:

He has used 3 black tiles and 11 white ones. By adding more tiles he can make the row longer.

(a) Complete this table:

No. of black tiles	1	2	3	4	5	6	25	
No. of white tiles			11					101

(b) Write down a formula for the number, W,

of white tiles needed with *B* black tiles:

$W =$

(7 marks)

5

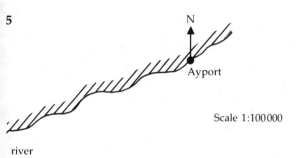

Scale 1:100 000

river

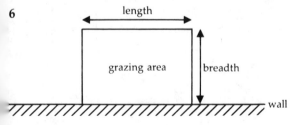

A ferry sets out from Ayport to sail directly to Bidock on the South side of the river. It sails on a bearing of 195° from Ayport.

(a) Show the position of Bidock on the map.

(b) By making an accurate measurement, find the actual distance between Ayport and Bidock, to the nearest tenth of a kilometre.

(c) Calculate the time the journey will take, at a steady speed of 6.5 km per hour. Answer to the nearest minute.

(10 marks)

6

Mr Archer wants to enclose an area of land next to a wall for his sheep to graze in. He has 120 m of fencing and wants to use it to enclose as large an area as possible.

(a) He decides first to enclose a rectangular area; and tries various possible values for the breadth.

Example: The breadth could be 10 m.
Then the length must be 100 m (*check*).
So the area will be 1000 m².
He records his calculations in a table.

(i) Complete this table:

Breadth (m)	10	20	30	40	50	60
Length (m)	100					
Area (m²)	1000					

(ii) What dimensions give the rectangle of greatest area?

(b) Next, he decides to make his 120 m of fencing into a semicircle:

(i) Calculate the radius of the semicircle.

(ii) Calculate the area enclosed.

(iii) Is the semicircle a better shape than a rectangle for this situation?

(Assume $\pi = 3.14$ and give your answers correct to 3 significant figures)

(10 marks)

TACKLING INVESTIGATIONS

Number of triangles	1	2	3	4	5	6	7
Number of matches	3	5	7	9	11	13	15

Each time you add another triangle you add *two more* matches

It's 3 + (one less than the number of triangles) × 2 matches

n triangles have $3 + (n - 1) \times 2$ or $2n + 1$ matches.

The pattern for the cubes made from matches is

12, 20, 28, 36, 44 . . .

ie $12 + (n - 1) \times 8$ or $8n + 4$ matches

SECTION 1 **Basic Computation**

1.1 Number patterns

1 10; they go up in twos

2 11; you add two more each time

3 25; there are 5 × 5 squares in the next shape

4 $1 = 1$
$1 + 3 = 4$
$1 + 3 + 5 = 9$
$1 + 3 + 5 + 7 = 16$
$1 + 3 + 5 + 7 + 9 = 25$
square numbers

5 (a) 36 (b) 100

6 (a) 15; you add five more dots (or one more dot than last time)
(b) 21; 28

7 (a) 21; 28
(b) $1 + 3 = 4$
$3 + 6 = 9$
$6 + 10 = 16$
$10 + 15 = 25$
square numbers
(c) yes; $15 + 21 = 36$; $21 + 28 = 49$

8 $1 = 1$
$1 + 2 + 1 = 4$
$1 + 2 + 3 + 2 + 1 = 9$
$1 + 2 + 3 + 4 + 3 + 2 + 1 = 16$
square numbers

9 (a) 25 (b) $25 = 5 \times 5$ (c) 36; 6×6

10 (a) (i) 24 (ii) 54 (b) (i) 8 (ii) 27 (c) 96 (d) 64

11 13; you add three more each time

12 (a) 24, 28 (b) 19, 22 (c) 29, 34

13 (a) 32; 25; 39
(b) 40; 31; 49
(c) 80; 61; 99
(d) $4n$; $3n + 1$; $5n - 1$

14 (a) 36, 42 (b) 54, 63 (c) 23, 27 (d) 20, 23

15 (a) 48; 72; 31; 26
(b) 60; 90; 39; 32
(c) 120; 180; 79; 62
(d) $6n$; $9n$; $4n - 1$; $3n + 2$

16 (a) 96, 192 (b) 243, 729 (c) 4, 2 (d) 103, 98

17 (a) 384; 2187; 1; 93
(b) 1536; 19683; 0.25($\frac{1}{4}$); 83

18 (a) 5, 10, 15, 20, 25
(b) 7, 14, 21, 28, 35
(c) 2, 9, 16, 23, 30
(d) 4, 9, 16, 25, 36
(e) 1, 3, 6, 10, 15, 21

19 (a) $\frac{1}{16}, \frac{1}{32}$ (b) 1.3, 1.6 (c) 0.01, 0.001 (d) $5\frac{1}{3}, 10\frac{2}{3}$
(e) 3.5, 4.2 (f) 8.3, 9.5

20 (a) 32

(b)
Generations back	0	1	2	3	4	5	6
Number in generation	1	2	4	8	16	32	64

(c) 1024

21 (a) $1^3 = 1$
$1^3 + 2^3 = 9$
$1^3 + 2^3 + 3^3 = 36$
$1^3 + 2^3 + 3^3 + 4^3 = 100$
(b) square numbers: $1^2, 3^2, 6^2, 10^2$
(c) $15^2 = 225$

22 (a) 32 (b) 16 (c) 2

23 (a)
Number of teams	2	3	4	5	6
Number of matches	1	3	6	10	15

23 (b) Yes; with 6 teams,
the first team plays 5 matches
the second team plays 4 other matches
the third team plays 3 other matches
so the number of matches is $5 + 4 + 3 + 2 + 1 = 15$

1.2 Multiples, factors and primes

1 (a) (i) 4, 8, 12, 16, 20, 24, 28, 32, 36, 40, 44, 48, 52, 56, 60, 64, 68, 72, 76, 80, 84, 88, 92, 96, 100
  (ii) 6, 12, 18, 24, 30, 36, 42, 48, 54, 60, 66, 72, 78, 84, 90, 96
  (iii) 7, 14, 21, 28, 35, 42, 49, 56, 63, 70, 77, 84, 91, 98
 (b) (i) 12, 24, 36, 48, 60, 72, 84, 96
  (ii) 28, 56, 84 (iii) 42, 84
 (c) (i) 12 (ii) 28 (iii) 42
 (d) 84

2 (a)

Whole numbers from 1 to 75

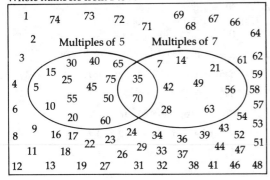

 (b) 35, 70   (c) 35

3 (a) 144   (b) 720

4 (a) 5, 10, 15, 20, 25, 30
 (b) they are either 0 or 5

5 (a) 4, 8, 12, 16, 20, 24, 28, 32, 36, 40, 44, 48
 (b) they go 4, 8, 2, 6, 0, 4, 8, 2, 6, 0, 4, 8 . . .

6 (a) 18, 27, 36, 45 . . .
 (b) 9
 (c) yes up to 90;
   then 99 → 18 → 9;
   189 → 18 → 9 etc

7 Saturday

8 7 weeks

9 12 days

10 after 24 working days; i.e. 4 weeks and 5 days on Friday

11 after 9 laps

12 after 8 laps

13 (a) (i) 1, 2, 4, 8 (ii) 1, 2, 3, 4, 6, 12
  (iii) 1, 2, 3, 6, 9, 18
 (b) (i) 1, 2, 4 (ii) 1, 2 (iii) 1, 2, 3, 6
 (c) (i) 4 (ii) 2 (iii) 6
 (d) 2

14 (a)

Whole numbers from 1 to 48

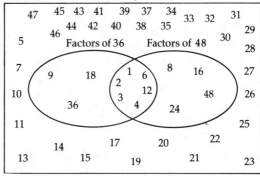

 (b) 1, 2, 3, 4, 6, 12
 (c) 12

15 (a) 24 (b) 12

16 (a) 1, 2, 4, 8, 16, 32, 64, 128, 256
 (b) 1, 3, 9, 27, 81, 243, 729, 2187
 (c) 1, 2, 3, 4, 6, 8, 9, 12, 16, 18, 24, 32, 36, 48, 64, 72, 96, 144, 192, 288, 576
 (d) 1, 2, 3, 4, 5, 6, 9, 10, 12, 15, 18, 20, 25, 27, 30, 36, 45, 50, 54, 60, 75, 90, 100, 108, 135, 150, 180, 225, 270, 300, 450, 540, 675, 900, 1350, 2700

17 (a) 1, 5 (b) 1, 11 (c) 1, 29 (d) 1, 37 (e) 1, 47

18

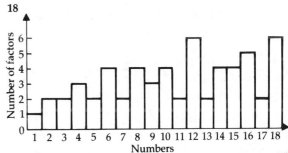

19 (a) (i) 2, 3, 5, 7, 11, 13, 17, 19, 23, 29, 31, 37, 41, 43, 47, 53, 59
  (ii) 6, 12, 18, 24, 30, 36, 42, 48, 54, 60
 (b) apart from 2 and 3 all the other prime numbers are one more or one less than a multiple of 6
 (c) yes

20 (a) (i) 4 (ii) 4 (iii) 6 (b) a £20

21 (a) 1, 2, 3, 6 (b) 6 (c) 1, 2, 4, 8; 8

22 (a) 1, 2, 3, 4, 5, 6, 8, 9, 10, 12, 15, 18, 20, 24, 30, 36, 40, 45, 60, 72, 90, 120, 180, 360
 (b) they are all factors of 360

1.3 Place value, addition and subtraction

1 (a) fifty (b) five hundred
 (c) five hundred (d) five thousand

2 (a) four hundred and sixty-five
(b) five hundred and six
(c) three thousand, two hundred and seventy
(d) four thousand and forty-two
(e) nine thousand and three

3 (a) 325   (b) 206   (c) 5002   (d) 1203   (e) 12100

4 (a) +50   (b) −200   (c) +100
(d) −7   (e) +150   (f) −900

5 (a) 100   (b) 100   (c) 10
(d) 100   (e) 1000

6 (a) 2530   (b) 7600   (c) 437   (d) 770

7 (a) ×10   (b) ÷100   (c) ×10
(d) ÷10   (e) ×10   (f) ÷1000

8 (a) 49   (b) 101   (c) 629   (d) 743   (e) 1392

9 (a) 26   (b) 127   (c) 1398

10 (a) 19 (7 + 3 = 10)   (b) 59 (37 + 13 = 50)
(c) 109 (77 + 23 = 100)   (d) 145 (88 + 12 = 100)
(e) 200 (85 + 15 = 100, 37 + 63 = 100)

11 690

12 315

13 (a) 5493   (b) 1

14 (a) 89977   (b) 18238; 7547; 4238

15 (a) 404000   (b) 96000

16 (a) 23   (b) 19   (c) 422   (d) 117   (e) 469

17 (a) 52   (b) 34   (c) 213   (d) 205   (e) 172   (f) 266

18 (a) 75   (b) 750   (c) 27   (d) 573   (e) 400   (f) 800

19 (a) 131   (b) 35   (c) 129   (d) 625

20 (a) 27   (b) 21   (c) 42   (d) 8

21 24

22 61

23 184

24 574

25 11638

26 12371

27 2534

28 6380

29 3620

30 £18.15

31 (a)   (b)

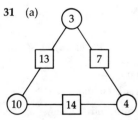   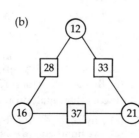

1.4   Multiplication and division

1 (a) 106   (b) 183   (c) 148   (d) 140   (e) 102

2 (a) 333   (b) 546   (c) 1377   (d) 1032   (e) 10373

3 (a) 848   (b) 1575   (c) 11748   (d) 8192

4 (a) true   (b) true   (c) true   (d) true

5 (21 × 3) × 10; (21 × 10) × 3; (21 × 6) × 5; (21 × 5) × 6;
(21 × 2) × 15; (21 × 15) × 2, etc

6 (a) true   (b) true   (c) true   (d) true

7 12 × 70; 6 × 140; (8 × 3) × 35; 24 × (5 × 7); 120 × 7, etc

8 (a) 180   (b) 240   (c) 210   (d) 390   (e) 1470

9 168 hours

10 (a) £56   (b) £168   (c) £525

11 (a) (i) 100 (ii) 300 (iii) 375
(b) (i) £1.20 (ii) £3.60 (iii) £4.50

12 (a) (i) 30 (ii) 72 (iii) 144
(b) (i) 54 (ii) 144 (iii) 450
(c) (i) £72 (ii) £240 (iii) £600

13 (a) 222 pints   (b) 555 pints   (c) 962 pints

14 (a) £55.50   (b) £277.50   (c) £481.00

15 (a) 342   (b) 1368   (c) 3420

16 (a) £10.26   (b) £205.20   (c) £492.48

17 (a) 16   (b) 4   (c) 27   (d) 52   (e) 61

18 (a) 33   (b) 26   (c) 32   (d) 123

19 (a) 53   (b) 53   (c) 75   (d) 132

20 (a) true   (b) true   (c) true   (d) true

21 (288 ÷ 6) ÷ 2; (288 ÷ 2) ÷ 6; (288 ÷ 3) ÷ 4; (288 ÷ 4) ÷ 3

22 (a) true   (b) true   (c) true   (d) true

23 (490 ÷ 7) ÷ 5; (490 ÷ 5) ÷ 7; 70 ÷ 5; 98 ÷ 7; 980 ÷ 70

24 (a) 47   (b) 18   (c) 23   (d) 26

25 (a) 43 weeks   (b) 8 days

26 (a) 8   (b) 32

27 (a) 20   (b) 500

28 19

29 26

30 (a) 322 pints   (b) 46 pints

31 13241

32 (a) 235   (b) (i) £592.20 (ii) £28.20

33 (a) 25   (b) £625

1.5   Using arithmetic

1 838 miles

2 594

3 5580

4 31

5 (a) 42 (b) 7

6 1743

7 2964 miles

8 (a) 220 miles (b) 277 miles

9 £177.84

10 1461 days

11 1960 lb

12 20

13 (a) 504 (b) 31
(c) (i) £3.75; £3.15; £2.85; £4.15; £3.55; £4.45; £3.30
(ii) £25.20

14 (a) 1272 (b) 128 (c) 282
(d) (i) £16 350; £8 400; £24 600; £14 250 (ii) £63 600
(e) 53

15 (a) 2567 (b) 19 686 (c) 314 (d) 3281
(e) (i) £3.25$\frac{1}{2}$; £17.02$\frac{1}{2}$; 0; 0; £2.37; £25.84$\frac{1}{2}$ (ii) £48.49$\frac{1}{2}$

16 (a) £2.97 (b) £4.95

17

No of articles	1	2	3	4	5	6	7	8	9	10
Cost in £	0.99	1.98	2.97	3.96	4.95	5.94	6.93	7.92	8.91	9.90

18 (a)

	1	2	3	4	5	6	7	8	9	10
£1.99	1.99	3.98	5.97	7.96	9.95	11.94	13.93	15.92	17.91	19.90
£2.99	2.99	5.98	8.97	11.96	14.95	17.94	20.93	23.92	26.91	29.90
£3.99	3.99	7.98	11.97	15.96	19.95	23.94	27.93	31.92	35.91	39.90
£4.99	4.99	9.98	14.97	19.96	24.95	29.94	34.93	39.92	44.91	49.90
£5.99	5.99	11.98	17.97	23.96	29.95	35.94	41.93	47.92	53.91	59.90
£6.99	6.99	13.98	20.97	27.96	34.95	41.94	48.93	55.92	62.91	69.90
£7.99	7.99	15.98	23.97	31.96	39.95	47.94	55.93	63.92	71.91	79.90
£8.99	8.99	17.98	26.97	34.96	44.95	53.94	62.93	71.92	80.91	89.90
£9.99	9.99	19.98	29.97	39.96	49.95	59.94	69.93	79.92	89.91	99.90

(b) (i) £23.97 (ii) £20.93 (iii) £24.95 (iv) £17.91

19 (a) (i) 3 (ii) 5 (iii) 8
(b) (i) 3p short of £18 (ii) 5p short of £30
(iii) 8p short of £48

20

	£0	£1	£2	£3	£4	£5	£6	£7	£8	£9
£0	–	1.45	2.90	4.35	5.80	7.25	8.70	10.15	11.60	13.05
£10	14.50	15.95	17.40	18.85	20.30	21.75	23.20	24.65	26.10	27.55
£20	29.00	30.45	31.90	33.35	34.80	36.25	37.70	39.15	40.60	42.05
£30	43.50	44.95	46.40	47.85	49.30	50.75	52.20	53.65	55.10	56.55
£40	58.00	59.45	60.90	62.35	63.80	65.25	66.70	68.15	69.60	71.05
£50	72.50	73.95	75.40	76.85	78.30	79.75	81.20	82.65	84.10	85.55
£60	87.00	88.45	89.90	91.35	92.80	94.25	95.70	97.15	98.60	100.05
£70	101.50	102.95	104.40	105.85	107.30	108.75	110.20	111.65	113.10	114.55
£80	116.00	117.45	118.90	120.35	121.80	123.25	124.70	126.15	127.60	129.05
£90	130.50	131.95	133.40	134.85	136.30	137.75	139.20	140.65	142.10	143.55
£100	145.00	146.45	147.90	149.35	150.80	152.25	153.70	155.15	156.60	158.05

(b) (i) $72.50 (ii) $116.00 (iii) $59.45
(iv) $110.20 (v) $178.35
(c) (i) £40 (ii) £50 (iii) £31 (iv) £57

21 (a)

Cheese	£2.80
Butter	£1.68
Yoghurt	£1.02
Gold	£0.38
Milk	£4.20
Eggs	£1.44

(b) £11.52
(c) £46.08

1.6 Fractions

1 (a) (i) $\frac{3}{4}$ (ii) $\frac{2}{3}$ (iii) $\frac{4}{6}$
   (b) (i) $\frac{1}{4}$ (ii) $\frac{1}{3}$ (iii) $\frac{2}{6}$

2 (a) (i) $\frac{3}{8}$ (ii) $\frac{3}{9}$ (iii) $\frac{2}{6}$
   (b) (i) $\frac{2}{8}$ (ii) $\frac{4}{9}$ (iii) $\frac{3}{6}$
   (c) (i) $\frac{3}{8}$ (ii) $\frac{2}{9}$ (iii) $\frac{1}{6}$

3 (b)

(i) (ii) (iii)

4 (a) (i) $\frac{4}{6}$ (ii) $\frac{6}{9}$ (iii) $\frac{8}{12}$
   (b) (i) $\frac{2}{6}; \frac{1}{3}$ (ii) $\frac{3}{9}; \frac{1}{3}$ (iii) $\frac{4}{12}; \frac{2}{6}; \frac{1}{3}$

5 (a) $\frac{4}{6}$ (b) $\frac{9}{12}$ (c) $\frac{5}{40}$ (d) $\frac{8}{14}$

6 (a) $\frac{1}{4}$ (b) $\frac{2}{3}$ (c) $\frac{3}{4}$ (d) $\frac{3}{5}$ (e) $\frac{1}{5}$ (f) $\frac{9}{20}$

7 (a) $\frac{5}{7}$ (b) $\frac{2}{8} = \frac{1}{4}$
   (c) $\frac{6}{4} = \frac{3}{2}$ (d) $\frac{16}{10} = \frac{8}{5}$

8 (a) $\frac{3}{4}$ (b) $\frac{3}{6} = \frac{1}{2}$ (c) $\frac{1}{8}$ (d) $\frac{7}{10}$

9 (a) $\frac{5}{6}$ (b) $\frac{17}{12}$ (c) $\frac{1}{20}$ (d) $\frac{11}{35}$

10 (a) $\frac{3}{2}$ (b) $\frac{7}{3}$ (c) $\frac{13}{8}$ (d) $\frac{15}{4}$

11 (a) $4\frac{1}{4}$ (b) $1\frac{5}{6}$ (c) $3\frac{29}{40}$ (d) $1\frac{7}{8}$

12 2

13 $2\frac{1}{9}$

14 (a) 2 (b) $\frac{9}{4} = 2\frac{1}{4}$ (c) 2 (d) $\frac{8}{3} = 2\frac{2}{3}$ (e) $\frac{35}{8} = 4\frac{3}{8}$

15 (b) (c) 3

16 (a) 5 (b) 15

17 (a) 7 (b) 21 (c) 35 (d) 49

18 (a) 2 (b) 8 (c) $\frac{15}{4} = 3\frac{3}{4}$ (d) $\frac{18}{5} = 3\frac{3}{5}$ (e) 30

19 (a) $\frac{1}{8}$ (b) $\frac{1}{10}$ (c) (i) $\frac{3}{8}$ (ii) $\frac{1}{5}$ (iii) $\frac{4}{10}$

20 (a), (b)

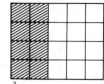 (c)

(d) $\frac{6}{20} = \frac{3}{10}$

21 (a) $\frac{3}{8}$ (b) $\frac{3}{10}$ (c) $\frac{3}{10}$ (d) $\frac{8}{15}$

22 (a) 54 (b) 34 (c) $1\frac{1}{3}$ (d) $\frac{20}{7} = 2\frac{6}{7}$ (e) $4\frac{3}{8}$

23 (a) 6 (b) 20 (c) 8 (d) 8

24 (a) 6 (b) 20 (c) 8 (d) 8 (e) $\frac{25}{2} = 12\frac{1}{2}$

25 (a) (i) true (ii) false (iii) false
   (b) (ii) $\frac{15}{56}$ (iii) $\frac{32}{9}$

1.7 Using fractions

1 $\frac{2}{7}$

2 (a) $\frac{24}{32} = \frac{3}{4}$ (b) $\frac{1}{4}$

3 $\frac{2}{5}$

4 (a) (i) $\frac{15}{36}$ (ii) $\frac{21}{36}$ (b) (i) $\frac{5}{12}$ (ii) $\frac{7}{12}$

5 (a) (i) $\frac{8}{32}$ (ii) $\frac{12}{32}$ (iii) $\frac{4}{32}$ (iv) $\frac{8}{32}$
   (b) (i) $\frac{1}{4}$ (ii) $\frac{3}{8}$ (iii) $\frac{1}{8}$ (iv) $\frac{1}{4}$
   (c) 1

6 (a) $\frac{3}{8}$ (b) $\frac{5}{8}$ kg

7 (a) $\frac{11}{8} = 1\frac{3}{8}$ kg (b) $\frac{7}{8}$ kg (c) $\frac{1}{8}$ kg

8 (a) 8 kg (b) $11\frac{1}{5}$ kg (c) 146 kg

9 (a) $31\frac{1}{2}$ km (b) 126 km (c) $1642\frac{1}{2}$ kg

10 (a) (i) $\frac{5}{6}$ (ii) $\frac{1}{6}$ (b) 6 weeks

11 (a) $\frac{1}{20}$ (b) (i) $\frac{1}{4}$ mile (ii) 1 mile (iii) $3\frac{3}{4}$ miles

12 (a) $\frac{4}{15}$ (b) (i) Anne £1200; Beni £1000; Carla £800

13 (a) $\frac{3}{5}$ (b) $\frac{1}{10}$

14 (a) rent £33; food £44; travel £22; savings £11;
   others £22
   (b) $\frac{1}{6}$

15 $\frac{1}{6}$

16 (a) $\frac{6}{20} = \frac{3}{10}$ (b) £18

17 $\frac{2}{15}$

18 (a) 120 (b) 80 (c) 40

19 (a) $22\frac{1}{2}$ cm (b) 160

20 (a) 72 cm (b) 36

21 (a) $1\frac{3}{8}$ kg (b) 13

22 £125

23 (a) 24 kg (b) 20 kg

24 (a) $2\frac{1}{4}$ litres (b) $\frac{9}{16}$ litre

25 (a) 33 metres (b) 14

26 (a) $\frac{13}{60}$ (b) $\frac{1}{20}$ (c) $\frac{9}{20}$ (d) $1\frac{9}{20}$

1.8 Decimals

1 (a) $\frac{3}{10}$ (b) $1\frac{7}{10}$ (or $\frac{17}{10}$) (c) $\frac{3}{100}$ (d) $3\frac{7}{100}$ (or $\frac{307}{100}$)
(e) $5\frac{43}{100}$ (or $\frac{543}{100}$)

2 (a) 0.3 (b) 0.07 (c) 0.23 (d) 7.3 (e) 2.07

3 (a) $\frac{8}{10} = \frac{4}{5}$ (b) $\frac{6}{100} = \frac{3}{50}$ (c) $\frac{15}{100} = \frac{3}{20}$ (d) $\frac{25}{100} = \frac{1}{4}$
(e) $2\frac{16}{100} = 2\frac{4}{25}$

4 (a) 0.2 (b) 0.8 (c) 0.35 (d) 0.65
(e) 0.12 (f) 1.92

5 (a) $\frac{5}{10}$ (b) $\frac{7}{100}$ (c) $\frac{3}{10}$ (d) $\frac{3}{100}$

6 (a) forty-six and five tenths
(b) five and six hundredths
(c) thirty-two and seven tenths
(d) ninety and six hundredths

7 (a) +5 (b) +0.2 (c) +0.05 (d) −0.07 (e) +0.5
(f) +0.06

8 (a) 100 (b) 10 (c) 10 (d) 100 (e) 1000

9 (a) 253 (b) 760 (c) 4.3 (d) 0.77

10 (a) ×10 (b) ÷10 (c) ÷100 (d) ÷10
(e) ÷1000 (f) ×1000

11 (a) 4.9 (b) 10.1 (c) 6.29 (d) 7.43 (e) 0.92

12 (a) 12.7 (b) 14.36

13 (a) 2.5 (b) 0.5 (c) 0.05 (d) 3.09 (e) 0.34

14 (a) 2.1 (b) 0.01

15 (a) £18.90 (b) £1.10

16 (a) £11.26 (b) £3.68

17 (a) 7.4 cm (b) 7.7 cm (c) 8.3 cm

18 (a) 19.3 cm (b) 0.4 cm

19 (a) 6.3 cm (b) 16.8 cm (c) 8.5 cm

20 (a) 8.6 (b) 15.3 (c) 29.2 (d) 48.5

21 (a) 1.1 cm (b) 2.1 cm (c) 3.2 cm (d) 3.3 cm

22 (a) 1.1 cm (b) 2.3 cm (c) 4.4 cm (d) 4.9 cm

23 (a) 1.2 (b) 2.4 (c) 1.9 (d) 4.2

24 (a) (i) 96p (ii) 72p (iii) £3.40
(b) (i) 0.96 (ii) 0.72 (iii) 3.4

25 (a) 2 (b) 1.92 (c) 1.08

26 (a) (i) £22.50 (ii) £32.60 (iii) £6.75
(iv) £83.10 (v) £174.65
(b) £319.60

27 (a) £15.42 (b) £107.94

28 (a) 714 mm^2 (b) 7.14 cm^2

29 (a) 2491 mm^2 (b) 24.91 cm^2

30 (a) 2.76 cm^2 (b) 8.88 cm^2

31 (a) (i) 2491 (ii) 4088 (iii) 7462
(b) (i) 24.91 (ii) 40.88 (iii) 74.62

1.9 Using decimals

1 (a) 59.7 sec (b) 0.3 sec

2 (a) 320.1 cm (b) 7.5 cm

3 (a) £8.15 (b) £1.85

4 (a) 2.49 kg (b) 0.01 kg

5 (a) 1.23 litres (b) 0.27 litres

6 (a) 50.3 cm (b) 16.4 cm

7 (a) 9.2 sec (b) 212.6 sec or 3 min 32.6 sec
(c) 7.4 sec (d) 2.6 sec

8 (a) 4.9 kg; 18.3 kg (b) 66.5 kg (c) 60.3 kg
(d) 39.7 kg (e) 89.7 kg

9 (a) 13.25 m (b) 6.75 m

10 (a) 0.7 l (b) 2.29 l (c) 2.71 l

11 (a) 3 l (b) 3.96 l (c) 5 l (d) 3.96 l
(e) 1 l (f) 37.5 l

12 (a)

Biro	£3.60
Crayon	£3.00
Pencil	£1.08
Pen	£22.00
Rubber	£2.80
Ruler	£2.60
Sellotape	£0.86

(b) £35.94

13 (a) 17 cm (b) 11.3 cm (c) 17.5 cm (d) 16.04 cm

14 (a) 144 mm (b) 14.4 cm

15 (a) 144 mm^2 (b) 1.44 cm^2

16 (a) 198 mm (b) 19.8 cm

17 (a) 1600 mm^2 (b) 16 cm^2

18 7.6 cm

19 156.25 cm^2

20 £148.75

21 183.05 miles

22 342.925 miles

23 53.4 m.p.g.

24 £109.46

25 (a) 372.06 m^2 (b) 31 bags (c) £134.85

26 (a) (i) \$32.4 (ii) \$194.4 (iii) \$591.3 (iv) \$12.15
(b) (i) £12.35 (ii) £74.07 (iii) £450 (iv) £3.50

27 (a) £267.80 (b) £17.80

28 (a) 35.77 cm^2; 29.89 cm^2; 44.53 cm^2
(b) 220.38 cm^2 (c) 218.197 cm^3

29 (a) 73.5 cm^3 (b) 42.875 cm^3

1.10   Using your calculator

1 (a) 1232   (c) 245

2 (a) £11.04   (b) £8.96

3 (a) £35.85   (b) £4.15

4 (a) $5 - 7 = -2$

5 $3 + (4 \times 5) = 23; (3 + 4) \times 5 = 35$

6 (b) $(8 + 6) \div 2 = 7; 8 + (6 \div 2) = 11$

7 (a) £7.60   (b) £7.60; 7.6

8 (a) £13.40   (b) £13.40; 13.4

9 (a) 0.666 66 . . .

10 (a) $1 \div 3 \times 3$ should be 1

11 (a) 0.666 66 . . .

12 (a) 6

13 (a) 25

14 (a) 3

15 $3^2 = 9; 9 + 4 = 13; 13^2 = 169$
$8 + (7 \div 2) + 3 = 8 + 3.5 + 3 = 14.5$

16 (a) 1089   (b) 1089

17 (a) 1089
(b) all results are 1089 except if result from subtraction is less than 100

18 (a) 731   (b) 731   (c) 93 522; 8502; 654
(d) result is first three digits of original number
(e) yes

19 (a) (i) 1001   (ii) 654 654
(b) 1001 × any three digit number gives the three digits repeated

20 (a) 333 333 333   (b) 444 444 444   (c) 12 345679 × 18
679 × 18 × 18   (d) 999 999 999

21 (a) (i)   1 000 000   (ii)   100 000 000
1–06   1–08
(iii) 1 000 000 000   (iv) 10 000 000 000
1–09   1–10
(b) the number at the right tells you how many zero's there are
(c) 1–12 (ie 12 zeros)

1.11   Estimation

1 (a) the number of people was between 35 000 and 45 000
(b) the number of people was between 42 500 and 43 500

2 (a) (i) 40 000 (ii) 44 000 (iii) 44 400 (iv) 44 380
(b) Andy was correct

3 (a) (i) 2000 (ii) 1600 (iii) 1630   (b) 25

4 (a) (i) 20 000 (ii) 18 000 (iii)17 500 (iv) 17 530
(b) 28

5 (a) (i) 80 (ii) 50 (iii) 90 (iv) 280 (v) 1000
(b) (i) 100 (ii) 600 (iii) 1000 (iv) 5700

6 (a) (i) $80 \times 50 = 4000$ (ii) $50 \times 80 = 4000$
(iii) $50 \times 30 = 1500$ (iv) $20 \times 70 = 1400$
(b) (i) 3952 (ii) 4056 (iii) 1457 (iv) 1311

7 (a) (i) $100 \times 400 = 40 000$ (ii) $600 \times 800 = 480 000$
(iii) $1000 \times 500 = 500 000$
(b) (i) 54 432 (ii) 456 876 (iii) 517 514

8 (a) (i) $400 \div 100 = 4$ (ii) $900 \div 200 = 4.5$
(iii) $600 \div 100 = 6$ (iv) $900 \div 300 = 3$
(v) $800 \div 200 = 4$ (vi) $1000 \div 200 = 5$
(b) (i) 3 (ii) 4 (iii) 5 (iv) 3 (v) 4 (vi) 4

9 (a) (i) £3 + £6 + £7 + £2 = £18
(ii) $5 \times £1 + 2 \times £2 = £9$
(b) (i) £18.22 (ii) £9.45

10 (a) (i) $50 \times 10 = 500$ (ii) $70 \times 20 = 1400$
(iii) $50 \times 50 = 2500$ (iv) $70 \times 30 = 2100$
(b) (i) 576 (ii) 1728 (iii) 2496 (iv) 2176

11 (a) £8 + £2 = £10; £120   (b) £119.76

12 (a) $30 \times 60 = 1800$ (b) $1728 \, m^2$

13 (a) $1200 \div 30 = 40$ (b) 44 tins

14 (a) (i) $20 + 30 + 30 + 60 + 60 = 200$
(ii) $600 + 800 - 200 - 500 = 700$
(iii) $2000 + 8000 + 6000 + 5000 = 21 000$
(iv) $4000 - 3000 + 10 000 - 7000 = 4000$
(b) (i) 197 (ii) 695 (iii) 21 262 (iv) 3628

15 (a) $50 \times 20 + 700 = 1700; 1761$
(b) $(60 + 30) \times 70 = 6300; 6052$
(c) $80 \times (200 + 400) = 48 000; 46 123$
(d) $600 + 70 \times 20 = 2000; 1997$

16 (a) $(70 + 50) \div 6 = 20; 20$
(b) $(300 + 700) \div 4 = 250; 238$
(c) $(900 - 400) \div 20 = 25; 18$
(d) $(900 + 900) \div 30 = 60; 62$

17 (a) $600 \div (20 + 40) = 10; 9.8$
(b) $3000 \div (600 - 200) = 7.5; 7.63$
(c) $\dfrac{(400 + 800)}{(10 + 20)} = 40; 44.7$ (d) $\dfrac{(800 + 1000)}{(40 + 20)} = 30; 27$

18 (a) $50 \times 50 = 2500; 2601$
(b) $30 \times 30 \times 30 = 27 000; 24 389$
(c) $20 \times 20 \times 20 \times 20 = 160 000; 130 321$
(d) $100 \times 100 \times 100 = 1 000 000; 1 442 897$
(e) $1000 \times 1000 = 1 000 000; 942 841$

1.12   Checking

1 938; 1961

2 (a) 133

3 (a) 45 322

4 (a) 2260   (b) 32 184

5 (a) 15 668

6 (a) 2 264 922

7 (a) £12.65

8 (a) 121

9 (a) the second column is the first in the reverse order
(b) £26.32

10 (a) (i) 1836 cm³ (ii) 37.835 cm³ (iii) 18.9 cm³
(b) multiply the numbers in the reverse order

11 (a) (i) 43 200 sec (ii) 10 080 min
(iii) 43824 hours with one leap year
(b) multiply the numbers in the reverse order

12 (a) 1701
(b) subtract the numbers from 12 000 in reverse order

13 (a) 144
(b) divide 36 288 by 12 and then by 21 or vice versa

15 (a) 246

16 (a) 1217

17 (a) 189

18 (a) 89

19 (a) 753

20 (a) 7051

21 (a) 4674

22 (a) 351 556

23 (a) (i) 8 (ii) 47
(b) do multiplications and divisions in reverse order

24 (a) £62.94

25 (a) £29.90

26 Jane £92.16; Bill £92.16; they are both the same

27 (a) 27 264 (b) Find 27 264 ÷ 2 ÷ 2 ÷ 2 ÷ 2 ÷ 2 ÷ 2

28 (a) (i) 13 405 (iii) 3168 (iv) 3
(b) (i) the 3 and 4 have been interchanged
(iii) the brackets were left out (ie $61 + 27 \times 36$)

(iv) the 42 was multiplied instead of divided
(ie $\dfrac{6678}{53} \times 42$)

29 (a) (i) 32×34 (ii) 1234×56
(b) (i) 782 (ii) 74 144

1.13 Approximations

1 (a) 2.3 (b) 2.4 (c) 2.4 (d) 8.4 (e) 8.3

2 (a) 2.3 (b) 2.4 (c) 2.4 (d) 8.3

3 (a) 5.24 (b) 5.36 (c) 5.41 (d) 7.10

4 (a) 0.46 (b) 0.47 (c) 0.46 (d) 0.05

5 (a) 0.462 (b) 0.467 (c) 0.465 (d) 0.050

6 (a) (i) 7.44 (ii) 46.56 (iii) 4.38 (iv) 245.85
(b) (i) 7.4 (ii) 46.6 (iii) 4.4 (iv) 245.9

7 (a) (i) 3.696 (ii) 36.975 (iii) 9.792 (iv) 0.0945
(b) (i) 3.70 (ii) 36.98 (iii) 9.79 (iv) 0.09

8 (a) (i) 10.35 cm² (ii) 65.25 cm² (iii) 41.28 cm²
(b) (i) 10.4 cm² (ii) 65.3 cm² (iii) 41.3 cm²

9 (a) 9.9 sec (b) 9.92 sec

10 (a) 0.7 mm (b) 0.74 mm (c) 0.737 mm

11 (a) 1.0 mm (b) 1.01 mm

12 (a) 25.225 mm (b) (i) 25.2 mm (ii) 25.23 mm

13 (a) 3.375 m² (b) (i) 3.4 m² (ii) 3.38 m²

14 (a) £16.503 75 (b) (i) £16.5 (ii) £16.50 (iii) £16.504
(c) £16.50

15 (a) 0.3 m (b) 0.29 m

16 (a) 60 (b) 70 (c) 500 (d) 600 (e) 4000

17 (a) 2 (b) 4 (c) 0.2 (d) 0.05 (e) 0.04

18 (a) 760 (b) 730 (c) 3600 (d) 7100 (e) 7000

19 (a) 0.46 (b) 0.43 (c) 0.094 (d) 0.10

20 (a) (i) 3283 (ii) 37.96 (iii) 27 260 (iv) 0.49
(b) (i) 3000 (ii) 40 (iii) 30 000 (iv) 0.5

21 (a) (i) 74.52 cm³ (ii) 593.775 cm³ (iii) 516 cm³
(b) (i) 70 cm³ (ii) 600 cm³ (iii) 500 cm³

22 31 536 000 sec; (a) 30 000 000 (b) 32 000 000
(c) 31 500 000 (d) 31 540 000

23 (a) 20 000 (b) 19 000 (c) 19 200

24 (a) 20 000 (b) 20 000 (c) 19 700

25 (a) 10 000 (b) 9600 (c) 9620

26 (a) 10 000 (b) 9700 (c) 9670

27 (a) (i) 5000; 5000; 5000 (ii) 4700; 5200; 5000
(iii) 4650; 5220; 4970 (b) 14 841; 15 000

28 (a) 274 875; 275 000 (b) £14 568 375; £14 600 000

1.14 Accuracy & sensible answers

1 £4.42

2 27p

3 32p

4 50p

5 8p per bottle

6 £2.86 per hour

7 21p

8 68 sec

9 (a) 10.11 sec (b) 11.56 sec (c) 14.74 sec

10 (a) 29.12 sec (b) 31.01 sec (c) 31.18 sec
(d) 34.09 sec

11 (a) 19.41 sec (b) 20.68 sec (c) 20.79 sec
(d) 22.72 sec

12 (a) 1.33 m or 133 cm (b) 1.67 m or 167 cm
(c) 0.82 m or 82 cm (d) 1.22 m or 122 cm
(e) 1.42 m or 142 cm

13 (a) 3.99 m (b) 5.01 m (c) 2.46 m (d) 3.66 m
    (e) 4.26 m

14 100.5 m; 99.5 m

15 100.05 m; 99.95 m

16 (a) 396.5 cm or 3 m 96.5 cm
    (b) 383.5 cm or 3 m 83.5 cm

17 (a) (i) 1.05 sec (ii) 0.95 sec
    (b) (i) 3 sec (ii) 180 sec (iii) 4320 sec (iv) 129 600 sec
        (v) 1 576 800

18 (a) 48.75 cm²; 35.75 cm²
    (b) 42.6525 cm²; 41.3525 cm²
    (c) 42.065 025 cm²; 41.935 025 cm²

19 (a) (i)   49 cm²; 36 cm²
            43 cm²; 41 cm²
            42 cm²; 42 cm²
       (ii)   48.8 cm²; 35.8 cm²
            42.7 cm²; 41.4 cm²
            42.1 cm²; 41.9 cm²
       (iii) 48.75 cm²; 35.75 cm²
            42.65 cm²; 41.35 cm²
            42.07 cm²; 41.94 cm²
    (b) (i)   6 cm² too large or 7 cm² too small
            1 cm² too large or too small
            same as actual area
       (ii)   6.2 cm² too large or 6.8 cm² too small
            0.6 cm² too large or 0.7 cm² too small
            0.1 cm² too large or too small
       (iii) 6.25 cm² too large or 6.75 cm² too small
            0.65 cm² too large or too small
            0.06 cm² too large or 0.07 cm² too small

20 (a) 40.572 cm³
    (b) 2.35 cm; 3.65 cm; 4.95 cm
    (c) 42.46 cm³
    (d) 2.25 cm; 3.55 cm; 4.85 cm
    (e) 38.74 cm³
    (f) 1.83 cm³ too large or 1.89 cm³ too small

1.15 Squares and square roots

1 (a) 64 (b) 81 (c) 100

2 (a)

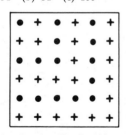

$1 + 3 + 5 + 7 + 9 + 11$

    (b) 36
    (c) (i) 49 (ii) 100

3 (a) 4 cm² (b) 16 cm² (c) 9 cm²

4 (a) 144 (b) 625 (c) 1.44 (d) 6.25

5 (a) 144 cm² (b) 625 cm² (c) 1.44 cm² (d) 6.25 cm²

6 (a) 289 (b) 320.41 (c) 2.89 (d) 320.41

7 (a) 841 cm² (b) 7921 cm² (c) 8.41 cm² (d) 79.21 cm²

8 (a) 100 (b) 10 000 (c) 1 000 000 (d) 0.01

9 (a) 6400 (b) 0.64

10 (a) 8100 (b) 0.81

11 (a) 2.25 (b) 22 500

12 $11^2 = 121$ $1.1^2 = 1.21$ $16^2 = 256$ $1.6^2 = 2.56$
    $12^2 = 144$ $1.2^2 = 1.44$ $17^2 = 289$ $1.7^2 = 2.89$
    $13^2 = 169$ $1.3^2 = 1.69$ $18^2 = 324$ $1.8^2 = 3.24$
    $14^2 = 196$ $1.4^2 = 1.96$ $19^2 = 361$ $1.9^2 = 3.61$
    $15^2 = 225$ $1.5^2 = 2.25$ $20^2 = 400$ $2.0^2 = 4.00$

13 256; 529; (a) 2.56 (b) 5.29 (c) 25 600 (d) 52 900

14 961; 3249; (a) 9.61 (b) 96 100 (c) 32.49 (d) 324 900

15

$16 \times$ $4 \times$ $9 \times$

    so $16 + 9 + 4 + 1 = 30$ altogether

16 (a) 16 m² (b) £80

17 225

18 100

19 156

20 (a) 5 (b) 6 (c) 9 (d) 12 (e) 15

21 (a) 5 (b) 6 (c) 9 (d) 12 (e) 15

22 (a) 5 (b) 9 (c) 10 (d) 11 (e) 13

23 (a) 4 (b) 20 (c) 14 (d) 16 (e) 18

24 (a) 1.1 (b) 1.3 (c) 1.4 (d) 1.6

25 17; 6.32; (a) 1.7 (b) 170 (c) 63.2 (d) 0.632

26 (a) 4 cm (b) 12 cm (c) 17.32 cm

27 (a) 19 cm (b) 2.2 cm (c) 8.3 cm

28 (a) 22 cm (b) 14.14 cm

29 3.5 m by 3.5 m

30 (a) 17.32 m (b) 70 (c) £84

1.16 Directed numbers

1 (a) 20°C (b) 50°C (c) 10°C (d) 30°C

2 (a) 25°C (b) 17°C (c) 10°C (d) 13°C

3 (a) 18°C (b) 39°C (c) 22°C (d) 25°C: so (b)

4 (a) 26°C (b) 25°C (c) 41°C (d) 38°C: so (c)

5 (a) 24°C (b) 41°C

6 90 feet below sea level

7 (a) ⁻22°C (b) fell by 16°C

8 (a) 0 (b) 0 (c) ⁻4
   ⁻1   1   ⁻3
   ⁻2   2   ⁻2
   ⁻3   3   ⁻1
   ⁻4   4   0

9 (a) ⁻4 (b) 3 (c) ⁻4 (d) ⁻10

10 (a) 3 (b) ⁻3 (c) ⁻6 (d) ⁻11

11 (a) 7 (b) 11 (c) ⁻3 (d) 5

12 (a) ⁻6 (b) ⁻12 (c) ⁻21 (d) ⁻20

13 yes

14 (a) ⁻15 (b) ⁻18 (c) ⁻56 (d) ⁻15 (e) ⁻18 (f) ⁻56

15 (a) ⁻4 (b) 6

16 (a) £25 (b) ⁻£50 (c) £75

17 (a) (i) ⁻15°C (ii) ⁻36°C (iii) ⁻66°C
   (b) (i) ⁻10°C (ii) ⁻16°C (iii) ⁻8°C
   (c) ⁻33°C and ⁻11°C

18 (a) 2   (b) ⁻4 (c) ⁻3
        0   ⁻2   0
       ⁻2   0   3
       ⁻4   2   6
       ⁻6   4   9

19 (a) ⁻32 (b) ⁻15 (c) 18 (d) 56 (e) 40 (f) 84

20 (a) ⁻70 (b) 84 (c) 40 (d) ⁻24

21 (a) 2   (b) 2   (c) −1
         1   4   −3
         0   −4   3
        −1   −2   1
        −2   −1 $4\frac{1}{2}$

22 (a) ⁻3 (b) ⁻2 (c) ⁻4 (d) ⁻5 (e) 5 (f) 12

23 (a) ⁻12 (b) ⁻28

24 (a) true (b) false (⁻12) (c) true
   (d) true (e) true (f) true

25 (a) ⁻97 (b) ⁻30 (c) 4 (d) 25

26 (a) ⁻3 (b) ⁻3 (c) ⁻8

27 (a) ⁻37 (b) 20

1.17 Indices

1 (a) $2 \times 2 \times 2 \times 2 \times 2 = 32$
  (b) $3 \times 3 \times 3 \times 3 = 81$
  (c) $4 \times 4 \times 4 = 64$
  (d) $5 \times 5 = 25$

2 (a) $3 \times 3 = 3^2$
  (b) $2 \times 2 \times 2 \times 2 = 2^4$
  (c) $3 \times 3 \times 3 = 3^3$
  (d) $5 \times 5 \times 5 = 5^3$

3 (a) 3^2 (b) 4^3 (c) 2^5 (d) 6^4

4 (a) five squared (b) eight cubed
  (c) two to the power four

(d) three to the power five
(e) six to the power seven

5 (a) 32 (b) 81 (c) 64 (d) 625 (e) 216

6 (a) 784 (b) 1849 (c) 4913 (d) 1728 (e) 15 625

7 (a) 204.49 cm² (b) 438.976 cm³

8 (a) 100 (b) 1000 (c) 10 000 (d) 10 000
  (e) 1 000 000

9 (a) 100 (b) 1000 (c) 10^2; 10^3

10 (a) 10 000 (b) 1 000 000 (c) 10^4; 10^6

11 (a) 1 000 000 (b) (i) 1 000 000 (ii) 1 000 000 000 000
   (c) (i) 1 000 000 000 (ii) 1 000 000 000 000 000 000
   (d) 10^6; 10^6; 10^{12}; 10^9; 10^{18}

12 (a) $(10 \times 10) \times (10 \times 10 \times 10) = 10^5$
   (b) $(10 \times 10 \times 10 \times 10) \times (10 \times 10 \times 10) = 10^7$
   (c) $(10 \times 10) \times (10 \times 10 \times 10 \times 10 \times 10) = 10^7$
   (d) you add the indices to get the index of the result

13 (a) $(3 \times 3 \times 3 \times 3) \times (3 \times 3 \times 3 \times 3 \times 3) = 3^9$
   (b) $(7 \times 7) \times (7 \times 7 \times 7) \times (7 \times 7 \times 7 \times 7) = 7^9$

14 (a) false (2^5) (b) true (c) true (d) true

15 (a) 5^{10} (b) 4^9 (c) 6^{10} (d) 3^9 (e) 7^{16} (f) 8^{15}

16 (a) $\dfrac{10 \times 10 \times 10}{10 \times 10} = 10^1$

   (b) $\dfrac{10 \times 10 \times 10 \times 10 \times 10}{10 \times 10 \times 10} = 10^2$

   (c) $\dfrac{10 \times 10 \times 10 \times 10}{10 \times 10} = 10^2$

   (d) you subtract the indices to get the index of the result

17 (a) $\dfrac{3 \times 3 \times 3 \times 3 \times 3 \times 3}{3 \times 3 \times 3 \times 3} = 3^2$ (b) $\dfrac{7 \times 7 \times 7 \times 7 \times 7}{7 \times 7 \times 7} = 7^2$

18 (a) true (b) true (c) false (4^4) (d) true

19 (a) 5^2 (b) 4^2 (c) 6^6 (d) 3^4 (e) 7^1 (f) 8^6

20 (a) (i) $1 = 5^0$ (ii) $1 = 3^0$ (iii) $1 = 2^0$
   (b) (i) 1 (ii) 1 (iii) 1
   (c) yes

21 (a) $\dfrac{1}{5 \times 5} = \dfrac{1}{5^2}$ (b) $\dfrac{1}{7 \times 7} = \dfrac{1}{7^2}$

   (c) $\dfrac{1}{3 \times 3 \times 3 \times 3} = \dfrac{1}{3^4}$ (d) $\dfrac{1}{2 \times 2 \times 2} = \dfrac{1}{2^3}$

22 (a) true (b) false ($\frac{1}{16}$) (c) true (d) true

23 (a) $\frac{1}{64}$ (b) $\frac{1}{243}$ (c) $\frac{1}{128}$ (d) $\frac{1}{81}$ (e) $\frac{1}{16}$ (f) $\frac{1}{125}$

24 (a) 2^7 (b) 3^2 (c) 3^4 (d) 5^{-3}
   (e) 2^7 (f) 3^{-1} (g) 4^{-8} (h) 7^0

25 (a)

Powers of 2	2^0	2^1	2^2	2^3	2^4	2^5	2^6	2^7
	1	2	4	8	16	32	64	128
Differences		1	2	4	8	16	32	64

(b) the differences are also powers of 2

(c)

Powers of 3	3^0	3^1	3^2	3^3	3^4	3^5	3^6	3^7
	1	3	9	27	81	243	729	2187
Differences		2	6	18	54	162	486	1458

the differences are two times each power of 3
(d) the differences are three times each power of 4

1.18   Problems and investigations: some hints

1 (a) Try 654 × 321, 643 × 521, 632 × 541 etc
(b) Try 1 × 23 456, 2 × 13 456 , 23 × 1456 etc
(c) Try 42 × 31, 41 × 32, 8654 × 7321, 8321 × 7654, etc

2 Look at the sizes of the numbers.
Does this help you to decide when to +, −, × or ÷?

3 (a), (b), (c), (d) Can you change the problem round to use the two given numbers?
(e), (f), (g), (h), (i), (j) Look at the units figures – does this help?
Look at the largest figures – does this help?

4 Why can't B + C be more than 9?
Try using 1 + 6 or 2 + 5 rather than 1 + 2 or 8 + 9.
What is the sum of two odd numbers, two even numbers, one odd and one even?

5 Look at the sizes of the numbers.

6 Look at the sizes of the numbers.
What about the units figures?
Can you simplify the equation?

7 Try finding the top ○ number first.

8 Try halving the 168 to put in each of the left hand circles. What happens when you increase one of these numbers?

9 (b) Try being systematic
4 + 4 + 4, 4 + 4 − 4, 4 + 4 × 4, 4 + 4 ÷ 4
Try using brackets 4 + 4 × 4, (4 + 4) × 4
(c) Try being systematic 4 + 4 + 4 + 4, 4 + 4 + 4 − 4 etc
Try using brackets
(4 + 4) × (4 + 4), (4 + 4 + 4) ÷ 4 etc
Is it easier to find a particular number or to see what number you can make?
What about 44 + 4 + 4 or (44 ÷ 4) + 4?

10 Write the powers of 2 beside each result. Does this help?

11 Make a table for your results.
Compare these with the powers of 2.
If you know how to move 3 discs can you use this to move the first three discs on your pile of 4 discs?

12 Make a table for your results.
Try two teams of 1 frog each
two teams of 2 frogs each
two teams of 3 frogs each etc.
Compare your results with the powers of 2.
Does it make a difference whether there are an even number of frogs in the team?

13 Write the two sets of results for (a) and (b) alongside each other. How are they related?

14 Try some parallel lines.
Try making some of the lines pass through the same point.
Try adding a fourth line in different ways to your results for three lines.
Make a table to show the maximum number of intersection points for each set of lines.

15 Compare the number of regions with the powers of 2.
Take great care with 6 points on the circle!!
Make sure you don't have three diagonals meeting at the same point inside the circle.
Check that you have the same number of diagonals leaving each point on the circle.

1.19   Oral test

1 36

2 16

3 7, 14, 21

4 12, 24, 36

5 1, 3, 5, 15

6 1, 2, 3, 4, 6, 12

7 38 shirts

8 44 tapes

9 63 badges

10 4 cakes

11 18 cakes

12 14 tins

13 $\frac{1}{6}$

14 $\frac{1}{4}$

15 16

16 23

17 2.5

18 2.7

19 800

20 200

21 500 (10 × 50)

22 49

23 4900

24 9

25 30

26 23°C

27 13°C

28 27

29 16

30 five squared (5²)

SECTION 2 Everyday arithmetic

2.1 Times and Timetables

1 10 15

2 (a) 11 35 (b) 07 35 (c) 07 30 (d) 14 35
   (e) 20 30 (f) 23 30

3 (a) 10.35 a.m. (b) 5.31 a.m. (c) 9.00 a.m.
   (d) 0.41 a.m. (e) 3.35 p.m. (f) 12.01 p.m.
   (g) 8.30 p.m. (h) 11.59 p.m.

4 (a) 11.03 a.m. (b) 4.38 p.m. (c) 12.30 p.m.
   (d) 10 06 (e) 08 28 (f) 23 48

5 (a) 12.15 p.m. (b) 7.30 p.m. (c) 0.42 a.m.
   (d) 11 48 (e) 11 03 (f) 00 10

6 (a) 1 hour 30 min (b) 1 hour 22 min
   (c) 9 hours 38 min

7 (a) 8 hours 50 min (b) 3 hours 54 min

8 (a) 8.52 a.m. (b) 8.45 a.m.

9 (a) 38 min (b) 18 03 (6.03 p.m.)

10 (a) (i) 38 min (ii) 37 min (b) 17 min

11 1.32 pm (13 32)

12 (a) 1 hour 35 min (b) 2 hours 25 min
   (c) 2 hours 25 min

13 (a) 6 hours 30 min (b) 1 hour 20 min
   (c) (i) 2 hours 20 min (ii) 2 hours
   (d) 4 hours 40 min

14 (a) 217 (b) S20 (c) 07 50(×20) (d) S20
   (e) 2 hours 59 min (f) 1 hour 15 min

15 (a) SX – Saturdays excepted; FO – Friday only;
   SO – Saturdays only
   (b) 2 hours 53 min
   (c) London
   (d) 2 hours 49 min
   (e) 17 55 from London; 19 min

16 (a) 6 hours 53 min (b) 11 hours 44 min

17 (a) 14 20 (b) 17 min (c) 21 13

18 (a) (i) 2 hours (ii) 9.50 a.m.
   (b) 1 hour 50 min; 2 hours 10 min

2.2 Reading dials, charts and tables

1 (a) 1001 (b) £350.35

2 (a) 3164 (b) 730 units (c)

4   2   9   8

3 (a) 112 (b) 187 (c) £19.17

4 (a) £1.50 (b) £5.50 (c) £9.75

5 (a) £10.50 (b) £6.50 (c) £47.50 (d) £325

6 £19.49

7 (a) £38.50 (b) £28.00 (c) £87.50 (d) £24

8 (a) (i) 193 miles (ii) 311 km
   (b) (i) 194 miles (ii) 312 km
   (c) (i) 304 miles (ii) 242 miles

9 (a) 385 (b) 315 (700–385)

10 (a) £273 (b) £633 (c) £597 (d) 2 × £273 + £180 = £726

2.3 Changing units of measure

1 (a) 300 mm (b) 0.3 m

2 0.25 kg

3 (a) 0.2 l (b) 200 ml

4 (a) 0.25 km (b) 25 000 cm

5 (a) 0.8 g (b) 0.0008 kg

6 (a) (i) 76 cm (ii) 12.2 cm
   (b) (i) 760 mm; 122 mm (ii) 0.76 m; 0.122 m

7 (a) 2.58 l (b) (i) 258 cl (ii) 2580 ml

8 (a) 1651 g (b) (i) 1.651 kg (ii) 1 651 000 mg

9 (a) (i) 300 cm² (ii) 10 cm²
   (b) (i) 30 000 mm²; 1000 mm² (ii) 0.03 m²; 0.001 m²

10 (a) (i) 160 cm^3 (ii) 12 cm^3
(b) (i) 160 000 mm^3; 12000 mm^3
(ii) 0.000 16 m^3; 0.000 012 m^3

11 (a) 11.82 inches (b) 30.48 cm

12 (a) 39.4 inches (b) 91.44 cm

13 (a) (i) 10 miles (ii) 45 miles (iii) 75 miles
(b) (i) 40 km (ii) 96 km (iii) 152 km

14 (a) 0.22 gallons (b) 4.55 l (c) 0.45 l
(d) 50 gallons ≈ 227 l (e) 87.5 pints (f) 4.57 l

15 (a)

Kilograms	0.5	1	1.5	2	2.5	3
Pounds	1.1	2.2	3.3	4.4	5.5	6.6

(b)

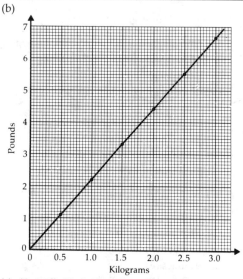

(c) (i) 1.7 lb (ii) 2.8 lb (iii) 5.0 lb
(d) (i) 0.45 kg (ii) 1.8 kg (iii) 1.1 kg

16

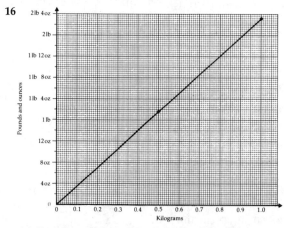

(a) 7 ounces (b) 1 pound 5 ounces (c) 0.57 kg

2.4 Measures in the home

1 (a) 1200 g (b) 1.2 kg

2 (a) 26p (b) 36p (c) 70p

3 (a) 300 kg (b) 12.5 kg (c) (i) £54 (ii) £51.75

4 (a) 1292 cm (b) 12.92 m

5 (a) 17.6 cm (b) 27.4 cm (c) 176 mm; 274 mm

6 (a) 72 m (b) 48 (c) 6

7 (a) 225 m^2 (b) 0.0225 ha

8 (a) (i) 15 200 cm^2 (ii) 1.52 m^2

9 (a) (i) 2640 cm^2 (ii) 0.264 m^2 (b) 50

10 2.208 m^3

11 (a) 24 000 000 cm^3 (b) 24 m^3

13 24.4 cm; 18.8 cm; 1.4 cm; 694848 mm^3;
694.848 m^3 (these values may vary slightly)

14 12

15 (a) 50 (b) 8 days plus 2 extra doses

16 0.125 or $\frac{1}{8}$

17 20.6 ml

18 (a) 4.75 l

(b) 1.5 l; 1 l + 0.5 l; 0.75 l + 0.75 l; 0.33 l + 0.67 l + 0.5 l; 4
(c) 0.83 l; 1.08 l; 1.17 l; 1.25 l; 1.33 l; 1.42 l; 1.67 l; 1.83 l;
2.0 l; 2.17 l; 2.25 l; 2.5 l (using 2 bottles);
1.58 l; 1.92 l; 2.17 l; 2.25 l; 2.33 l; 2.67 l; 2.83 l; 3.0 l;
3.17 l; 3.25 l (using 3 bottles) etc
(d) Fill the 0.75 l bottle and pour it into the 0.5 l bottle.
There is 0.25 l left
(e) 0.08 l (0.75–0.67); 0.17 l (0.5–0.33); 0.25 l; 0.34 l;
0.42 l etc

2.5 Bills and money calculations

1 58p

2 (a) £1.32 (b) £3.68

3 (a) £1.96 (b) £3.04

4 £1.35

5 £1.48

6 (a) £2.10 (b) £1.75

7 5

8 16; 12p

9 2 × 13p + 3 × 18p

10 (a) £1.60 (b) 12 (c) 15; 20p

11 (a) £18.20 (b) £1.80 (c) £4.55

12 (a) £2.35 (b) £2.00 (c) £5.55

13 Potatoes 35p
Carrots 36p
Sprouts 15p
‾‾‾‾
86p

14 (a) £5.90 (b) £5.55 (c) £9.00

15 (a) £10.88 (b) £6.80 (c) £4.76

16 (a) £13.14 (b) £6.57 (c) £11.68

17 (a) (i) £1.07 (ii) £1.24 (iii) £1.50
(b) the largest size (c) £0.45

18 (a) (i) £1.55 (33333) (ii) £1.54(6) (iii) £1.53 (33333)
(b) the third store (c) 15p (d) 11p

19 (a) the smaller packet
(b) yes, £1.07 for 550 g is £1.94(5) per kg
49p for 250 g is £1.96 per kg

20 (a) £11.92 (b) £34.00 (c) £47.80 (d) £72.64

21 (a) 20p (b) £1.00 (c) £1.50 (d) £2.40

22 (a) £45.60 (b) £197.60 (c) £292.60 (d) £463.60

23 (a) Each bill would be reduced by £1.17
(i.e. $\frac{2}{13}$ of £7.60)
(b) Each bill would be increased by £0.58
(i.e. $\frac{1}{13}$ of £7.60)

24 (a) £21.45 (b) £50.05 (c) £59.29 (d) £72.36

25 The last will be cheaper by 21p

2.6 Ready reckoners and conversion graphs

1

Litres	5	10	15	20	25	30	35	40	45
Gallons	1.1	2.2	3.3	4.4	5.5	6.6	7.7	8.8	9.9

(a) 23 litres (b) 32 litres (c) 41 litres

2

×	9	19	29	39	49	59	69	79	89	99
1	9	19	29	39	49	59	69	79	89	99
2	18	38	58	78	98	118	138	158	178	198
3	27	57	87	117	147	177	207	237	267	297
4	36	76	116	156	196	236	276	316	356	396
5	45	95	145	195	245	295	345	395	445	495
6	54	114	174	234	294	354	414	474	534	594
7	63	133	203	273	343	413	483	553	623	693
8	72	152	232	312	392	472	552	632	712	792
9	81	171	261	351	441	531	621	711	801	891
10	90	190	290	390	490	590	690	790	890	990
20	180	380	580	780	980	1180	1380	1580	1780	1980

3 (a) £6.23 (b) £15.13 (c) £7.77 (d) £33.67

4 (a) £1.16 (b) £2.03 (c) £2.61 (d) £4.35

5 (a) £1.45 (b) £14.16 (c) (i) £1.52 (ii) £3.12
(iii) £6.32

6 (a) (i) 13.6 l (ii) 34.1 l
(b) (i) 4.4 gallons (ii) 7.7 gallons

7 (a) (i) £7.28 (ii) £12.74 (b) (i) £6.62 (ii) £14.06

8 (a)

Number of gallons	1	2	3	4	5
Number of litres	4.5	9.1	13.6	18.2	22.7

(b)

Number of litres	5	10	15	20	25
Cost of £	1.79	3.57	5.36	7.14	8.93

(c)

Number of gallons	1	2	3	4	5
Cost in £	1.62	3.24	4.87	6.49	8.11

9 (a) 50% (b) 25% (c) 75% (d) 67% (e) 83%

10 (a)

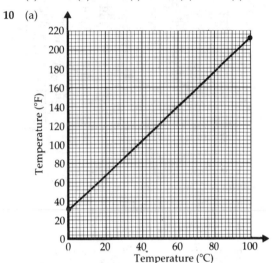

(c) (i) 95°F (ii) 10°C

11 (a)

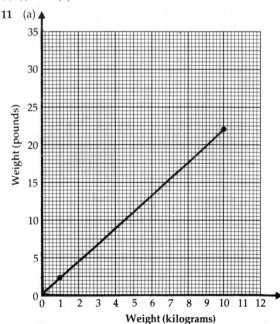

(c) (i) 11.44 lb (ii) 8.18 kg

12 (a) (i) 99p (ii) £2.38 (iii) £1.33 (iv) 63p
(b) (i) 57p (ii) £1.70 (iii) 85p (iv) £2.56

13 (a)

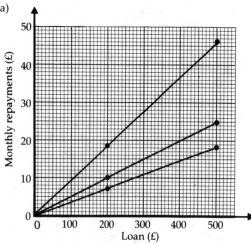

(b) (i) £11.50 (ii) £21.09 (iii) £8.33

2.7 Percentages, decimals and fractions

1 (a) $\frac{21}{100}$ (b) $\frac{7}{100}$ (c) $\frac{25}{100}$ or $\frac{1}{4}$

2 (a) 21% (b) 7% (c) 25%

3 (a)

(b)

(c)

(d)

4 (a) $\frac{50}{100} = \frac{1}{2}$ (b) $\frac{25}{100} = \frac{1}{4}$ (c) $\frac{75}{100} = \frac{3}{4}$
(d) $\frac{20}{100} = \frac{1}{5}$ (e) $\frac{80}{100} = \frac{4}{5}$
(f) $\frac{15}{100} = \frac{3}{20}$ (g) $\frac{44}{100} = \frac{11}{25}$ (h) $\frac{65}{100} = \frac{13}{20}$
(i) $\frac{5}{100} = \frac{1}{20}$ (j) $\frac{150}{100} = 1\frac{1}{2}$

5 (a) $\frac{8}{100} = \frac{2}{25}$ (b) $\frac{12}{100} = \frac{3}{25}$ (c) $\frac{40}{100} = \frac{2}{5}$ (d) $\frac{97}{100}$

6 (a) 3% (b) 15% (c) 60% (d) 75% (e) 6%
(f) 50% (g) 25% (h) 12.5% (i) 37.5% (j) 0.5%

7 (a) $\frac{1}{4}$ (b) $\frac{2}{5}$ (c) $\frac{6}{25}$

8 (a) 25% (b) 40% (c) 24%

9 (a) 0.19 (b) 0.47 (c) 0.85 (d) 0.07 (e) 1.09

10 (a) 13% (b) 65% (c) 9% (d) 70% (e) 12.5%

11 (a) £8 (b) £12 (c) £60 (d) £1.50 (e) £15 (f) £4

12 (a) 7p (b) 15p (c) 48p (d) 48p (e) 27p (f) £1.20

13 (a) £8.84 (b) £12.92 (c) £20.93
(d) £12.42 (e) £37.35 (f) £8.76

14 (a) 7.31 m (b) 16.56 kg (c) 25.65 l
(d) 28.35 km (e) 91.25 gallons (f) £2160

15 (a) 4 m (b) 3.5 km (c) 33 l (d) 1 kg (e) 3.28 g
(f) 2.8 cm (g) 130.79 tons (h) 69.75 cl (i) £42.00
(j) £1262.25

16 (a) 20% (b) £2.00 (c) 64p

17 1128

18 (a) 80% (b) 20% (c) £16.80

19 (a) £16 (b) £2

20 (a) 51p (b) £3.91

21 (a) £744 (b) £5456

22 (a) £1134 (b) 73%

23 (a) £327 (b) £5123

24 (a) £27 (b) £10.90 (c) the first shop (d) 10p

25 (a) £100 − 15% × £100 = £85:
£85 − 30% × £85 = £59.50
£100 − 30% × £100 = £70: £70 − 15% × £70 = £59.50,
so no difference

2.8 Percentages and percentage change

1 (a) 14% (b) 86%

2 (a) 86% (b) 14%

3 (a) 36% (b) 8% (c) 8% (d) 52%
(e) 45% (f) 95% (g) 10% (h) 90%

4 (a) 50% (b) 50% (c) 50% (d) 50%
(e) 75% (f) 75% (g) 20% (h) 70%

5 65%

6 (a) 75% (b) 90% (c) 75% (d) 65%

7 (a) 50% (b) 75% (c) 62%
(d) 25% (e) 40% (f) 65%

8 (a) 50% (b) 25% (c) 38%
(d) 75% (e) 60% (f) 35%

9 (a) 55% (b) 62.5% (c) 62.5% (d) 25%
(e) 68.75% (f) 20% (g) 12.5% (h) 37.5%

10 (a) 95%; 92.5%; 88.75%; 53.75%
(b) 82.5%
(c) 82.5%

11 (a) 50p (b) 25%

12 (a) £5   (b) 20%

13 (a) 15°F increase   (b) 20%

14 (a) 15°F decrease   (b) 25%

15 In question **14** the 15°F is then found as a percentage of 60°F, rather than 75°F so the percentage change is larger

16 16%

17 16%

18 54

19 (a) 20   (b) 100   (c) 12.5%

20 (a) 10   (b) 40   (c) 24%

21 (a) 472   (b) 2% (now 20%)

22 (a) £265   (b) 5.6%

2.9   VAT, discount, profit and loss

1 (a) £15; £7.50; £30; £45; £22.50; £37.50
(b) £115; £57.50; £230; £345; £172.50; £287.50

2 (a) 15p   (b) 30p   (c) 75p   (d) 18p   (e) 66p

3 (a) £3.00   (b) £23.00   (c) £4.60

4 (a) 54p   (b) £4.14

5 £46

6 (a) £12   (b) £108

7 £600

8 (a) £376   (b) In the first shop she will save £1

9 (a) (i) £47 (ii) £46 (iii) £46.92
(b) If he buys 50 slabs he will save £2
(c) With a 3% discount 50 slabs would cost £48.50, so it would be cheaper to buy the 48 slabs

10 £12 with a 10% discount is £10.80, £11.50 with a 6% discount is £10.81

11 (a) £1.40   (b) £21.40   (c) 5%

12 (a) £56   (b) 14%

13 (a) 12%   (b) £282.50

14 (a) 60p   (b) £1.80

15 (a) 25%   (b) 20%

16 (a) £2400   (b) £2160   (c) £160   (d) 8%

17 (a) £3.52   (b) 12.5%   (c) £3.04

18 (a) £6.75   (b) 27p   (c) 4%   (d) 6.4%

19 (a) £82.08   (b) none

2.10   Interest, loans and hire purchase

1 (a) £22   (b) £66

2 (a) £222   (b) £24.42

3 (a) £120   (b) £600

4 (a) £10   (b) £180

5 (a) £24; £25; £26   (b) £10

6 £12

7 £900

8 (a) £12.60   (b) £54   (c) £328.50

9 (a) £107   (b) £1.07   (c) £114.49   (d) £122.50

10 (a) £106   (b) £112.36   (c) £119.10

11 (a) £218   (b) £237.62   (c) £259.01

12 (a) £24   (b) £224

13 (a) £12   (b) £236

14 (a) £16   (b) £216   (c) £18

15 (a) £36   (b) £336   (c) £14

16 £2700

17 (a) £270   (b) £2070   (c) £57.50

18 £405

19 (a) £100.80   (b) £520.80   (c) £21.70

20 £100.80 − £67.20 = £33.60

2.11   Wages and taxes

1 £5760

2 £4368

3 (a) £36.00   (b) £1872

4 (a) £4.70   (b) £28.20   (c) £30.30   (d) £121.20

5 (a) £5.88   (b) £305.76

6 (a) £43.20   (b) £518.40

7 £10.32

8 (a) (i) £416 (ii) £96   (b) £349.44   (c) £107.33

9 (a) £104   (b) £3.90   (c) £15.60   (d) £119.60

10 (a) £153.30   (b) £58.40   (c) £211.70

11 £21.78

12 (a) 40   (b) £180   (c) £36

13 (a) £46.63   (b) £202.08

14 (a) £72.98   (b) £316.25

15 (a) £2575   (b) £2775   (c) £3575

16 (a) £1205   (b) £1405   (c) £2405

17 (a) £324   (b) £378   (c) £540   (d) £594

18 (a) £1755
(b) £162 per month (£1944 per year)
(c) £40.50 per week (£2106 per year)

19 £540

20 £1620

21 (a) £2375   (b) (i) £641.25 (ii) £12.33

22 (a) (i) £336 (ii) £6.46
(b) (i) £3168 (ii) £60.92

23 (a) £3005   (b) £811.35

24 (a) £4833   (b) £1000   (c) £2070

25 (a) £2700   (b) £5673   (c) £10 383

2.12 Household finance

1 (a) yes; he has £1.25 left   (b) no

2 (a) £31.15   (b) £6.65 or 95p per day

3 (a) £66.75   (b) £10 per week

4 (a) £145   (b) £468

5 Longleat costs £17.50; Alton Towers costs £17.25.
The second is cheaper

6 (a) 12 m^2 (b) £45

7 (a) 14 m   (b) 28 widths   (c) 7 rolls   (d) £20.75

8 (a) 625 m^2 (b) 5 cans   (c) £13.45

9 (a) 100 m   (b) 300 plants   (c) £5.40

10 (a) (i) £25 (ii) £20; the second
(b) (i) £18 (ii) £18.60; the first
(c) 75 miles

11 (a) 208 hours   (b) £137.40
(c) £12.60 (£137.40 − £124.80)

12 (a) yes; it will cost him £4732 per year   (b) £1.31

13 (a) 200 gallons   (b) £350   (c) £720   (d) 12p

14 (a) £59.75   (b) £8.50   (c) £238.50

15 (a) 55 gallons   (b) 250 litres
(c) 1162.5 francs (or £116.25)

16 (a) £354.75   (b) yes; this would cost £350
(c) £414.25

17 £1010

18 (a) £366.67   (b) £74.25

2.13 Ratio and proportion

1 (a) 4 : 5   (b) 2 : 7   (c) 8 : 1

2 7 : 3

3 (a) 15 : 16   (b)160 : 171

4 15:14

5 (a) 3 : 10   (b) 10 : 3

6 (a) 4 : 5   (b) 16 : 25

7 8 : 27

8 (a) 40   (b) 15

9 It is a square

10 (a) (i) 3 cm (ii) 6 cm (iii) 15 cm
(b) (i) 1 cm (ii) 2 cm (iii) 4 cm

11 (a) (i) 4 girls (ii) 8 girls (iii) 20 girls
(b) (i) 3 boys (ii) 6 boys (iii) 9 boys

12 (a) 2 : 6   (b) 2 : 10   (c) 4 : 6   (d) 4 : 10

13 (a) 1 : 3   (b) 4 : 1   (c) 2 : 3   (d) 3 : 2

14 (a) 2 : 12   (b) 3 : 21   (c) 2 : 5   (d) 3 : 4

15 1 cm and 4 cm

16 2 cm and 8 cm

17 (a) $\frac{1}{5}$ and $\frac{4}{5}$ (b) £4 and £16

18 (a) $\frac{1}{5}$ and $\frac{4}{5}$ (b) £12 and £48

19 100 g and 150 g

20 30 cm and 50 cm

21 $\frac{1}{3}$l (0.33 l) and $\frac{2}{3}$l (0.67 l)

22 (a) 1 : 6 : 9   (b) 2 : 3 : 11   (c) 3 : 4 : 9

23 (a) 20 cm and 70 cm   (b) 3 cm and 6 cm

24 (a) $\frac{1}{6}$, $\frac{2}{6}$ and $\frac{3}{6}$ (b) £10, £20 and £30

25 (a) 36 kg of sand and 72 kg of gravel   (b) 30 kg

26 (a) 50 g, 150 g   (b) 300 g, 150 g

2.14 Maps and scale drawing

1 (a) 1.5 cm   (b) 2.5 cm   (c) 1 : 2

2 6 cm; 2:5

3 6 cm; 4:15

4 (a) 192 cm   (b) 84 cm   (c) 48 cm   (d) 16

5 (a) 50 cm   (b) 15 cm   (c) 56 cm

6 (a) (i) 1.5 cm by 2.0 cm (ii) 1.5 cm by 1.5 cm
(b) (i) 300 cm by 400 cm (ii) 300 cm by 300 cm
(c) 1 cm by 0.5 cm

7 (a) (i) 8 cm by 12 cm (ii) 12 cm by 16 cm
(iii) 6 cm by 10 cm

9 (a) 12 cm by 6 cm   (b) 1 cm by 1 cm

10 (a) 3 cm by 4 cm   (b) 108 cm^2 (c) 12 cm^2
(d) ratio of areas is 1 : 9

11 (a) 100 000 cm   (b) 1000 m   (c) 1 km

12 (a) 600 000   (b) 30 cm

13 (a) (i) 8 km (ii) 10 km (iii) 24 km
(b) (i) 2 cm (ii) 2.5 cm (iii) 6 cm
(c) 4 km^2 (d) 4 cm^2

14 (a) (i) 20 000 cm (0.2 km) (ii) 60 000 cm (0.6 km)
(iii) 50 000 cm (0.5 km)
(b) 0.01 km^2 (10 000 m^2)

2.15 Similarity and enlargement

1 (a) 2 : 3   (b) 3 cm

2 (a) 3 : 4   (b) 3 cm, 4.5 cm; 4 cm, 2$\frac{2}{3}$ cm (2.67 cm)

3 (a) 4 : 5   (b) 4.4 cm; 2.5 cm

4 (a) 8 cm (b) 3.75 cm

5 6 cm; 7.5 cm, 6 cm

6 (a) 6 cm by 4 cm (b) 9 cm by 6 cm
  (c) 4.5 cm by 3 cm (d) 3.75 cm by 2.5 cm

7 (a) 5 cm (b) 12 cm

8 (a) 72 cm (b) 60 cm (c) 20 cm (d) 19.2 cm

9 (a) 1 : 2 (b) yes (c)

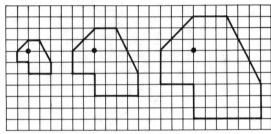

10 (a) yes (b) yes (c) yes (d) 1 : 3 (e) yes

11

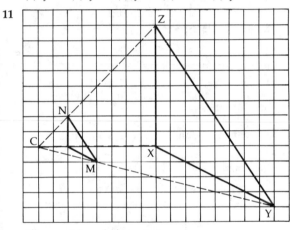

  (f) yes (g) scale factor is 4

12 (a) (i) 25 cm^2 (ii) 100 cm^2
  (b) The area is multiplied by 4 (i.e. 2^2)

13 (a) (i) 4 cm^2 (ii) 36 cm^2
  (b) The area is multiplied by 4 (i.e. 2^2)

14 (a) (i) 12 cm^2 (ii) 48 cm^2
  (b) the area is multiplied by 4 (i.e. 2^2)

15 (a) (i) 30 cm^2 (ii) 270 cm^2
  (b) The area is multiplied by 9 (i.e. 3^2)

16 (a) and (b)

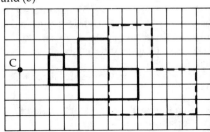

(c) 3 square units (d) 12 square units; 27 square units
(e) When the scale factor is 2 the area is multiplied by 2^2
  When the scale factor is 3 the area is multiplied by 3^2

17 (a) (i) 125 cm^3 (ii) 1000 cm^3
  (b) The volume of the cube is multiplied by 8 (i.e. 2^3)

18 (a) (i) 8 cm^3 (ii) 216 cm^3
  (b) The volume of the cube is multiplied by 27 (i.e. 3^3)

2.16 Speed, distance and time

1 (a) 40 m.p.h. (b) 32 m.p.h. (c) 64 m.p.h.
  (d) 48 m.p.h. (e) 45 m.p.h. (f) 46 m.p.h.

2 (a) 9.42 a.m. (b) 48 m.p.h.

3 (a) 42 m.p.h. (b) 68 m.p.h. (c) 55 m.p.h.

4 (a) 14 m.p.h. (b) 15 m.p.h. (c) 1 m.p.h.
  (d) 1 mile

5 (a) 56 m.p.h. (b) 224 miles

6 (a) 24 miles (b) 60 miles (c) 7 miles (d) 4 miles
  (e) 6 miles (f) 30 miles

7 (a) 10.26 a.m. (b) 70 miles

8 (a) 48 miles; 57 miles (b) 6 miles

9 21 km

10 (a) (i) 100 miles (ii) 96 miles
  (b) 196 miles (c) 56 m.p.h.

11 (a) 2 hours (b) 4 hours (c) half an hour
  (d) quarter of an hour (e) 1½ hours (f) 1¼ hours

12 (a) 3 hours (b) 7.30 p.m.

13 (a) 6 hours (b) 4½ hours
  (c) the father by half an hour

14 16 minutes

15 10 minutes

16 50 minutes

17 (a) 30 minutes (b) 54.55 km per hour

18 (a) 7½ minutes (b) 7½ minutes (c) 30 minutes
  (d) 5 minutes

19 (a) 22500 m (i.e. 22.5 km)
  (b) 22.5 km per hour (c) 2 min 8 sec

20 (a) (i) 600 m (ii) 36 000 m (i.e. 36 km)
  (b) 36 km per hour

21 (a) (i) 533.33 m (ii) 32 000 m (i.e 32 km)
  (b) 32 km per hour

22 The section is 75 km; 75 km in 50 min is 90 km per
  hour

2.17 Direct and inverse proportion

1 (a) (i) £2.65 (ii) £6.36 (b) (i) £5.30 (ii) £3.18
  (c) 30 bottles

2 (a) (i) £1.68 (ii) £6.30 (b) (i) £5.04 (ii) £2.10
  (c) 24 bags

3 (a) (i) £1.95 (ii) 13p (iii) 39p (b) 30lb

4 (a) 40p (b) £2.00

5 (a) £1.25 (b) £6.25

6 (a) £11.25 (b) £24

7 £29.95

8 120 g flour; 96 g sugar; 54 g butter; 750 g fruit;
6 eggs; 3 tablespoons milk

9 160 miles

10 7.5 gallons

11 (a) (i) 495 miles (ii) 378 miles
(b) 2.94 gallons
(c) 26 miles

12 (a) 126 miles (b) 504 miles (c) 661.5 miles

13 (a) 20 minutes (b) 1 hour 20 minutes
(c) 30 minutes

14 (a) 96 m.p.h. (b) 32 m.p.h. (c) 64 m.p.h.

15 (a) 6 days (b) 1½ days (c) 2 days

16 10 hours

17 3 hours

18 8 cars

19 (a) (i) 24 men (ii) 16 men
(b) (i) 180 racquets (ii) 300 racquets
(iii) 500 racquets
(c) (i) 5 weeks (ii) 3 weeks

20 (a)

Number of gallons	0	1	2	3	4	5	6
Miles covered	0	35	70	105	140	175	210

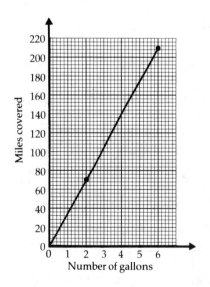

(d) (i) 122.5 miles (ii) 2.29 gallons

21 (a)

Time in hours	1	2	3	4	5	6
Speed in m.p.h.	3000	1500	1000	750	600	500

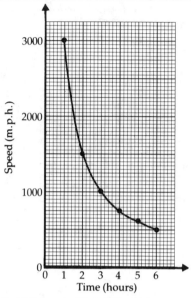

(d) (i) 1200 m.p.h. (ii) 3.75 hours

22 (a) 1 hour (b) 1500 letters (c) 36 students
(d) 600 letters

2.18 Problems and investigations: some hints

1 Try finding the times for each film from one part of
the program and then use these to fill in the gaps.

2 Use a piece of squared paper, and the four carpet
sizes. Find the total area needed and also the area
wasted.

3 Draw the corridor on squared paper.
Try drawing lines in various positions to find the
maximum length.

4 Draw a line for each tariff on the same piece of graph
paper. Find where these lines intersect.

5 Draw three lines on the same graph to represent the
three wage settlements. Find where these lines
intersect.

6 (a) Make up a table for a line of 2, 3, 4, 5 tiles in a row.
Is there a pattern?
(b) Make up a table for different rectangles e.g.
$2 \times 3, 2 \times 4, 3 \times 4, 2 \times 5, 3 \times 5$ etc. Is there a pattern?
Is the number of tiles in the border connected to the
perimeter of the rectangle?
(c) Make up a table for different └ shapes, see (b)
above.
(d) Make up a table for different ⊔ shapes. See (b)
above.

7 Write down the results from your calculator.
Look at the sets of numbers used.
Look at the order of these numbers.

8 Write down the results from your calculator.
Look at the sets of numbers used.
Look at the order of these numbers.
Do the 'thirteenths' fall into two groups?

9 Draw a diagram to show which days they have off.
Look at the multiples. Does this help?
Draw diagrams for the different possible starting days.

10 It may help you to draw a diagram to show the number of handshakes.
What about using a table like this?

```
      a  b  c  d
a     x  ✓  ✓  ✓
b        x  ✓  ✓
c           x  ✓
d              x
```

Are the results connected with the triangle numbers?

11 The lines joining the corresponding vertices for any pair of squares should meet at a point.
Look at the ratio of the distances of this point to each of the corresponding vertices.
Look at the ratio of the sides of the two squares.

12 Use your calculator and write down the results.
(ii) add 6% on twice   (iii) add 3% on four times
(iv) add 1% on twelve times.

2.19   Oral test

1 £160

2 £7.92

3 £3.60

4 8p

5 40p

6 36p

7 4⅔ cakes

8 £4; £14

9 17p

10 169 years (to 1989)

11 1902 (to 1989)

12 25 minutes

13 14 minutes

14 14°C

15 £4.50

16 £24

17 £17.60

18 £50

19 16 minutes

20 10 minutes

21 50 minutes

SECTION 3   Geometry and measures

3.1   Lines and angles

1 (a) 6 cm   (b) 63 mm or 6 cm 3 mm

2 (a) (i) 4 cm (ii) 6 cm (iii) 7 cm
  (b) (i) 4 cm (40 mm) (ii) 5.8 cm (58 mm)
     (iii) 7.2 cm (72 mm)

3 (a) (i) 10 cm (ii) 11 cm
  (b) 1.5 cm

5 16.5 cm

6 (a) (i) acute (ii) reflex (iii) right-angled (iv) obtuse
  (b) (i) 30° (ii) 240° (iii) 90° (iv) 120°

8 AC is 6.4 cm; BC is 5.4 cm

9 (a) 220°   (b) 110°   (c) 50°   (d) 150°

10 (a) 40°   (b) 70°   (c) 30°   (d) 122°

11 (a) 60°; 120°   (b) 90°; 65°   (c) 48°; 42°
   (d) 30°; 120°; 30°

12 (a) 58°   (b) 34°   (c) 42°   (d) 133°

13 (a) $p = r$ (corresponding); $q = s$ (alternate)
   (b) $p = r$ (vertically opposite); $r = s$ (alternate);
      $q = t$ (corresponding)

14 (a) $p = 65°$ angle on straight line with 115°
      $q = 65°$ alternate angle to p
   (b) $p = 72°$ alternate angle with 72°
      $q = 72°$ corresponding angle to p
      $r = 72°$ alternate angle to q
      $s = 72°$ vertically opposite to 72°

3.2   Symmetry and reflection

1 (a)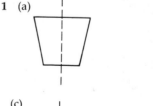

  (b)

  (c)

  (d)

2 (a)

  (b)

  (c)

  (d)

3 (a)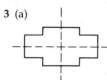

(b)

(e)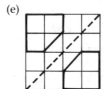

(f)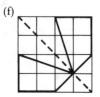

4 (a)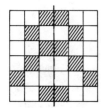

(b)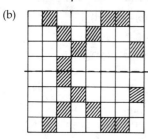

10

(a)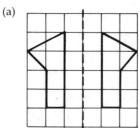

(b)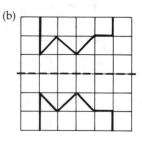

5 (a)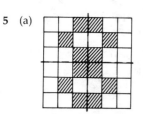

(b)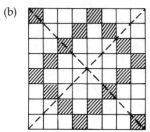

(c)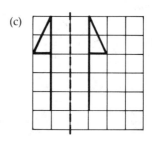

(d)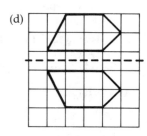

(e)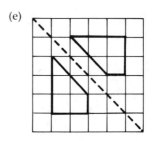

(f)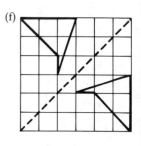

8 (a) the right eye (b) clockwise

11

9 (a) 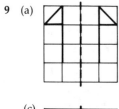 (b)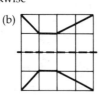

(c)   (d)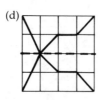

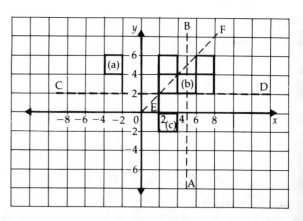

12

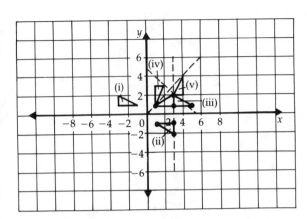

13 (a) (i) (7, 2) (ii) (6, 4) (iii) (5, 1) (iv) (8, 5)
(b) You find how far the first coordinate is to the left of 4 and then add this to 4; the second coordinate stays the same

14 (a) (i) (0, 10) (ii) (2, 8) (iii) (3, 11) (iv) (0, 7)
You find how far the second coordinate is below 6 and then add this to 6. The first coordinate stays the same
(b) (i) (2, 1) (ii) (4, 2) (iii) (1, 3) (iv) (5, 0)
You swap the coordinates

3.3 Symmetry and rotation

1 (a)

order 4

(b)

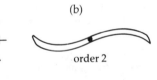

order 2

(c)

order 2

(d)

order 5

2 (a)

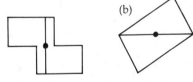

(b)

(c)

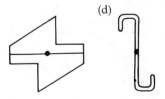

(d)

3 (a)

(b)

4 (a)

(b)

5 (a)

(b)

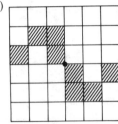

6 (a)

(b)

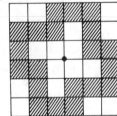

9 (a)

(b)

(c)

(d)

10 (a)

(b)

11 (a)

(b)

12

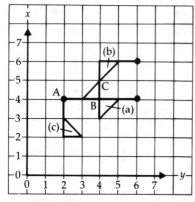

13

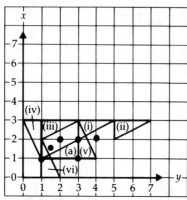

14 (a) (i) $(-1, 0)$ (ii) $(0, -4)$ (iii) $(-3, -1)$ (iv) $(-2, -5)$
(b) You change the sign of both coordinates

15 (a) (i) $(0, 1)$ (ii) $(-4, 0)$ (iii) $(-1, 3)$ (iv) $(-5, 2)$
(b) The first coordinate becomes the second and the second coordinate becomes the first with a change of sign

3.4 Triangles

1 (a) $70°$ (b) $90°$ (c) $120°$ (d) $40°$

2 (a) isosceles; acute-angled
(b) scalene; right-angled
(c) isosceles; obtuse-angled
(d) scalene; obtuse-angled

3 (a) (b)

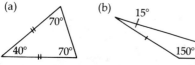

4 (a) $60°$ (d) (e) order 3

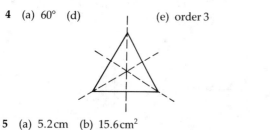

5 (a) 5.2 cm (b) 15.6 cm^2

6 (a)

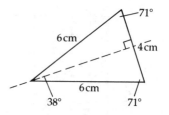

(b) $71°; 71°; 38°$ (d) 5.7 cm (e) 11.4 cm^2

7 (a)

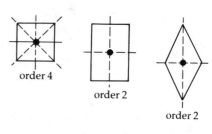

(b) $37°; 53°; 90°$; (c) right-angled (d) 13.5 cm^2

8 (b) 60 cm^2

9 (a) $65°$

10 (b) 8.7 cm (c) $30°; 90°$ (d) 21.7 cm^2

11 (b) $90°; 6.6$ cm; 4.6 cm

13 (c) 6.8 cm; 34 cm^2

14 (b) You should get the same area each time
(c) The three perpendiculars should meet at the same point

15 (b) The perpendicular bisectors of each side should meet at the same point

3.5 Quadrilaterals and polygons

1

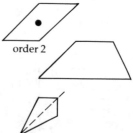

2 (b) The properties should include statements about equal sides, equal angles, and the diagonals

3 (c) The properties should include statements about equal sides, equal angles and the diagonals

4

	S	R	Rh	P	T	K
4 equal sides	✓	x	✓	x	x	x
4 equal angles	✓	✓	x	x	x	x
2 pairs equal sides	x	✓	x	✓	x	✓
Opposite sides equal	✓	✓	✓	✓	x	x
Adjacent sides equal	✓	x	✓	x	x	✓
Opposite angles equal	✓	✓	✓	✓	x	x
Diagonals equal	✓	✓	x	x	x	x
Diagonals bisect	✓	✓	✓	✓	x	x
Diagonals at 90°	✓	x	✓	x	x	✓

5

(a)

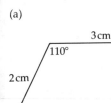

(b)

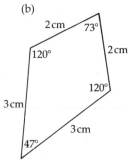

6 (a) 150° (b) 110°

7 (a) 60°; 60°; 120°; 120°

8

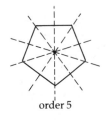

order 5

order 6

order 8

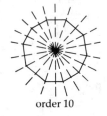

order 10

9 (a)

	Number of sides	Lines of symmetry	Order of rot. sym
Triangle	3	3	3
Square	4	4	4
Pentagon	5	5	5
Hexagon	6	6	6
Octagon	8	8	8
Decagon	10	10	10
N-agon	n	n	n

(b) The number of lines of symmetry and the order of rotational symmetry is the same as the number of sides of the polygon

10 (a)

Number of sides	3	4	5	6	8	10
Number of diagonals	0	2	5	9	20	35

(b) (i) 54 (ii) 119 (iii) 1224 (iv) $\dfrac{n \times (n-3)}{2}$

11, 12

Number of sides	3	4	5	6	8	10
Sum of angles (°)	180	360	540	720	1080	1440
Each angle (°)	60	90	108	120	135	144

13 (a) (i) 60° (ii) 45°
(b) (i) 360° (ii) 360°; yes

3.6 Circles and loci

1 (c) angle APB = 90° (d) angle AQB = 90°
(e) all the angles are 90°

2 (e) angle CRD = angle CSD

3 All the angles are the same size

4 All the angles are the same size
Each of these plus each of the angles in Question **3** add up to 180°

5 yes

6 (b) yes

7

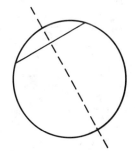

The two halves fit on top of each other
The line of symmetry cuts the chord in half
The line of symmetry is at right-angles to the chord

8 (a) yes   (b) yes

9 (d) yes

10 There are 2 points 5cm from P and 6cm from Q

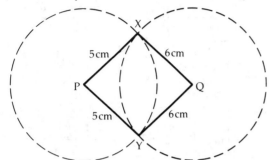

11 (c) All the points lie on a straight line which bisects AB and is at right-angles to AB

12

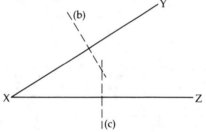

(d) There is only one point which is the same distance from X, Y and Z

13

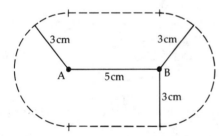

(c) Use a piece of rope attached first to A and then to B to mark out the semicircular ends. Then mark the straights by joining the ends of the semicircles

14 (a)

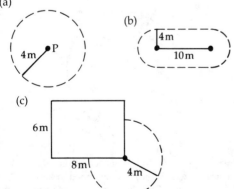

(b)

(c)

15 (a) $\pi \times 4^2 \, m^2 = 50.2 \, m^2$
(b) $(50.2 + 80) \, m^2 = 130.2 \, m^2$
(c) $\frac{3}{4}$ of $(\pi \times 4^2) = 37.7 \, m^2$

16

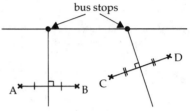

Draw the perpendicular bisectors of AB and CD. Find where these cross the road

17

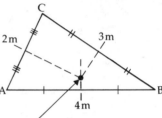

distribution box

(a) Draw the perpendicular bisectors of AB, BC and AC.
The box should be put where these meet.
(b) 6.2km

3.7   Perimeter and area: rectangles

1 (a) 18 cm   (b) 20 cm   (c) 12 cm   (d) 30 cm

2 (a) four times   (b) three times
(c) six times   (d) (number of sides) times

3 Add the lengths of the four sides
Add the length and width and double the result
Double the length and the width and then add the results

4 $(3.5 + 5.5 + 3.5 + 5.5) \, cm$; $2 \times (3.5 + 5.5) \, cm$; $(2 \times 3.5 + 2 \times 5.5) \, cm$

5 $(p + q + p + q) \, cm$; $2 \times (p + q) \, cm$; $(2 \times p + 2 \times q) \, cm$

6 (a) 12 cm^2 (b) 84 cm^2 (c) 8.2 cm^2 (d) 9.43 cm^2

7 (a) multiply the length by the width
(b) divide the area by the known side

8 (a) 6cm   (b) 12cm   (c) 3.1cm   (d) 16cm^2

9 (a) (i) 7.5cm^2 (ii) 6cm^2 (b) (i) 13cm (ii) 12cm

10 (a) 11cm^2 (b) 11cm^2

11 (a) 44cm^2 (b) 99cm^2 (c) 176cm^2 (d) 1100cm^2

12 When you double the grid size you multiply the area by 4
When you treble the grid size you multiply the area by 9
When you multiply the grid size by n you multiply the area by n^2.

13 (a) 36cm^2 (b) 144cm^2 (c) 4.41mm^2 (d) 39.69cm^2

14 (a) multiplied by 4　(b) multiplied by 9

15 (a) 3cm　(b) 8cm　(c) 20cm　(d) 2.5cm

16 (a) 66 slabs　(b) £117.40

17 (a) (i) 10m^2 (ii) 8m^2 (iii) 9.5m^2 (iv) 10.5m^2
(b) 19 litres　(c) £27.80

18 (a) 8m^2
(b) 200 tiles assuming they can be cut exactly in half
(c) 6 packs; £28.50

3.8　Area; triangles and parallelograms

1 (a) 4.5cm^2　(b) 6cm^2　(c) 6cm^2　(d) 6cm^2

2 (a) (i) 30cm^2 (ii) 70cm^2 (iii) 20cm^2 (iv) 60cm^2
(b) (i) 15cm^2 (ii) 35cm^2 (iii) 10cm^2 (iv) 30cm^2

3 (a) 30cm^2　(b) 27cm^2

4 (a) 12cm^2　(b) 35cm^2　(c) 12cm^2　(d) 6cm^2

5 (a) 6cm^2　(b) 30cm^2　(c) 42.5cm^2
(d) 7.2cm^2　(e) 24.82cm^2

6 (a) 4cm　(b) 5cm　(c) 3cm　(d) 8cm

7 (a) 14cm^2; 28cm^2　(b) 15cm^2; 30cm^2
(c) 12cm^2; 24cm^2　(d) 42cm^2; 84cm^2

8 (a) 45cm^2　(b) 52cm^2　(c) 25cm^2　(d) 16cm^2

9 (a) 8cm　(b) 2cm　(c) 2.1cm　(d) 4cm

10 (a) 8cm　(b) 12cm　(c) 6cm　(d) 21cm

11 (a) 45cm^2　(b) 18cm^2　(c) 66.5cm^2　(d) 69cm^2

12 (a) 5m^2　(b) 22.7m^2　(c) £36.32

13 8m

3.9　Perimeter and area: circles

1 (a) 31.4cm　(b) 18.84cm　(c) 25.12cm　(d) 50.24cm

2 (a) 6.28cm　(b) 15.7cm　(c) 37.68cm　(d) 7.85cm

3 (a) 18.84cm　(b) 50.24cm　(c) 87.92cm　(d) 15.7cm

4 (a) 2cm　(b) 4cm　(c) 5cm　(d) 7cm

5 (a) You divide the circumference by 2π or 6.28
(b) (i) 1cm (ii) 2cm (iii) 3.25cm

6 (a) 12.56cm　(b) 6280cm or 62.8m　(c) 8 times

7 (a) 514m　(b) 114m

8 63.7m

9 (a) 31.4cm　(b) 101.4cm or 1.014m

10 (a) 28.26cm^2　(b) 50.24cm^2　(c) 314cm^2
(d) 153.86cm^2

11 (a) 12.56cm^2　(b) 78.5cm^2　(c) 452.16cm^2
(d) 19.625cm^2

12 (a) You divide the diameter by 2 to get the radius
Then square the radius and multiply by π
(b) (i) 12.56cm^2 (ii) 78.5cm^2 (iii) 452.16cm^2
(iv) 13.8474cm^2

13 (a) 39.25cm^2　(b) 12.56cm^2　(c) 9.42cm^2
(d) 9.42cm^2

14 (a) 3.16cm　(b) 2cm　(c) 5cm

15 (a) 14.13cm^2　(b) 100.48cm^2

16 3cm

17 (a) 3.44cm^2　(b) 10.75cm^2　(c) 13.76cm^2
(d) 18.84cm^2

18 (a) 105.12cm^2　(b) 29.76cm^2

19 2.3925cm^2

20 1.7712cm^2

3.10　3-D shapes and their nets

1 (a)

	Number of edges	Number of faces	Number of vertices
Cube	12	6	8
Cuboid	12	6	8
Prism	18	8	12
Pyramid	8	5	5

(b) 14; 14; 20; 10
(c) The number of faces plus the number of vertices is two more than the number of edges

2 (a) no　(b) yes　(c) yes　(d) yes

3

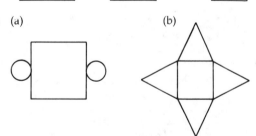

4 (a) 　　　　　　　　　　(b)

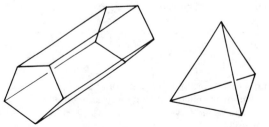

5 (a) (i) A prism with each end a pentagon
(ii) A pyramid with a triangular base
(b) (i) 　　　　　　　　　　(ii)

7

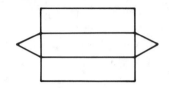

8 (a) (i) 8 cm²; 6 cm²; 12 cm²
　　　(ii) 20 cm²; 12 cm²; 60 cm²
　　(b) (i) 52 cm² (ii) 184 cm²

9 (a) (i) 30 cm²; 40 cm²; 50 cm²
　　　(ii) 72 cm²; 96 cm²; 120 cm²
　　(b) (i) 6 cm² (ii) 24 cm²
　　(c) (i) 132 cm² (ii) 336 cm²

10 (a) 96 cm² (b) 160 cm² (c) 88 cm² (d) 72 cm²

11 (a) (i) 12.56 cm　(ii) 31.4 cm
　　(b) (i) 125.6 cm² (ii) 94.2 cm²
　　(c) (i) 12.56 cm² (ii) 78.5 cm²
　　(d) (i) 150.72 cm² (ii) 251.2 cm²

12 (a) 294 cm² (b) 900 cm²
　　(c) 1152 cm² (d) 140.8 cm²

3.11 Volume: cubes and cuboids

1 (a) 5 (b) 9 (c) 4

2 (a) (i) 8 cm³ (ii) 27 cm³ (iii) 64 cm³
　　(b) You cube the length i.e. length × length × length
　　(c) (i) 125 cm³ (ii) 216 cm³ (iii) 1000 cm³

3 (a) (i) 12 (ii) 6
　　(b) (i) 3 (ii) 8
　　(c) (i) 36 (ii) 48

4 (a) 16; 5; 80; 80 cm³
　　(b) 30; 3; 90; 90 cm³
　　(c) 16; 10; 160; 160 cm³
　　(d) 30; 2; 60; 60 cm³

5 You multiply the length by the breadth by the width

6 (a) 84 cm³ (b) 100 cm³
　　(c) 180 cm³ (d) 216 cm³

7 (a) 60 cm³ (b) 112 cm³
　　(c) 135 cm³ (d) 216 cm³

8 (a) 64 cm³ (b) 100 cm³
　　(c) 48 cm³ (d) 32 cm³

9 cube: 3375 cm³; cuboid 3315 cm³; yes

10 64

11 1000

12 50

13 (a) On the bottom layer you would put 2 packets along the 60 cm, and 5 packets along the 1 m. You would have 5 layers in all
　　(b) no

(c)

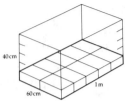

14 (a) 260 cm³ (b) 300 cm³

15 1400 cm³

16 2800 cm³

17 (a) 375 cm³ (b) 1800 cm³ (c) 2400 cm³

18 432 cm³

19 153.6 cm³

3.12 Volume: cylinders and prisms

1 (a) (i) 12.56 cm² (ii) 78.5 cm²
　　(b) (i) 125.6 cm³ (ii) 628 cm³

2 (a) 50.24 cm²; 251.2 cm³
　　(b) 28.26 cm²; 565.2 cm³
　　(c) 113.04 cm²; 1356.48 cm³
　　(d) 200.96 cm²; 2411.52 cm³

3 You first find the area of one end by squaring the radius and then multiplying the result by π (3.14). You then multiply this result by the length

4 (a) 125.6 cm³ (b) 226.08 cm³ (c) 942 cm³
　　(d) 861.616 cm³　(e) 38.4336 cm³

5 (a) 1570 cm³ (b) 785 cm³; the first

6 (a) (i) 4521.6 cm³ (ii) 4710 cm³ (iii) 4019.2 cm³
　　　　(iv) 4019.2 cm³; 2nd
　　(b) (i) 1657.9 cm² (ii) 1570 cm² (iii) 1406.7 cm²
　　　　(iv) 2110.1 cm²; 3rd

7 (a) 42 cm³ (b) 24 cm³ (c) 942 cm³ (d) 62.8 cm³

8 (a) 375 cm³ (b) 225 cm³ (c) 706.5 cm³ (d) 176.6 cm³

9 23.864 m³

10 (a) 94 200 m³ (b) 52 987.5 m³ (c) 20 887.5 m³
　　(d) 38 550 m³

11 9043.2 cm³

12 70 650 cm³

13 (a) You divide the volume by π (3.14) and then by half the diameter and then by half the diameter again

$$\text{i.e. } V \div \pi \div \left(\frac{D}{2}\right) \div \left(\frac{D}{2}\right)$$

　　(b) You divide the volume by π (3.14) and then by the length and then find the square root of the result
　　　i.e. $V \div \pi \div l$ and then find $\sqrt{\ }$

14 (a) 1.6 cm (b) 4 cm

15 11.7 cm

3.13 Bearings

1 (a) E (b) W (c) SE (d) SW

2 (a) Tower (b) Station (c) Cinema (d) Bank

5 (a) E (b) N (c) SW (d) NW

6 (a) 045° (b) 180° (c) 225° (d) 315°

7 (a) Q (b) P (c) R (d) S

8 (a) (i) 060° (ii) 250°
    (b) (i) 240° (ii) 070°

9 (a) 050° (b) 090° (c) 270° (d) 230° (e) 130°
    (f) 310°

10 230°

11 (a)   (b) 045°

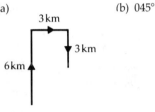

12 (a) 118.5 km (b) 2.4 km

13 091°; 118.5 km

14 (a)   (b) 4.9 miles (c) 245°

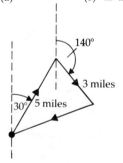

15 none

16 (a)   (b) 17 km (c) 180°

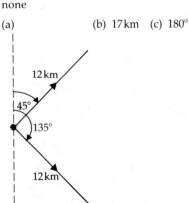

17 (a)

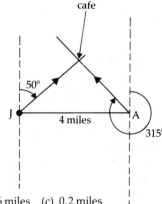

(b) 2.6 miles (c) 0.2 miles

18 (b) 2.8 km; 3.9 km (c) 52 minutes

19 (a)

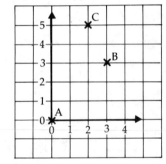

(b) 045° (c) 068° (d) 333° (e) 2.24

3.14 Pythagoras

1 (a) 169 cm^2 (b) 25 cm^2; 144 cm^2 (c) yes

2 (a) $6^2 + 8^2 = 10^2$ (b) $8^2 + 10^2 = 164$ but $12^2 = 144$
    (c) $7^2 + 24^2 = 25^2$

3 (a) no (b) yes (c) yes

4 (a) 3.53 cm^2; 6.28 cm^2; 9.81 cm^2
    (b) $3.53 + 6.28 = 9.81$

5 (a) 10 cm (b) 15 cm (c) 26 cm

6 20 cm

7 3.16 m

8 130 m

9 7.5 miles

10 (a) 6 cm (b) 9 cm (c) 10 cm

11 12 cm

12 2.83 m

13 150 m

14 (a) 25 cm (b) 41 cm

15 (a) 5 cm (b) 13 cm

16 (a) 7.07 cm (b) 5.66 cm

17 (a) 12 cm (b) 14 cm (c) no

18 (a) 16 m (b) 24 m

3.15 Trigonometry: tangent

1 (a) 0.364 (b) 0.700 (c) 4.705
(d) 0.885 (e) 2.379 (f) 2.921

2 (a) 35° (b) 42° (c) 14° (d) 45°
(e) 58° (f) 32.5° (g) 37.2° (h) 61.4°

3 (a) (i) 0.75 (ii) 0.4 (iii) 1 (iv) 2
(b) (i) 36.9° (ii) 21.8° (iii) 45° (iv) 63.4°

4

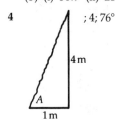

; 4; 76°

5 (a) 1.73 cm (b) 2.38 cm (c) 4.29 cm (d) 2.33 cm

6 (a)

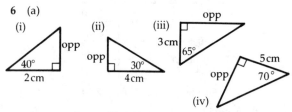

(b) (i) 1.68 cm (ii) 2.31 cm (iii) 6.43 cm (iv) 10.7 cm

7 (a) 4.77 cm (b) 6.49 cm (c) 13.74 cm (d) 8.58 cm

8 (a) 3.12 cm (b) 4.28 cm (c) 4.77 cm (d) 4.66 cm

9 47.1 m

10 38.7°

11 (a) 6.25 m (b) 32°

12 (a) 171 m (b) 236 m (c) 54°

3.16 Trigonometry: sine and cosine

1 (a) 0.342 (b) 0.574 (c) 0.978
(d) 0.663 (e) 0.922 (f) 0.946

2 (a) 30° (b) 65° (c) 60° (d) 2°
(e) 36.9° (f) 39.6° (g) 49.4° (h) 25.7°

3 (a) (i) 0.75 (ii) 0.4 (iii) 0.8 (iv) 0.5
(b) (i) 48.6° (ii) 23.6° (iii) 53.1° (iv) 30°

4

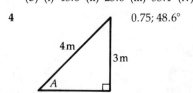

0.75; 48.6°

5 (a) 1 cm (b) 2.3 cm

6 (a)

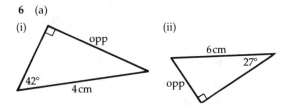

(b) (i) 2.68 cm (ii) 2.72 cm

7 (a) 3.77 cm (b) 4.41 cm

8 (a) 4.23 cm (b) 4.77 cm

9 (a) 0.940 (b) 0.819 (c) 0.208
(d) 0.749 (e) 0.388 (f) 0.324

10 (a) 60° (b) 25° (c) 30° (d) 88°
(e) 53.1° (f) 50.4° (g) 40.6° (h) 64.3°

11 (a) (i) 0.5 (ii) 0.8 (iii) 0.75 (iv) 0.2
(b) (i) 60° (ii) 36.9° (iii) 41.4° (iv) 78.5°

12

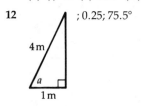

; 0.25; 75.5°

13 41.4°

14 (a) 4.64 cm (b) 7.73 cm

15 (a) (i)

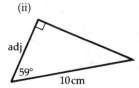

(b) (i) 3.88 cm (ii) 5.15 cm

16 (a) 6.52 cm (b) 4.56 cm

17 (a) 5.59 cm (b) 3.98 cm

3.17 Trigonometry: applications

1 (a)

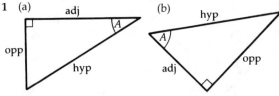

2 (a) $\tan A = \frac{5}{2}$; 68.2° (b) $\cos B = \frac{5}{10}$; 60°
(c) $\sin C = \frac{1}{4}$; 30° (d) $\sin D = \frac{4}{5}$; 53.1°

3 (a) $a = 4 \sin 65°$; 3.63 cm (b) $b = 5 \tan 25°$; 2.33
(c) $C \sin 70° = 10$; 10.64 cm (d) $d \tan 75° = 8$; 2.14 cm

4 (a) 4.62 cm (b) 1.92 cm (c) 11.08 cm (d) 13 cm
 (e) 169 cm^2 (f) $(25 + 144)$ cm^2 = 169 cm^2
 (g) yes (h) a right-angled triangle

5 (a) 38.7 km (b) 80.9 km

6 037°

7 032°

8 (a)

(b) 027°

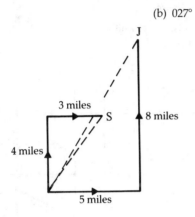

9 (a) 0.34 m (b) 0.06 m

10 63.4°

11 (a) 3.35 m (b) 12.5 m (c) 37.5 m (d) 37.5 m

12 61°; 30°; 44.4°

13 (a) 84.5 m
 (b) It would be easier to cycle up the dotted line, since the angle this makes with level ground is smaller

14 (a) 1.81 m (b) 0.85 m

15 (a)   C ; 36.7°   (b) Q   ; 45°

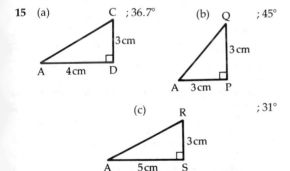

16 (a) AC = 5 cm; AQ = 4.24 cm; AR = 5.83 cm
 (b) AC^2 = 3^2 + 4^2 = 25; AQ^2 = 3^2 + 3^2 = 18;
  AR^2 = 5^2 + 3^2 = 34

3.18 Problems and investigations: some hints

1 Try putting each pair of edges together in turn. Turn one of the triangles over and repeat.

2 See **1**.

3 Try to be systematic in finding your triangles. Fix two points and vary the third point. Check that you haven't already got the triangle in a different position or as a reflection. There are less than 10 possibilities. Look at the lengths of the sides and the angles.

4 Try to make all the shapes of one type e.g. squares, parallelograms, trapeziums etc. before moving onto another shape. There are less than 20 possibilities.

5 Make part of a shape and then rotate the grid through 90° or 180° to see where the shape moves to. Make part of a shape and then reflect this in a line.

6 Try making all the triangles on the top row, then the bottom row, then the left hand column then the right hand column. Use tracing paper to help you find rotations.

7 Try making one shape and then move one corner. Does this increase or decrease the number of sides?

8 Recording several results is important. Try moving one corner. What does this do to the area? What does this do to the number of dots inside? How is (dots on edge ÷ 2) + dots inside connected to the area?

9 Try and be systematic. Use four squares in a line, then three etc. Move one square at a time. Try to imagine the squares folded in. If necessary cut out the shape and fold it up. Try to identify the top or the bottom of the box.

10 See **9**.

11 Always draw the diagonals from one corner to make the triangles. How many triangles are there for a given polygon? How do you get the sum of the angles? How do you get each angle? How many diagonals will there be from each vertex? How many vertices? Don't count the diagonals twice.

12 Find combinations of angles which add up to 360° using:
 (a) one angle e.g. 90° + 90° + 90° + 90°
 (b) two angles e.g. 120° + 60° + 120° + 60°
 (c) three angles e.g. 150° + 90° + 120°

3.19 Oral test

1 360°

2 180°

3 100°

4 55°

5 (a) one (b) none

6 (a) two (b) two

7 28 cm

8 26 cm

9 84 cm^2

10 60 cm^3

11 (a) 5 cm^2 (b) 12 cm

12 (a) 6 (b) 12 (c) 8

13 (a) 5 (b) 5 (c) 8

14 (a) 32 cm (b) 63 cm^2

15 (a) 15 cm^2 (b) 20 cm^2

16 (a) 24 m (b) 48 m^2

17 12 metres

18 4 miles due east of where I started

19 5 miles

20 13 cm

21 (a) 3 cm (b) $\frac{3}{5}$ or 0.6

22 (a) $\sqrt{32}$ cm (5.66 cm) (b) 45°, 45°, 90° (c) 1

SECTION 4 Graphs, algebra and statistics

4.1 Coordinate and graphs

1 (a) (1, 2); (1, 1); they are the same
(b) (−1, −1); (2, −1); they are the same

2, 3

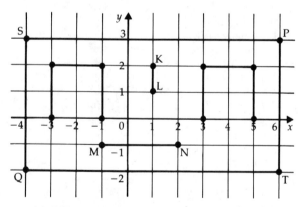

2 (c) PSQT is a rectangle

3 (a) a square (b) a square

4 (2, 3); (4, 2); (4, 0); (3, −2)

5 (a) yes
(b) (−4, 2); (−4, 0); (−3, −2)
(c) The first coordinate changes sign, the second coordinate stays the same
(d) an octagon

6 (a) yes
(b) (4, −2); (3, 2)
(c) The first coordinate stays the same, the second coordinate changes sign
(d) It is the same point (4, 0)

7 (a) 2 units (b) 1 unit (c) (4, 1); (10, 2); (8, 4); (2, 5)
(d) (−4, 1); (−10, 2); (−8, 4); (−2, 5)

8 (a)

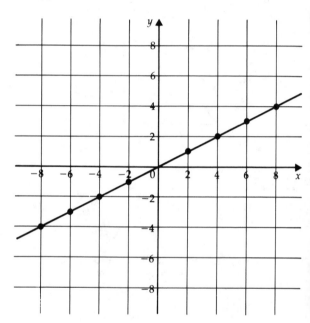

(b) In each the first coordinate is double the second coordinate
(d) (i) (5, 2.5) (ii) (3, 1.5) (iii) (1, 0.5)

9 (a) The second coordinate is one more than the first coordinate
(b) The second coordinate is two less than the first coordinate
(c) The second coordinate is double the first coordinate
(d) The second coordinate is three times the first coordinate
(e) The second coordinate is two times the first coordinate plus three
(f) The second coordinate is the square of the first coordinate

10 (a) (1.6, 0.8); (3.0, 1.5)
(b) the second coordinate is half of the first coordinate
(c) yes
(d) (1.0, 0.5); (2.0, 1.0); (4.0, 2.0)
(e) yes

11 (a)

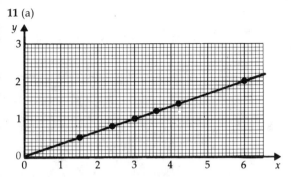

(b) The points all lie on a straight line
(c) The second coordinate is one third of the first coordinate
(d) (4.5, 1.5); (4.8, 1.6); (1.2, 0.4); (5.4, 1.8)

12 (a)

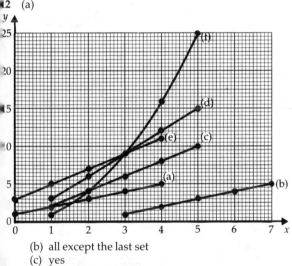

(b) all except the last set
(c) yes

4.2 Reading and drawing graphs

1 (a) 50 miles   (b) 80 miles
(c) (i) 9.30 a.m. (ii) 10.51 a.m.   (d) 30 miles
(e) She stopped to shop or have a cup of coffee

2 (a) (i) 70°F (ii) 65°F (iii) 70°F
(b) (i) 7.00 a.m. (ii) 7.30 a.m.; 5.00 p.m.
(iii) 9.00 a.m.; 12.30 p.m.; 3.30 p.m.
(c) 80°F   (d) 2.00 p.m.

3 (a) (i) 4 gallons (ii) 7 gallons (iii) 3 gallons
(iv) 8 gallons
(b) (i) 25 miles (ii) 50 miles

4 (a) (i) £18 (ii) £33 (iii) £20.50 (iv) £41.75
(b) (i) 800 units (ii) 1000 units
(iii) 300 units (iv) 750 units
(c) There is a standing charge of £8 for the service
(d) £2.50
(e) 2.5p per unit

5 (a) (i) 1 hour (ii) 160 miles

(b)

Time in hours	1	2	3	4	5	6	7	8
Distance in miles	80	160	240	320	400	480	560	640

(c)

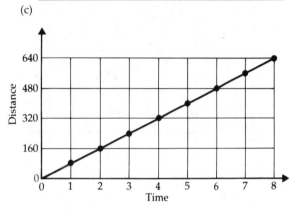

(d) (i) 280 miles (ii) 5.25 hours

6 (a)

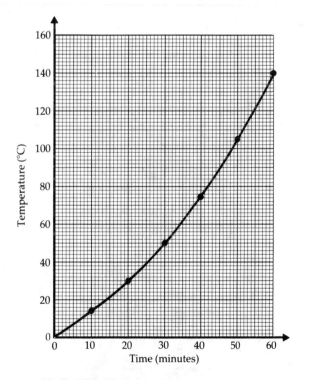

(c) (i) 90°C (ii) 48 minutes
(d) 25°C, 30°C, 35°C
(e) 62.5 minutes

7   (a)

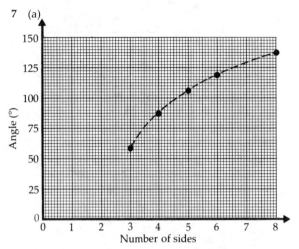

(d)

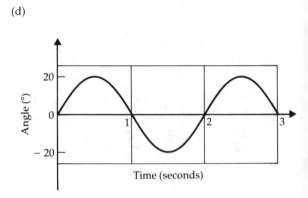

(c) Only those corresponding to a whole number of sides

(d) You can't have a polygon with less than three sides

(e) 129°

(f) As the number of sides increases the angle increases up to 180°

4.3   Graphs and formulae

1   (a)

Number of days	1	2	3	4	5	6	7
Number of papers	50	100	150	200	250	300	350

(b) 50 times

(c)

8   (a)

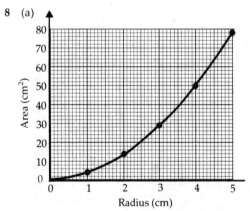

(c) yes, for each radius there will be a corresponding area

(d) (i) 20 cm² (ii) 3.5 cm

9   (a)

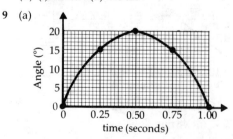

(c) (i) 18.5° (ii) 0.15 sec and 0.85 sec

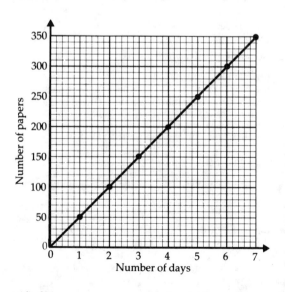

(d) £14

2 (a) You divide the 60 miles by the speed

(b)

Speed in m.p.h.	2	4	6	8	10	12
Time in hours	30	15	10	7.5	6	5

(c)

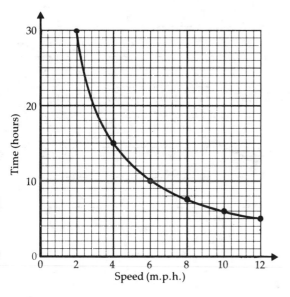

(d) yes

3 (a)

Side in cm	1	2	3	4	5	6	7	8
Area in cm²	1	4	9	16	25	36	49	64

(b) the side multiplied by the side

(c)

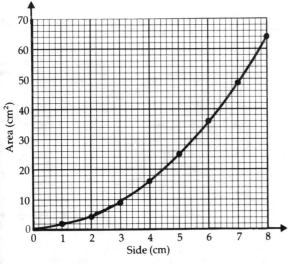

(d) (i) 20cm² (ii) 6.3cm

4 (a)

Number of weeks	4	8	12	16	20	24
Cost of hiring	28	36	44	52	60	68

(b) There is a minimum charge of £20 plus a weekly charge

(c) (i) £32 (ii) £50 (iii) £124

(d) Multiply the number of weeks by 2 and add 20

5 (a) You find the distance in miles by multiplying the number of hours by 120

(b)

Number of hours	1	2	3	4	5
Number of miles	120	240	360	480	600

(c)

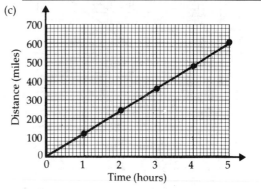

6 (a) You multiply the length by itself and itself again

(b)

Edge in cm	1	2	3	4	5	6	7	8	9	10
Volume in cm³	1	8	27	64	125	216	343	512	729	1000

(c)

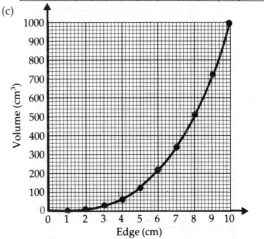

(d) (i) 90cm³ (ii) 5.9cm

7 (a)

Number of miles	20	40	60	80	100
Cost of hiring (£)	16	17	18	19	20

(b) (i) £17.50 (ii) £18.50 (iii) £19.25
(c) £15
(d) 5p per mile
(e) You multiply 5p by the number of miles and then add on £15

8

Number of miles	10	20	30	40	50
Cost of hiring in £	2	3.5	5	6.5	8

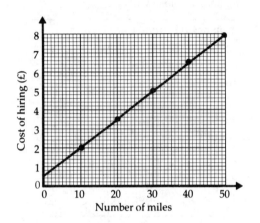

4.4 x–y graphs

1 (a)

x	0	1	2	3	4	5	6	7	8
$x + 3$	3	4	5	6	7	8	9	10	11

(b)

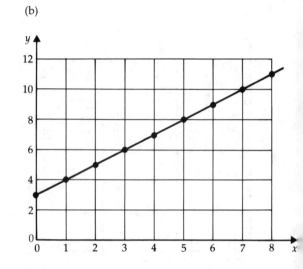

(c) yes
(d) (0, 3)

9 (a)

Weight of letter in g	50	75	100	125	175
Cost of letter in pence	13	20	20	24	30

(b) (i) 13p (ii) 20p (iii) 30p
(c) over 60 g up to 100 g
(d) 2 × 55 g cost 26p; 1 × 110 g costs 24p

10 (a) (i) 18p (ii) 26p (iii) 40p
(b) Over 100 g up to 150 g
(c)

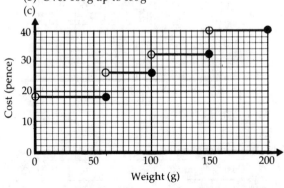

2 (a)

x	0	1	2	3	4	5	6	7	8
$x + 1$	1	2	3	4	5	6	7	8	9

(b), (c)

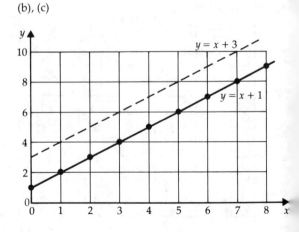

(d) This line is parallel to the line in question **1**
(e) (0, 1)

3 (a)

x	0	1	2	3	4	5	6	7	8
$x + 2$	2	3	4	5	6	7	8	9	10

(b), (c)

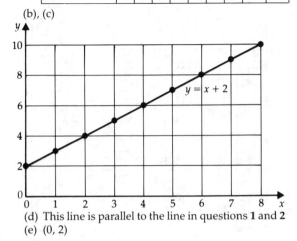

(d) This line is parallel to the line in questions **1** and **2**
(e) (0, 2)

4 The line $y = x + 4$ will be parallel to the other three lines and pass through (0, 4) on the y-axis

5 (a)

x	0	1	2	3	4	5	6	7	8
$4x$	0	4	8	12	16	20	24	28	32

(b), (c)

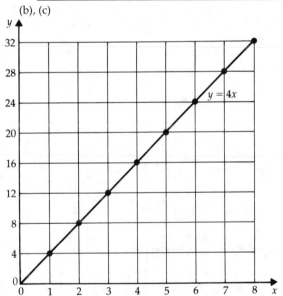

6 (a)

x	0	1	2	3	4	5	6	7	8
$2x$	0	2	4	6	8	10	12	14	16

(b)

x	0	1	2	3	4	5	6	7	8
$3x$	0	3	6	9	12	15	18	21	24

(c)

x	0	1	2	3	4	5	6	7	8
$\frac{1}{2}x$	0	0.5	1	1.5	2	2.5	3	3.5	4

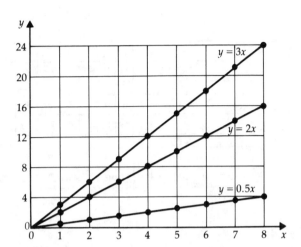

7 The graph of $y = 5x$ will pass through (0, 0) but will be steeper than the graphs in question **6**

8 (a)

x	0	1	2	3	4	5	6	7	8
$2x + 3$	3	5	7	9	11	13	15	17	19

(b), (c)

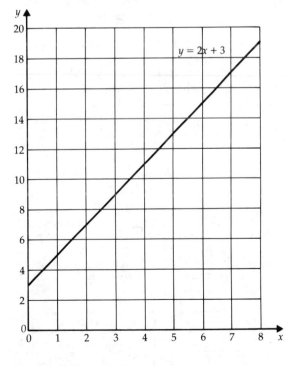

(d) (0, 3)

9 (a)

x	0	1	2	3	4	5	6	7	8
$2x + 5$	5	7	9	11	13	15	17	19	21

(b)

x	0	1	2	3	4	5	6	7	8
$3x + 2$	2	5	8	11	14	17	20	23	26

(c)

x	0	1	2	3	4	5	6	7	8
$4x - 1$	-1	3	7	11	15	19	23	27	31

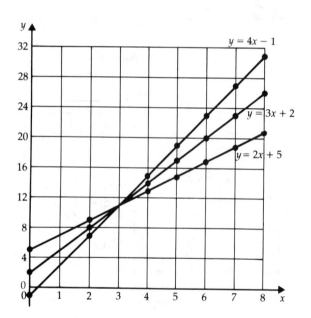

10 (a) yes; $y = 4x - 1$ is the steepest line
yes; in $y = 2x + 5$ the graph passes through $(0, 5)$
(b) $y = 3x + 5$ will be parallel to $y = 3x + 2$ and pass through $(0, 5)$
(c) The lines are all parallel but pass through different points on the y-axis i.e. $(0, 2)$, $(0, 4)$ and $(0, -2)$

11 (a)

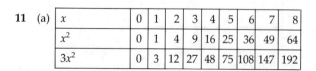

x	0	1	2	3	4	5	6	7	8
x^2	0	1	4	9	16	25	36	49	64
$3x^2$	0	3	12	27	48	75	108	147	192

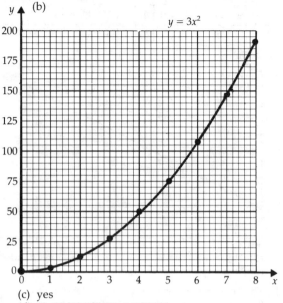

(b)

(c) yes
(d) (i) 61 (60.75) (ii) 5.8 (5.77)

12 (a)

x	0	1	2	3	4	5	6	7	8
x^2	0	1	4	9	16	25	36	49	64
$2x^2$	0	2	8	18	32	50	72	98	128

(b), (c)

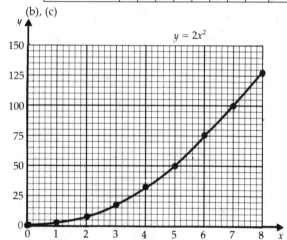

(d) Again the points lie on a smooth area but it is shallower than the curve for $y = 3x^2$
(e) (i) 41 (40.5) (ii) 7.1 (7.07)

13 (a)

x	0	1	2	3	4	5	6	7	8
x^2	0	1	4	9	16	25	36	49	64
$2x^2 + 1$	1	3	9	19	33	51	73	99	129

x	0	1	2	3	4	5	6	7	8
x^2	0	1	4	9	16	25	36	49	64
$3x^2 + 1$	1	4	13	28	49	76	109	148	193

(b), (c)

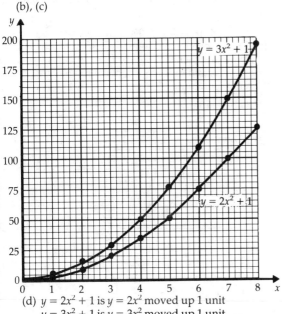

(d) $y = 2x^2 + 1$ is $y = 2x^2$ moved up 1 unit
$y = 3x^2 + 1$ is $y = 3x^2$ moved up 1 unit

(e) (i) 42 (41.5); 62 (61.75)   (ii) 7 (7.04); 5.7 (5.74)
The number multiplying the x^2 makes the graph steeper or shallower. Adding 1 moves the graph up 1 unit

14 (a)

x	-4	-3	-2	-1	0	1	2	3	4
x^2	16	9	4	1	0	1	4	9	16
x^2-3	13	6	1	-2	-3	-2	1	6	13

(b), (c)

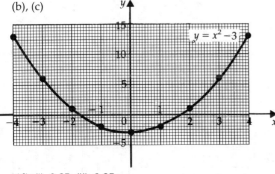

(d) (i) 3.25 (ii) 9.25
(e) $x = -1.73$ or $+1.73$

15 (a)

x	-4	-2	-1	-0.5	0.5	1	2	4
$1/x$	-0.25	-0.5	-1	-2	2	1	0.5	0.25

(b) This graph has two parts

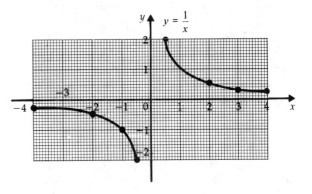

4.5   Gradients

1 (a)

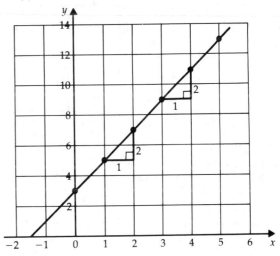

(b) (i) 5; 7 (ii) 9; 11
(c) yes
(d) yes

2 (a) (i) 4 (ii) 5
(b) If you are given the equation of the line as $y = mx + c$ the value of m is the gradient; **or** work out the value of y when $x = 1$ and when $x = 2$, the difference in these values of y is the gradient

3 (a) (i) 2 (ii) 3
(b) (i) 5 (ii) 4 (iii) 3 (iv) 1

4 (a) (i)   (ii)

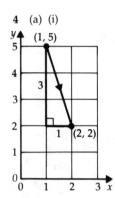
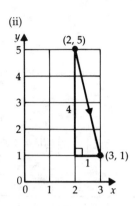

(iii)   (iv)

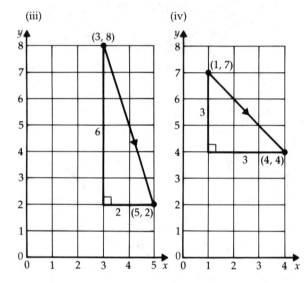

(b) (i) −3 (ii) −4 (iii) −3 (iv) −1

5 (a) (i) PQ; RS (ii) ST
(b) It is flat and has a zero gradient
(c) (i) 1 (ii) 1 (iii) −2

6 (a) 24 miles
(b) 2 hours
(c) 12 miles per hour
(d) 1½ hours
(e) 16 miles per hour

7 (a) 12; 0; −16
(b) The gradient corresponds to the speed
(c) A negative gradient indicates that the speed is in the opposite direction

8 (a) 4; 8; 0; −4; −8
(b) 4 m.p.h.; 8 m.p.h.; 0 m.p.h.; −4 m.p.h.; −8 m.p.h.
(c) between 7.30 p.m. and 8.00 p.m. and between 9.00 p.m. and 9.30 p.m.
(d) between 7.00 p.m. and 7.30 p.m. and between 8.30 p.m. and 9.00 p.m.
(e) between 8.00 p.m. and 8.30 p.m.

(f) He ran at 4 m.p.h. for half an hour, then at 8 m.p.h. for half an hour then he rested for half an hour. On his way back he ran at 4 m.p.h. for half an hour and finally at 8 m.p.h. for the last half an hour

9 (a) SP: 30 m.p.h.; PQ: 120 m.p.h.; QF: 30 m.p.h.
(b) The car accelerated from 0 m.p.h. to 120 m.p.h. in 1 minute, then travelled at 120 m.p.h. for half a minute, and then slowed down to a halt in another minute

4.6 Number patterns

1 (a)

Number	1	2	3	4	5	6	7	8
Result	5	7	9	11	13	15	17	19

(b) The results are all odd numbers; they go up in twos
(c) (i) 23 (ii) 43 (iii) 203 (iv) $2n + 3$

2 (a)

Number	1	2	3	4	5	6	7	8
Result	2	7	12	17	22	27	32	37

(b) The results all end in a 2 or a 7; they go up in fives
(c) (i) 47 (ii) 97 (iii) 497 (iv) $5n − 3$

3 (a)

Number	1	2	3	4	5	6	7	8
Result	15	18	21	24	27	30	33	36

(b) The results are all multiples of three; they go up in threes
(c) (i) 42 (ii) 72 (iii) 312 (iv) $(n + 4) \times 3$ or $3(n + 4)$

4 (a) (i) yes (ii) yes (iii) yes

5 Multiply the number by four and then add seven

6 (a) (i)   (ii)   (iii)   (iv)

(b) (i) 12; 18; 24; 36 (ii) 20; 30; 40; 100
(iii) 200; 300; 400; 10000 (iv) $2n$; $3n$; $4n$; n^2

(c) (i) Double the number
(ii) Multiply the number by three
(iii) Multiply the number by four
(iv) Multiply the number by itself

7

Number n		1	2	3	4	5	6	7
Result $3n − 2$		1	4	7	10	13	16	19

8

Number n		1	2	3	4	5	6	7
Result $5(n + 1)$		10	15	20	25	30	35	40

9 Multiply the number by five and then add five

10 (i) (a) 18 (b) 20 (c) 40 (d) $2n$
(ii) (a) 19 (b) 21 (c) 41 (d) $2n + 1$
(iii) (a) 27 (b) 30 (c) 60 (d) $3n$
(iv) (a) 28 (b) 31 (c) 61 (d) $3n + 1$
(v) (a) 81 (b) 100 (c) 400 (d) $n \times n$ or n^2
(vi) (a) 82 (b) 101 (c) 401 (d) $n^2 + 1$

11 (a) (i) 30 (ii) 42 (iii) 110
(b) (i) The 20th shape will have 20 dots on one side
and 21 dots on the other
(i.e. $20 \times 21 = 420$ dots in all)
(ii) The nth shape will be a rectangle with n dots
on one side and one more dot on the other
side (i.e. $n \times (n + 1)$ dots in all)
(c) $n \times (n + 1)$ (or $n^2 + n$)

12 (a) (i) 16 (ii) 32 (iii) 512 or 2^9
(b) You keep on doubling the number of rectangles
(i) $2 \times 2 \times 2 \times \ldots \times 2$ or 2^{19}
$\leftarrow$ 19 2's $\rightarrow$
(ii) $2 \times 2 \times 2 \times \ldots \times 2$ or $2^{(n-1)}$
$\leftarrow (n-1)$ 2's $\rightarrow$

13 (a) $2^9 = 512$ (b) 2^{63}

14 (a) 1, 3, 5, 7, 9, 11, 13, 15
(b) (i) 19 (ii) $2n - 1$
(c) (i) 4 (ii) 9 (iii) n^2 or $n \times n$

15 (a) 1, 2, 3, 4, 5, 6, 7, 8
(i) 10 (ii) n
(i) 3 (ii) 6 (iii) $\dfrac{n \times (n + 1)}{2}$
(b) 0, 1, 2, 3, 4, 5, 6, 7
(i) 9 (ii) $n - 1$
(i) 1 (ii) 3 (iii) $\dfrac{(n - 1) \times n}{2}$

16 (a) 2; 1
(b) 4; 3

(c)

Number of folds	Number of small rectangles	Number of fold lines
1	2	1
2	4	3
3	8	7
4	16	15
5	$32 = 2^5$	$31 = 2^5 - 1$
(d) 20	$2^{20} = 1\,048\,576$	$2^{20} - 1 = 1\,048\,575$

4.7 Use of letters and formulae

1 (a) $p + p + p = 3p$ (b) $a + a + a + a + a = 5a$

2 (a) $3a$ (b) $5p$ (c) $8a$ (d) $11p$ (e) $6a$ (f) $15p$

3 (a) $l + w + l + 2w = 2l + 3w$
(b) $p + r + q + p + r + q = 2p + 2r + 2q$

4 (a) $2a + 2b$ (b) $2p + 3q$ (c) $3a + 2b$
(d) $10p + 4q$ (e) $4a + 2b$ (f) $10p + 7q$

5 (a) p (b) $3a$ (c) $5p$ (d) $2a - 3b$
(e) $4p + 3q$ (f) $3a + 3b$

6 (a) area is length $\times$ length, or length squared
(b) $A = l \times l$, or l^2

7 (a) volume is length $\times$ length $\times$ length, or length
cubed
(b) $V = l \times l \times l$, or l^3

8 (a) volume is length $\times$ width $\times$ height
(b) $V = l \times w \times h$, or lwh

9 (a) area is half base $\times$ height
(b) $A = \frac{1}{2}b \times h$ or $\frac{1}{2}bh$

10 (a) (i) 49 (ii) 144 (iii) 6.25 (iv) 9.61
(b) You find the square root of A
(c) (i) 6 (ii) 13 (iii) 21 (iv) 1.2

11 (a) (i) 35 (ii) 96 (iii) 14 (iv) 40
(b) You divide A by l
(c) (i) 8 (ii) 12

12 (a) (i) 343 (ii) 1728 (iii) 15.625 (iv) 29.791
(b) (i) 2 (ii) 3

13 (a) (i) 70 (ii) 480 (iii) 30 (iv) 156.8
(b) (i) 2 (ii) 0.5

14 (a) (i) 20 (ii) 48 (iii) 7 (iv) 20
(b) (i) 8 (ii) 12

15 (a) (i) 60 m.p.h. (ii) 51 m.p.h. (iii) 80 m.p.h.
(b) You divide the distance by the time
(c) (i) average speed is
distance travelled $\div$ time taken
(ii) $S = d \div t$ or $\dfrac{d}{t}$

16 (a) $d = s \times t$ or st
(b) (i) 240 miles (ii) 154 miles (iii) 108 miles

17 (a) (i) £23 (ii) £44 (iii) £72.20
(b) $T = n \times £0.15 + £8$
(c) 200 units

18 (a) (i) £540 (ii) £945 (iii) £1640.25
(b) $T = 27\%$ of £$(G - 2425)$ or £$[\frac{27}{100} \times (G - 2425)]$
(c) £3425

4.8 Formulae from tables

1 (a) (3,6),(6,12)
(b) 20
(c) 200
(d) $y = 2x$

2 (a) (2,5),(6,9)
(b) 13
(c) 103
(d) $y = x + 3$

3 (a) (1,5),(5,13),(8,19)
(b) 23
(c) 203
(d) $y = 2x + 3$

4 (a) (3,16),(7,36)
(b) 51
(c) 501
(d) $y = 5x + 1$

5 (a) (3,36),(5,60)
(b) You multiply the number of bars by 12
(c) £1.20
(d) $C = 12N$

6 (a) 13
(b) You divide the total cost by 12
(c) $N = C \div 12$

7 (a) (3,165),(5,275)
(b) You multiply the time in hours by 55
(c) 550 miles
(d) $D = 55T$

8 (a) 12 hours
(b) You divide the number of miles by 55
(c) $T = D \div 55$

9 (a)

Number of shape	1	2	3	4	5	6
Perimeter of shape	4	6	8	10	12	14

(b) You double the number of the shape and add 2
(c) 22
(d) $P = 2N + 2$

10 (a)

Number of shape	1	2	3	4	5	6
Perimeter of shape	6	8	10	12	14	16

(b) You double the number of the shape and add 4
(c) 24
(d) $P = 2N + 4$

11 (a) 175p
(b) 375p
(c) $C = 10N + 75$
(d) 575p

12 $C = 50N + 100$;

N	1	2	3	4	5	6	7	8	9	10
C	150	200	250	300	350	400	450	500	550	600

13 (a)

Number of pattern	1	2	3	4	5	6
Number of dots		2	5	10	17	26	37

(b) You square the number of the pattern and then add 1
(c) 101
(d) $N = P \times P + 1$ or $N = P^2 + 1$

14 (a) (1,8),(4,17),(7,26)
(b) 35
(c) 65
(d) You multiply x by 3 and add 5
(e) $y = 3x + 5$

15 (a) $C = 5N + 350$

Number of units	100	200	300	400	500	600
Cost in £	8.50	13.50	18.50	23.50	28.50	33.50

(b) £41

4.9 Linear equations

1 8; you subtract 4 from 12

2 15; you add 7 to 8

3 8; you divide 24 by 3

4 36; you multiply 9 by 4

5 (a) 2 (b) 5 (c) 34 (d) 4 (e) 14 (f) 72

6 (a) 6 (b) 9 (c) 7 (d) 4 (e) 120 (f) 360

7 (a) $x = 6$ (b) $y = 15$ (c) $a = 16$ (d) $x = 7$
(e) $b = 5$ (f) $p = 68$

8 (a) $x = 2$ (b) $y = 5$ (c) $p = 6$ (d) $q = 30$
(e) $s = 52$ (f) $t = 6$

9 10; you subtract 5 and then halve the result

10 11

12 (a) 10 (b) 3 (c) 7 (d) 7 (e) 7 (f) 4

13 (a) $2n + 9 = 17$ (b) 4

14 (a) $3n + 5 = 17$ (b) 4

15 (a) $4n - 5 = 19$ (b) 6

16 (a) $x = 15$ (b) $p = 7$ (c) $a = 8$ (d) $y = 7$
(e) $b = 4.5$ (f) $q = 7.5$

17 (a) (i) $p + 130 = 180$ (ii) $67 + q + 43 = 180$
(b) (i) $p = 50$ (ii) $q = 70$

18 (a) (i) $y + 72 + 48 = 180$ (ii) $x + x + 70 = 180$
or $2x + 70 = 180$
(b) (i) $y = 60$ (ii) $x = 55$

19 (a) (i) $l + l + 7 + 7 = 28$ or $2l + 14 = 28$
(ii) $w + w + 9 + 9 = 28$ or $2w + 18 = 28$
(b) (i) $l = 7$ (ii) $w = 5$

20 (a) (i) $x + 12 + 15 = 36$
(ii) $p + p + 8 = 36$ or $2p + 8 = 36$
(b) (i) $x = 9$ (ii) $p = 14$

21 (a) $(a + 5)$ years (b) $a + (a + 5) = 31$ (c) 13; 18

22 (a) $(a - 3)$ years (b) $a + (a - 3) = 25$ (c) 14; 11

23 (a) $(w + 3)$cm (b) $w + w + (w + 3) + (w + 3) = 42$
(c) 12 cm; 9 cm

24 (a) £$(2c + c + c + c)$ or £$5c$ (b) $5c = 3.50$
(c) 70p; £1.40

25 (a) $n + (n + 1) = 93$ (b) 46; 47

26 (a) $n + (n + 1) + (n + 2) = 93$ (b) 30; 31; 32

27 24; 21; 19

28 (a) 30°, 60°, 90° (b) right-angled

29 (a) 36°, 72°, 108°, 144°

4.10 Linear inequalities

1 11

2 6

3 7, 8

4 5, 6

5   4, 5, 6, 7, 8, 9

6   (a) 3, 6, 9, 12
    (b) 10, 11, 12
    (c) 0, 1, 2, 3, 4
    (d) 7, 8
    (e) 4, 5, 6, 7

7   (a) 1
    (b) 1, 2, 3, 4
    (c) 8, 9, 10
    (d) 1
    (e) 1, 2, 3, 4
    (f) 7, 8, 9, 10

8   (a) 10, 11
    (b) 10, 11, 12, 13, 14
    (c) 18, 19, 20
    (d) 10, 11
    (e) 10, 11, 12, 13, 14
    (f) 17, 18, 19, 20

9   (a) 1
    (b) 9, 10
    (c) 5, 6, 7, 8, 9, 10
    (d) 9, 10
    (e) 1, 2, 3, 4
    (f) 8, 9, 10

10 (a) 7, 8, 9, 10
    (b) 1, 2, 3, 4
    (c) 8, 9, 10
    (d) 8, 9, 10
    (e) 1, 2, 3, 4
    (f) 10

11 (a) 1, 2, 3
    (b) 8, 9, 10
    (c) 7, 8, 9, 10
    (d) 1, 2, 3, 4
    (e) 7, 8, 9, 10
    (f) 1, 2, 3, 4, 5, 6, 7, 8

12 (a) $x < 4$
    (b) $x > 8$
    (c) $x < 8$

13 (a) $x > 14$
    (b) $x < 22$
    (c) $x > 6$

14 (a) $x < 10$
    (b) $x > 3$
    (c) $x < 7$
    (d) $x > 7$
    (e) $x < 7$
    (f) $x > 4$

15 (a) $2n + 9 < 17$
    (b) $n < 4$

16 (a) $3n + 5 > 17$
    (b) $n > 4$

17 (a) $4n - 5 < 19$
    (b) $n < 6$

18 (a) $x < 9$
    (b) $p > 7$
    (c) $a < 8$
    (d) $y > 7$
    (e) $b < 4.5$
    (f) $q > 7.5$

19 -3

20 -3

21 -5

22 0

23 3

24 -2

25 $-1, 0, 1, 2$

26 $-4, -3$

27 (a) $-7, -6, -5, -4, -3, -2, -1$
    (b) $-7, -6, -5, -4, -3$
    (c) $-4, -3, -2, -1, 0, 1, 2, 3, 4, 5, 6, 7, 8$
    (d) $-3, -2, -1, 0, 1, 2$
    (e) $-5, -4, -3$
    (f) -4

28 (a) $-5, -4, -3, -2, -1, 0, 1$
    (b) $-5, -4, -3$
    (c) $-2, -1, 0, 1, 2, 3, 4, 5$
    (d) $-4, -3, -2, -1, 0, 1, 2, 3, 4, 5$
    (e) $-5, -4, -3, -2, -1, 0, 1, 2, 3$
    (f) $-2, -1, 0, 1, 2, 3, 4, 5$

29 (a) $-5, -4, -3, -2, -1, 0, 1$
    (b) $-2, -1, 0, 1, 2, 3, 4, 5$
    (c) $-5, -4, -3, -2, -1, 0, 1$
    (d) $-5, -4, -3, -2, -1, 0, 1$
    (e) $-5, -4, -3, -2, -1, 0, 1$
    (f) $-2, -1, 0, 1, 2, 3, 4, 5$

30 (a) $x < -2$
    (b) $x > -3$
    (c) $x < -2$

31 (a) $x > 3$
    (b) $x < -3$
    (c) $x > -3$

32 (a) $x < -3$
    (b) $x > -5$
    (c) $x < -2$
    (d) $x > -3$
    (e) $x < -2$
    (f) $x > -2$

33 (a) $2n + 7 < 3$
    (b) $n < -2$

34 (a) $3n + 4 > -2$
    (b) $n > -2$

35 1, 2, 3, 4 ; $n^2 < 20$

36 7, 8 ; $40 < n^2 < 80$

37 Up to 71 ; $7n \leqslant 500$

38 22; $30n \geqslant 640$

39 Between 7 and 25 ; $4n \geqslant 25$ and $n \leqslant 25$

4.11   Brackets and factors

1 (a) £13.96
(c) $2 \times £4.99 + 2 \times £1.99$ or $2 \times £(4.99 + 1.99)$

2 (a) (i) £14.97 (ii) £5.97
(b) £20.94   (c) £6.98   (d) £20.94

3 (a) $15 + 18 = 33; 3 \times 11 = 33$; yes
(b) $8 + 10 + 12 = 30; 2 \times (15) = 30$
(c) yes; the second

4 (a) £34.96   (b) £52.44

5 (a) $2p + 2q$ (b) $3x + 3y$ (c) $4r + 4s$
(d) $2x - 2y$ (e) $5a - 5b$ (f) $7p - 7q$

6 (a) $2p + 2q + 2r$ (b) $3x + 3y + 3z$ (c) $4r + 4s + 4t$
(d) $2x + 2y - 2z$ (e) $5a - 5b + 5c$ (f) $7p - 7q - 7r$

7 (a) $112 + 168 = 280 = 7 \times (40)$
(b) $234 + 66 = 300 = 3 \times (100)$

8 (a) 280   (b) 300   (c) 400   (d) 20

9 (a) $6 \times (47 + 13) = 6 \times 60 = 360$
(b) $3 \times (12 + 13 + 15) = 3 \times 40 = 120$
(c) $7 \times (2.4 + 1.6) = 7 \times 4 = 28$
(d) $4 \times (53 + 21 + 26) = 40 \times 100 = 400$

10 (a) $3(p + q)$ (b) $5(a + b)$ (c) $7(x + y + z)$
(d) $3(a - b)$ (e) $6(x - y)$ (f) $4(p + q - r)$

11 (a) $p(x + y)$ (b) $a(x + y)$ (c) $s(x + y + z)$
(d) $x(a - b)$ (e) $p(x - y)$ (f) $k(p + q - r)$

12 (a) $8x + 2y$ (b) $3p + 21q$ (c) $8a + 12b + 20c$
(d) $15a - 5b$ (e) $7x - 28y$ (f) $10p + 35q - 40r$

13 (a) $p(2x + y)$ (b) $3a(x + y)$ (c) $2s(2x + y)$
(d) $x(a - 2b)$ (e) $4p(x - y)$ (f) $3k(p + 2q)$

14 (a) $p(2x + y + 3z)$ (b) $3a(x + y + z)$
(c) $s(4x + 2y - z)$ (d) $3k(3x + 2y + z)$
(e) $2x(2a - b + 4c)$ (f) $2p(2x - 2y + 3z)$

15 (a) (i) $49 \, \text{cm}^2$ (ii) $p \times p \, \text{cm}^2$ or $p^2 \text{cm}^2$
(iii) $(2x) \times (2x) \, \text{cm}^2$ or $4x^2 \text{cm}^2$
(iv) $(3y) \times (3y) \, \text{cm}^2$ or $9y^2 \text{cm}^2$
(b) (i) $343 \, \text{cm}^3$ (ii) $p \times p \times p \, \text{cm}^3$ or $p^3 \text{cm}^3$
(iii) $(2x) \times (2x) \times (2x) \, \text{cm}^3$ or $8x^3 \text{cm}^3$
(iv) $(3y) \times (3y) \times (3y) \, \text{cm}^3$ or $27y^3 \text{cm}^3$

16 (a) 5^2 (b) 6^3 (c) 3^4 (d) 7^5 (e) n^2 (f) w^3
(g) y^4 (h) z^5

17 (a) 25   (b) 64   (c) 81   (d) 32   (e) 144

18 (a) p^5 (b) y^9 (c) $9p^2$ (d) $8y^3$

19 (a) $10p^2$ (b) $12y^2$ (c) $24n^3$

20 (a) $2xy + 6xz$ (b) $2p^3 + p^2q$ (c) $6y^4 + 18y^2z$

21 (a) $p^2(b + c)$ (b) $p^2(2b + 3c)$ (c) $p^2(p + y)$

22 (a) $p^2(2x + 3y + 4z)$ (b) $p^2(2b + 3c + p)$

23 (a) $p^2(b^2 + pc^2)$
(b) $pb(2p^2 + 3pb + b^2) = pb(2p + b)(p + b)$

24 (a)

20	8
15	6

(b)

ac	ad
bc	bd

25 (a)

x^2	$3x$
$2x$	6

(b)

y^2	$5y$
$4y$	20

26 (a) yes; yes   (b) $x^2 + 5x + 6; y^2 + 9y + 20$

27 (a) $2x^2 + 15x + 18$ (b) $6y^2 + 37y + 45$

4.12   Transformation of formulae

1 (a) £85   (b) £134

2 (a) $£(h - 25)$ (b) $£(w + 25)$

3 (a) yes   (b) yes

4 (a) yes   (b) yes

5 (a) $h = w - 4$ (b) $h = w + 7$
(c) $h = p - q$ (d) $h = \dfrac{w}{3}$
(e) $h = 6w$ (f) $h = \dfrac{A}{b}$

6 (a) Divide the area by the length
(b) $w = \dfrac{A}{l}$
(c) 8 cm

7 (a) Multiply the speed by the time
(b) $d = s \times t$
(c) 156 miles

8 (a) Divide the volume by the length and the width
(b) $h = \dfrac{V}{lw}$
(c) 3 cm

9 (a) $p = \dfrac{(h - 2)}{5}$ (b) $p = \dfrac{(h - 3)}{4}$ (c) $p = \dfrac{(w - b)}{5}$
(d) $p = \dfrac{(h + 3)}{5}$ (e) $p = \dfrac{(w + 4)}{7}$ (f) $p = \dfrac{(k + a)}{2}$

10 (a) $x = \dfrac{(y - 2)}{3}$ (b) $x = \dfrac{(y + 3)}{4}$ (c) $x = \dfrac{y}{5} - 2$
(d) $x = \dfrac{(w - u)}{5}$ (e) $x = \dfrac{(t + s)}{7}$ (f) $x = \dfrac{k}{2} + 3$

11 (a) (i) £1.95 (ii) £2.67
(b) yes
(c) $n = \dfrac{(C - 75)}{8}$
(d) (i) 25 (ii) 10 (iii) 20

12 (a) (i) £11 (ii) £15 (iii) £29
(b) $c = n \times 2 + 5$ or $c = 2n + 5$
(c) (i) 2 (ii) 4 (iii) 14
(d) $n = \dfrac{(c - 5)}{2}$

13 (a) (i) £22 (ii) £31.80 (iii) £$(p + 1) \times 4$
(b) $B = 4 \times (p + 1)$ **or** $B = 4(p + 1)$
(c) (i) £5 (ii) £3.50 (iii) £7.95
(d) $p = \dfrac{B}{4} - 1$

14 (a) (i) £$(5 \times n)$ (ii) £$(7.50 \times n)$ (iii) £$(p + 1) \times n$
(b) $B = n \times (p + 1)$ **or** $B = n(p + 1) = np + n$
(c) (i) $n = \dfrac{B}{(p + 1)}$ (ii) $p = \dfrac{B}{n} - 1$

15 (a) (i) 10°C (ii) 20°C (iii) 30°C (iv) 0°C
(b) Multiply the Centigrade temperature by $\frac{9}{5}$ and then add 32
(c) 41°F

16 (a) 41°F (b) 50°F (c) 77°F (d) 86°F (e) 32°F

17 (a) $x \to$ ☐$\times m$ $\to$ ☐$+c$ $\to y$ (b) 43 (c) $x = \dfrac{(y - c)}{m}$

18 (a) (i) 19 (ii) 43 (b) $u = v - at$ (c) $a = \dfrac{(v - u)}{t}$

4.13 Flow charts

1 (a) It's the same as the one you started with
(b) as part (a)
(c) Yes

2 (a) The results are the same (e.g. $10 \to 101$)
(b) Again the results are the same (e.g. $8 \to 81$)
(c) As in parts (a) and (b)
(d) The number is multiplied by 10 and 1 is added

3 (a) 19
(b) They are the odd numbers
(c) Change the first instruction to:

> Write down 2 on a piece of paper

4 (a) 10
(b) They are the powers of 2
(c) Change the second instruction to:

> Multiply the number you have just written down by 3

5 (a)

> Write down **right-angled**

(b) equilateral; isosceles; scalene; scalene right-angled

6 (a) 1, 3; 2, 4, 6, 8, 12, 24; 27;
5, 7, 10, 11, 13, 14, 15, 16, 17, 18, 19, 20, 21, 22, 23, 25, 26, 28
(b) 1 and 3
(c) 5, 7, 10, 11, 13, 14, 15, 16, 17, 18, 19, 20, 21, 22, 23, 25, 26, 28

7 (a) square; rhombus; rectangle; parallelogram; none; kite; none; trapezium
(b) Fifth branch: none unless you include

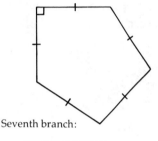

Seventh branch:

4.14 Bar charts and block graphs

1 (a) The chart shows the relative number of car sales by each of four companies in January.
(b) Ford: 8000 cars; Vauxhall: 6000 cars; Austin Rover: 5000 cars
(c) Vauxhall: 15 000 cars; Austin Rover: 12 500 cars; Peugeot Talbot: 5000 cars
(d) three times as many

Council expenditure in £1 millions

2

Schools	££££££££££££££££££££££££££
Police	££££££
Fire Services	£££
Other services	££££££££££££

3 (a) The amount each of three classes raised in a fund raising event
(b) Class 1 raised £500; class 2 raised £300; class 3 raised £400
(c) £1200
(d) Class 1: $\frac{5}{12}$; class 2: $\frac{3}{12}$ or $\frac{1}{4}$; class 3: $\frac{4}{12}$ or $\frac{1}{3}$

4 (a) The popularity of five common drinks
(b) Cocoa; 7
(c) Bovril; 1
(d) 20 people
(e) Bovril: $\frac{1}{20}$; Cocoa: $\frac{7}{20}$; Coffee: $\frac{4}{20}$ or $\frac{1}{5}$; Milk: $\frac{3}{20}$; Tea: $\frac{5}{20}$ or $\frac{1}{4}$

5 (a) The number of exam entries in each of five subjects
(b) English
(c) French

(d) 15

(e) 300

(f) English $\frac{90}{300}$ or $\frac{3}{10}$; French $\frac{30}{300}$ or $\frac{1}{10}$; History $\frac{60}{300}$ or $\frac{1}{5}$; Maths $\frac{75}{300}$ or $\frac{1}{4}$; Science $\frac{45}{300}$ or $\frac{3}{20}$

6 (a)

Vehicle	Frequency
Cars	17
Lorries	8
Buses	3
Bicycles	21

(b) 49

(c)

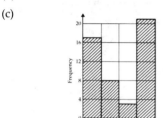

7 (a)

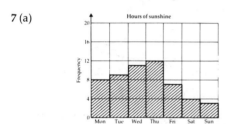

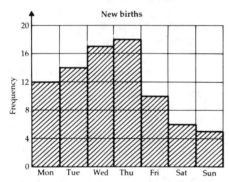

(b) In both graphs the greatest numbers occur on Wednesday and Thursday

(c) no

(d) Hospital staff want the weekends free!

8 (a) frequency

(b) a

(c), (d)

Length of word	1	2	3	4	5	6	7	8	9
Tally	ll	ʜʜ llll	ʜʜ ʜʜ l	ʜʜ ʜʜ ʜʜ ll	ʜʜ ll	ll	ll	ʜʜ	l
Frequency of word	2	9	11	17	7	2	2	5	1

(e)

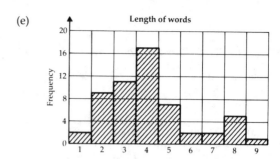

9 (a)

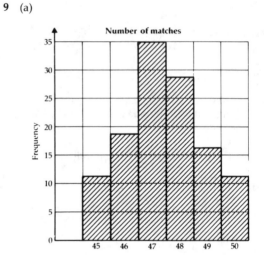

(b) 55 boxes

(c) $\frac{30}{120}$ or $\frac{1}{4}$

10 (a) The number of boys with various heights

(b) 40

(c) 160–169 cm

(d) It is somewhere between 190 and 199 cm

(e) 12

11 (a)

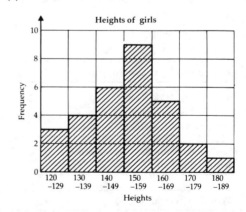

(b) It is somewhere between 180 and 189 cm

(c) 13

(d) The girls tend to be shorter, with the most common height in the range 150–159 cm rather than 160–169 cm

(e) 120–129; 150–159

4.15 Pie charts

1 (a) The number of cars sold by Ford and Vauxhall compared with other manufacturers

(b) (i) $\frac{1}{2}$ (ii) $\frac{1}{3}$ (iii) $\frac{1}{6}$

(c) (i) 1800 (ii) 1200 (iii) 600

(d) (i) 180° (ii) 120° (iii) 60°

2 (a) (i) $\frac{1}{3}$ (ii) $\frac{1}{4}$ (iii) $\frac{1}{6}$

(b) (i) 120° (ii) 90° (iii) 60°

(c)

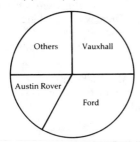

3 (a) 2 cars

(b)

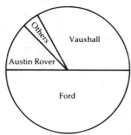

4 (a) The relative sizes of various crops

(b) 45°; 45°; 90°; 60°; 120°

(c) beans $\frac{1}{8}$; peas $\frac{1}{8}$; potatoes $\frac{1}{4}$; corn $\frac{1}{6}$; wheat $\frac{1}{3}$

(d) (i) 45; 45; 90; 60; 120

(ii) 90; 90; 180; 120; 240

(iii) 30; 30; 60; 40; 80

(iv) 112.5; 112.5; 225; 150; 300

5 (a) The relative amounts spent on various household expenses

(b) 90°; 60°; 90°; 75°; 45°

(c) $\frac{1}{4}$; $\frac{1}{6}$; $\frac{1}{4}$; $\frac{5}{24}$; $\frac{1}{8}$

(d) (i) £450; £300; £450; £375; £225

(ii) £300; £200; £300; £250; £150

(iii) £1200; £800; £1200; £1000; £600

6 (a) travel £1600; holidays £2400; food £2000; savings £1200

(b) rent £1440; travel £960; holidays £1440; savings £720

7 (a)

(b) 15°

(c) (i) 60° (ii) 75° (iii) 105°

8 (a)

Sleeping	120°
Eating	30°
Working	135°
Watching TV	30°

(b)

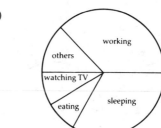

9 (a)

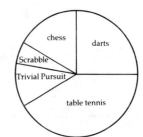

(b)

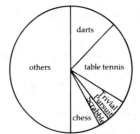

10 (a) 90°; 90°; 45°; 45°; 60°; 30°

(b) 480

(c) English 120; History 60; Geography 60; Science 80; French 40

11

12

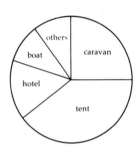

13 (a)   (b)

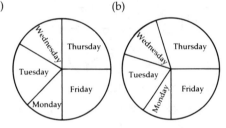

4.16 Information from tables

1 (a) £78
  (b) £57.50

2 (a)

New York	London	Moscow
7.00 p.m.	Midnight	3.00 a.m.
2.00 p.m.	7.00 p.m.	10.00 p.m.
11.00 p.m.	4.00 a.m.	7.00 a.m.

  (b) 7.00 a.m.
  (c) 3.00 p.m.
  (d) 6.00 a.m.

3 (a) 2 hours 31 minutes
  (b) 1728
  (c) 1200
  (d) The 1030; 4 hours 23 minutes

4 (a) (i) £10.02 (ii) 72p (iii) £8.35
  (b) (i) £8.65 (ii) £17.30 (iii) £1.00

5 (a) Self-catering; Bed and breakfast; Half-board
  (b) Starting on December 31st
  (c) £398
  (d) £960; on December 17th and January 14th

6 (a) Birmingham
  (b) Newcastle
  (c) Leeds
  (d) £56
  (e) 1815

7 (a) £7602
  (b) £8184
  (c) £9157

8 (a) (i) £6.09 (ii) £19.76 (iii) £12.83 (iv) £31.01
  (b) (i) £2000 (ii) £5000 (iii) £1500 (iv) £9000

4.17 Using statistics

1 (a) The profits were best between April and June, dropped between July and September and rose again between October and December
  (b) In each quarter the profits were better in 1987 than in 1986. There was a great increase in profits in the last quarter
  (c) April to June   (d) July to September
  (e) £6100   (f) 1500

2 (a) Eastenders   (b) That's Life
  (c) Five times as many people watched Eastenders.
  (d) 2500; 1000; 1000; 500

3 (a) 12 000   (b) 36 000   (c) Week 4
  (d) 24 000   (e) 42 000

4 (a) (i) 12.15 p.m. (ii) 11.00 a.m.
  (b) 12 miles
  (c) The cyclist from Coventry was travelling fastest between 10.30 a.m. and 11.30 a.m.
  The cyclist from Birmingham was travelling fastest between 11.45 a.m. and 12.15 p.m.

5 (a) 3.1 kg
  (b) $\frac{74}{21} \approx 3.52$ kg
  (c)

(histogram: Frequency against Weight (kg), with class intervals 2.5–2.9, 3.0–3.4, 3.5–3.9, 4.0–4.4, 4.5–5.0)

6 (a) £98 (b) £99 (c) £105

7 (a) (i) 50 (ii) 40
   (b) (i) 500 (ii) 400

8 (a) 3
   (b) 4 eggs
   (c) 121
   (d) 200
   (e) $\frac{121}{200}$
   (f) 3.025 eggs

9 (a) 200 (b) 1210

10 (b) $\frac{300}{1000} = \frac{3}{10}$; $\frac{250}{1000} = \frac{1}{4}$; $\frac{200}{1000} = \frac{1}{5}$; $\frac{100}{1000} = \frac{1}{10}$; $\frac{100}{1000} = \frac{1}{10}$; $\frac{50}{1000} = \frac{1}{20}$
   (c) yes; the numbers 1, 2, and 3 occur much more frequently than the numbers 4, 5 and 6

4.18 Mean, mode and median

1 (a) 3, 4, 5, 5, 5, 6, 7, 8, 11
   (b) 3; a flat
   (c) 11; an old Victorian house or a large detached house
   (d) 5; semi-detached houses on an estate

2 5 using the mode or 6 using the mean

3 5

4 5

5 6

6 7

7 (a) (i) 2 (ii) 2 (iii) 9 (iv) 4
   (b) (i) 2 (ii) 4 (iii) 8 (iv) 4
   (c) (i) 3 (ii) 4 (iii) 7 (iv) 4

8 (a) 127, 128, 133, 133, 136, 137, 142, 144, 153
   (b) (i) 127 cm (ii) 153 cm (iii) 136 cm

9 (a) 133 cm (b) 136 cm (c) 137 cm

10 (a)

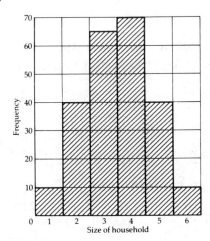

   (b) 4
   (c) 4

11 Three bedroomed houses; there are 70 households of size 4 but only 10 of size 1

12 (a)

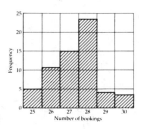

   (b) 28
   (c) 27

13 28 since this is the most frequent number of bookings

14 (a)

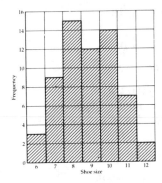

   (b) size 8
   (c) size 9
   (d) size 8

15 (a) Each graph shows the number of eggs laid on each day of the week
   (b) (i) 40; 40 (ii) 40; 40
   (c) 295; 295
   (d) £11.80; £11.80

16 (a) £50
   (b) £2.50

17 (a) 2
   (b) 5
   (c) The most likely number would be 2 goals

18 (a) 2 goals
   (b) 2 goals

19 (a) £100
   (b) £150
   (c) The second, since 10 × £150 = £1500

20 (a) 6; 7; 7.14
   (b) 8; 6.5; 6.17

4.19 Probability: equally likely events

1 (a) certain
  (b) impossible
  (c) certain
  (d) impossible

2

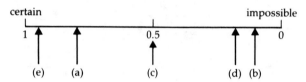

3 (a) 1; 0; 1; 0
  (b) 0.8; 0.1; 0.5; 0.17; 0.95

4 (a) a red bean
  (b) yes
  (c) 3 out of 10 i.e. 0.3

5 (a) 7
  (b) 7 out of 10 i.e. 0.7

6 (a) a red bean
  (b) a white bean
  (c) (i)   7 out of 16 i.e. 0.4375
      (ii) 3 out of 16 i.e. 0.1875
      (iii) 6 out of 16 i.e. 0.375

7 (a) 13
  (b) 13 out of 16 i.e. 0.8125
  (c) 9
  (d) 9 out of 16 i.e. 0.5625

8 (a) a girl
  (b) 3 out of 8 i.e. 0.375

9 (a) neither; each is equally likely
  (b) 4 out of 8 i.e. 0.5

10 (a) a number less than 4
   (b) (i) 2 out of 6 i.e. 0.33 (ii) 3 out of 6 i.e. 0.5

11 (a) neither; each is equally likely
   (b) (i) 1 out of 6 i.e. 0.17 (ii) 1 out of 6 i.e. 0.17

12 (a) an even number
   (b) (i)   3 out of 6 i.e. 0.5
       (ii) 3 out of 6 i.e. 0.5
       (iii) 2 out of 6 i.e. 0.33

13 (a) a 1
   (b) (i)   3 out of 6 i.e. 0.5
       (ii) 2 out of 6 i.e. 0.33
       (iii) 1 out of 6 i.e. 0.17

14 (a) an odd number
   (b) (i) 2 out of 6 i.e. 0.33
       (ii) 4 out of 6 i.e. 0.67

15 (a) $\frac{1}{8}$ (b) $\frac{3}{8}$ (c) $\frac{4}{8}$ (d) $\frac{2}{8}$ (e) $\frac{2}{8}$

16 (a) $\frac{1}{8}$ (b) $\frac{7}{8}$ (c) $\frac{3}{8}$

4.20 Probability: listing outcomes

1 (a) (i) 13 (ii) 26 (iii) 4
  (b) (i) $\frac{13}{52} = \frac{1}{4}$ (ii) $\frac{26}{52} = \frac{1}{2}$ (iii) $\frac{4}{52} = \frac{1}{13}$

2 (a) 13 (b) 3 (c) $\frac{3}{13}$

3 (a) 26 (b) 6 (c) $\frac{6}{26} = \frac{3}{13}$

4 (a) 12 (b) 2 (c) (i) $\frac{12}{52} = \frac{3}{13}$ (ii) $\frac{2}{52} = \frac{1}{26}$

5 (a) Monday, Tuesday, Wednesday, Thursday, Friday, Saturday, Sunday
  (b) 2
  (c) (i) $\frac{2}{7}$ (ii) $\frac{2}{7}$ (iii) $\frac{1}{7}$ (iv) $\frac{0}{7} = 0$
  (d) (i) $\frac{5}{7}$ (ii) $\frac{5}{7}$ (iii) $\frac{6}{7}$ (iv) $\frac{7}{7} = 1$

6 (a) 3
  (b) (i) $\frac{3}{12} = \frac{1}{4}$ (ii) $\frac{2}{12} = \frac{1}{6}$ (iii) $\frac{2}{12} = \frac{1}{6}$ (iv) $\frac{0}{12} = 0$
  (c) (i) $\frac{9}{12} = \frac{3}{4}$ (ii) $\frac{10}{12} = \frac{5}{6}$ (iii) $\frac{10}{12} = \frac{5}{6}$ (iv) $\frac{12}{12} = 1$

7 (a) 12
  (b) 4
  (c) (i) $\frac{4}{12} = \frac{1}{3}$ (ii) $\frac{6}{12} = \frac{1}{2}$ (iii) $\frac{3}{12} = \frac{1}{4}$ (iv) $\frac{3}{12} = \frac{1}{4}$

8 (a) 25
  (b) (i) $\frac{20}{25} = \frac{4}{5}$ (ii) $\frac{4}{25}$ (iii) $\frac{0}{25} = 0$

9 (a) 200 (b) 400

10 (a) 200 (b) 300

11 (a) $\frac{5}{100} = \frac{1}{20}$ (b) $\frac{95}{100} = \frac{19}{20}$

12 (a) 5 (b) 25 (c) 100 (d) 240

13 (a) $\frac{390}{1000} = \frac{39}{100}$ (b) $\frac{320}{1000} = \frac{8}{25}$ (c) $\frac{280}{1000} = \frac{7}{25}$

14 (a) 3900 (b) 3200 (c) 2800 (d) 100

15 (a) $\frac{12}{20} = \frac{3}{5}$ (b) $\frac{8}{20} = \frac{2}{5}$ (c) $\frac{4}{20} = \frac{1}{5}$ (d) $\frac{8}{20} = \frac{2}{5}$
   (e) $\frac{6}{20} = \frac{3}{10}$ (f) $\frac{2}{20} = \frac{1}{10}$

16 (a) 160
   (b) (i) $\frac{40}{160} = \frac{1}{4}$ (ii) $\frac{50}{160} = \frac{5}{16}$ (iii) $\frac{25}{160} = \frac{5}{32}$ (iv) $\frac{45}{160} = \frac{9}{32}$
   (c) (i) 2000; 2500; 1250; 2250

4.21 Probability: combined events

1 (a) head, head; head, tail; tail, head; tail, tail
  (b) 2
  (c) (i) $\frac{2}{4} = \frac{1}{2}$ (ii) $\frac{1}{4}$

2 (a)

2p	h	h	h	h	t	t	t	t
5p	h	h	t	t	h	h	t	t
10p	h	t	h	t	h	t	h	t

  (b) (i) 3 (ii) 3
  (c) (i) $\frac{3}{8}$ (ii) $\frac{3}{8}$ (iii) $\frac{3}{8}$ (iv) $\frac{1}{8}$

3 (a) (1, H), (2, H), (3, H), (4, H), (5, H), (6, H)
      (1, T), (2, T), (3, T), (4, T), (5, T), (6, T)
      12; Yes
  (b) 2
  (c) (i) $\frac{1}{12}$ (ii) $\frac{1}{12}$ (iii) $\frac{3}{12} = \frac{1}{4}$

4 (a)

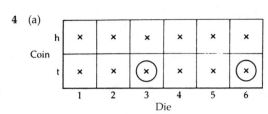

(b) $\frac{2}{12} = \frac{1}{6}$

5 (a) 36

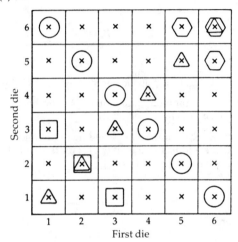

(i) ◯   total score of 7

(ii) ▢   a total score of 4

(iii) △   doubles

(iv) ⬡   a total score more than 10

6 (a) $\frac{6}{36} = \frac{1}{6}$ (b) $\frac{3}{36} = \frac{1}{12}$ (c) $\frac{6}{36} = \frac{1}{6}$ (d) $\frac{3}{36} = \frac{1}{12}$

7

(a) ◯   at least one 6

    ▢   an even number on each die.

(c) △   one 6 and an odd number

(d) all x's without a ◯ on

8 (a) $\frac{11}{36}$ (b) $\frac{9}{36} = \frac{1}{4}$ (c) $\frac{6}{36} = \frac{1}{6}$ (d) $\frac{25}{36}$

9 (a) yes; $\frac{1}{2} \times \frac{1}{2} = \frac{1}{4}$ (b) yes; $\frac{1}{2} \times \frac{1}{2} = \frac{1}{4}$

10 (a) (H, T) or (T, H) so $\frac{1}{4} + \frac{1}{4} = \frac{1}{2}$

     (b) The four outcomes represent all the possible events

11 (a) (i) $\frac{1}{6} \times \frac{1}{6} = \frac{1}{36}$ (ii) $\frac{1}{6} \times \frac{5}{6} + \frac{1}{6} \times \frac{5}{6} = \frac{10}{36}$ (iii) $\frac{5}{6} \times \frac{5}{6} = \frac{25}{36}$

     (b) (i) one possibility (ii) ten possibilities

         (iii) twenty-five possibilities

12 $\frac{1}{36} + \frac{10}{36} + \frac{25}{36} = \frac{36}{36} = 1$

These represent the 36 possible outcomes

13

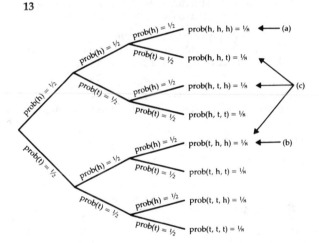

(c) $\frac{1}{8} + \frac{1}{8} + \frac{1}{8} = \frac{3}{8}$

14 (a) $\frac{1}{6} \times \frac{1}{6} \times \frac{1}{6} = \frac{1}{216}$

     (b) $\frac{1}{6} \times \frac{5}{6} \times \frac{5}{6} + \frac{5}{6} \times \frac{1}{6} \times \frac{5}{6} + \frac{5}{6} \times \frac{5}{6} \times \frac{1}{6} = \frac{75}{216}$

4.22 Problems and investigations: some hints

1 Look for routes using 3 cm, 4 cm and 5 cm once each.
Go from A to E then to G, Go from A to B then to G etc.
Don't forget you can come back to A providing you don't go over the same edge a second time.
Record each route. Be systematic.

2 Being systematic is vital for this. Start with r as first letter and find possibilities. Then use b as first letter etc.
With four colours remember you can fix the first and then rearrange the other three.

3 How many ways can you rearrange 2, 3 or 4 letters?

4 Be systematic! Use 1 number, then 2 numbers e.g. 3 + 1, 1 + 3 before moving on to other numbers. Are the answers powers of 2?

5 Be systematic! Pascal's triangle?

6 Be systematic! Change the position of the u. Keep one u fixed then move the other u.

7 Be systematic! Try going right as far as you can before going up. Then only go up one step before going right again if you can. Record your routes. Compare your answers with question 6. Pascal's triangle.

8 Powers of 2? Triangle numbers? Look for primes, multiples of 3, multiples of 5 etc.

9 How are these results related to Pascal's triangle? See if you can explain the result for $(11)^5$.

10 How many ways are there of getting (a) two heads (b) two tails (c) a head and a tail when you toss two coins? (Use a 2p and a 5p coin.)

11 How many ways are there of getting (a) three heads (b) two heads and a tail, etc when you toss three coins? (Use a 2p, 5p and 10p coin.)

12 How many ways are there of getting (a) two sixes (b) one six (c) no sixes when rolling two dice. (Use two different coloured dice.)

13 All up to 1000 finish at 1.

14 Some finish at 1; others do not.

15 4.30 a.m.

16 £60

17 £3.40

18 $3(x + y)$

19 $8(a - b)$

20 $8x + 2y$

21 $3x + 4y$

22 6

23 5

24 4

25 $-3, -2, -1, 0, 1, 2$

26 Find their sum and divide this by three; 10

27 The most common number in the set, or the number which occurs most frequently

28 Arrange the numbers in ascending order, and then take the middle number in the list (or the average of the two middle numbers)

29 7

30 20

31 $\frac{1}{3}$ (or $\frac{2}{6}$)

32 $\frac{1}{10}$

33 That the event is certain to happen.

34 That the event is impossible

35 36

36 $\frac{1}{36}$

4.23 Oral test

1 32

2 14

3 12

4 7

5 6

6 44

7 Multiply the length by the width

8 Multiply the length by the width and by the height

9 Divide the distance by the time

10 Multiply the speed by the time

11 The steepness of the line

12 10.12

13 12.47

14 1.00 p.m.

SECTION 5 **Practice exams**

GCSE

Paper 1

1 £1.87

2 (a) £5.20 (b) £3.70

3 (a) 3672 (b) 1245 (c) £71.24

4 (a) 860 m (b) 45 000 m² (c) 4.5 ha

5 (a) 23 miles (b) 55 minutes (c) 22.05 (d) £8.50

6 (a) £720 (b) £1080 (c) 9p

7 (a) 30 m² (b) 24 m² (c) 6 m²

8 (a) 78p (b) 78 minutes

9 (a) 6 (b) 12 (c) 900 cm³

10 16 km

11 length 32 m; breadth 12 m; area 384 m²

12 (a) 100° (b) £25 (c) £180 (d) $\frac{1}{3}$

13 157

14 (a) 58 km/h (b) 145 km

15 (a) (i) 40 km (ii) 17.5 miles (b) 75 miles

16 (b) 3000

17 (a) 50 (b)

Number of letters	1	2	3	4	5	6	7
Frequency	7	5	7	8	11	9	3

(d) 5

18 (a) £322 (b) £332.50 (c) £348

Paper 2

1 £5.20

2 70

3 (a) 92p (b) 11p (c) 2 kg size

4 (a) 6 cm (b) 6 cm^2 (c) 12 cm^2 (d) 4.5 cm (e) 95°

5 (a) 16° (b) −7°

6 (a) 12 years 9 months (b) 13 years 3 months

7 (a) 219.8 cm (b) 43.96 m (c) 454

8 (a) 10 minutes (b) 40 minutes (c) 15 minutes late

9 £73.60

10 AC = 14 cm; total distance is 29 km

11 (a) 125 (b) 30 (c) $\frac{2}{5}$, 0.4

12 (a) $\frac{1}{2}$ (b) $\frac{1}{5}$ (c) $\frac{7}{10}$ (d) $\frac{4}{9}$ (e) $\frac{5}{9}$

13 (a) length 72 cm; width 50 cm; height 50 cm
(b) 19 400 cm^2

14 (a) cube (b) 8 cm^3

15 (a) 29 (b) 16 cm (c) 15 cm (d) 10 handspans

16 (a) 9.50 a.m. (b) 12 minutes (c) 28 miles
(d) 10.39 a.m.

17 (a) £1760 (b) £1520

18 (a) $174 (b) £6

Paper 3

1 0.000 13

2 (a) 47 000 (b) 4.657×10^4

3 £1.20

4 (a) £1040 (b) £1123.20

5 (a) 86°F (b) 10°C (c) $\dfrac{5(F - 32)}{9}$

6 (a) £1.40 (b) £1.20; £1.20; £1.80

7 (a) −3 (b) $\frac{5}{3}$ (c) (0, 5) (d) $(\frac{8}{5}, 0)$

8 (a) 1.21 p.m. (b) 64 m.p.h. (c) £12.88

9 (a) 3 cm (b) 4 cm

10 (a) 63.4 cm (b) 20.56° (c) (i) 306° (ii) 3.41 km

11 (a) 33° (b) 73.5°

12 (a) 131.74 cm^3

13 (a) 4000 cm^2 (b) $\frac{1}{5}$ (c) 20%

14 (a) 19.625 m^2 (b) 6 packets

15 (a) $\frac{3}{20}$ (b) (i) $\frac{9}{25}$ (ii) $\frac{3}{20}$ (iii) $\frac{4}{25}$

16 (a) 17 cm (b) 21.9 cm

17 (a) £760 (b) $\frac{1}{50}x + 200$ (c) £32 000

18 (d) (i) 6.5 m (ii) 27 cm (iii) 9 cm, 4 cm

STANDARD GRADE

Foundation Level Paper 1

1 1470, 708

2 8th April

3 £1.10

4 525 cm

5 8.57

6

7

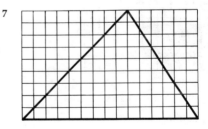

8 29 miles

9 9 gallons

10 126 miles

11 £50

12 £516.80

13 41°

14 Discount £2.55, Charge £14.45

15   80 tiles

16   £384.80

17   30.3 m

18   578

19

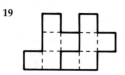

20   (a) 1244   (b) 46 minutes

21   480 m³

22   (a) 45 minutes   (b) 16 miles

Foundation Level Paper 2

1

Colour	Weight (grams)
Apricot (a)	10
Blue (b)	5
Crimson (c)	20
Dun (d)	15
Emerald (e)	35
Purple (p)	25

2

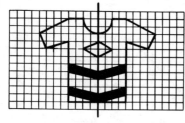

3   (a) 1800 m   (b) 2400 m   (c) 1800 m

4   (a) 6 gallons   (b) 8 gallons $7\frac{1}{2}$ gallons
  (d) $17\frac{1}{2}$ gallons

5   (a)

Anne	Billy
10	50
30	100
60	150
100	200
150	250

(b) £550   (c) £500

(d)

Carole	David
5	400
15	600
35	700
75	750
155	775

(e) £5115   (f) £797   (g) Carole

General Level Paper 1

1   2.5 mm

2   (a) 13 (b) 29%

3   24 000 tons

4 $7x + 32$

5   9th May – 15th June and 21st September – 2nd October

6 $\frac{3}{4}$ inch

7

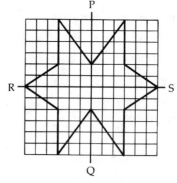

8   (a) 22p (b) Yes, 2p

9   (a) £16 875   (b) £18 984.38

10   110°

11   (a) 440 000 cm³ (b) 88 000 cm³

12 $y > 2$

13

4	−1	0
−3	1	5
2	3	−2

14   (a) 20 cm² (b) 58° (c) 40 cm

15   (a) 15 m

  (b) 0 m; It is at the same level as it was to begin with

General Level Paper 2

1 (a) £150; £120 (b) 200 miles; 300 miles

(c)

Distance (m)	Alpha (£)	Betta (£)
0	100	60
50	110	60
100	120	60
150	130	80
200	140	100
250	150	120
300	160	140

(d)

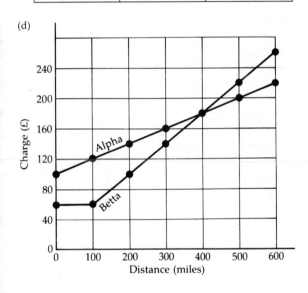

(e) 400 miles

2 (a) 150.5 m (b) 41.5° (c) 40.5 m

3

	Ann	**James**
Round 1	14	5
Round 2	9	15
Round 3	7	10
TOTAL	30	30

4 (a)

No. of black tiles	1	2	3	4	5	6	25	33
No. of white tiles	5	8	11	14	17	20	77	101

(b) $W = 3B + 2$

5 (b) 3.8 km (c) 35 minutes

6 (a) (i)

Breadth (m)	10	20	30	40	50	60
Length (m)	100	80	60	40	20	0
Area (m²)	1000	1600	1800	1600	1000	0

(ii) Breadth 30 m, length 60 m

(b) (i) 38.2 m (ii) 2290 m² (iii) Yes

SECTION 7 Mathematical vocabulary

The Scottish Standard Grade Regulations say that
the following terms may be used in the
examinations without explanation.

FOUNDATION LEVEL

Number

add
approximate
average
decimal (fraction)
divide
equals, equal to
even number
(vulgar) fraction
greater (than)
greatest
least
lesser
less (than)
more (than)
most
multiply
odd (number)
percentage
remainder
rounding off
square
subtract
total
to the nearest
   whole number

Money

basic wage
benefit
bill
bonus
commission
deductions (from income)
deposit
discount
double time
down payment
gross wage
hire purchase

GENERAL LEVEL

annum
decimal places,
   correct to
difference
digit
factor
index
integer
inverse
multiple
negative integer
number line
positive integer
power,
   raised to the power
prime number
product
quotient
ratio
square root
sum

exchange rates
premium (insurance)
principal
scientific notation
time-and-a-half

FOUNDATION LEVEL

income (tax)
instalment
invoice
loss
national insurance
net wage
overtime
pension
profit
receipt
rent
salary
simple interest
take home pay
union dues
unit (electricity)
VAT

Shape

acute
angle
area
arm of an angle
axis
axis of symmetry
bearing
centre
circle
circumference
cone
corner
cube, cuboid
cylinder
degree
diagonal
diameter
edge
face
horizontal
line
net
obtuse
origin
parallel
perimeter
plot
point

GENERAL LEVEL

adjacent
alternate (angles)
altitude
base
bilateral symmetry
bisect
centre of symmetry
complement (ary)
congruent
coordinates
corresponding angles
equilateral
gradient
hexagon
hypotenuse
intersect
isosceles
kite
opposite angles
parallelogram
pentagon
perpendicular
plane
(regular) polygon
prism
quadrilateral
reflection
reflex angle

FOUNDATION LEVEL

pyramid
radius
rectangle
revolution
right-angle
scale (drawing)
side
sphere
square
straight
tiling
triangle
vertical

Relationships

bar graph
direct proportion
find the value of
flow chart
formula
graph
line graph
pie chart
ready reckoner
scale

GENERAL LEVEL

rhombus
rotation
rotational symmetry
segment (of a line)
similar
supplement
tangent
trapezium
trigonometric ratios
    cosine, sine, tangent
vertically opposite
    angles

direct variation,
    varies (directly) as
equation
generalise
inequality
inverse proportion
linear
relationship
scale factor
simplify
sketch
solution
solve